Wilhelm Max Wundt, James Edwin Creighton, Edward Bradford Titchener

Lectures on human and animal psychology

Wilhelm Max Wundt, James Edwin Creighton, Edward Bradford Titchener

Lectures on human and animal psychology

ISBN/EAN: 9783743312005

Hergestellt in Europa, USA, Kanada, Australien, Japan

Cover: Foto ©berggeist007 / pixelio.de

Manufactured and distributed by brebook publishing software
(www.brebook.com)

Wilhelm Max Wundt, James Edwin Creighton, Edward Bradford Titchener

Lectures on human and animal psychology

LECTURES

ON

HUMAN AND ANIMAL
PSYCHOLOGY

BY

WILHELM WUNDT

Translated from the Second German Edition

BY

J. E. CREIGHTON & E. B. TITCHENER

London

SWAN SONNENSCHEIN & CO

NEW YORK : MACMILLAN & CO

1894

TRANSLATORS' PREFACE

THE present volume is the first of Professor Wundt's writings to be made generally accessible to the English-speaking public. Its comparatively popular and introductory character will, it is hoped, render it especially acceptable both to those beginning the study of psychology, to whom the technicalities of the author's *Grundzüge* would present very considerable difficulties, and to workers in other departments of science who may desire some knowledge of the methods and results of the new psychological movement.

The translators have endeavoured throughout to retain the oral form of the original Lectures. They have aimed, at the same time, to furnish a literal, as distinguished from a verbal, rendering of the German text. In view of the confusion which still obtains in English psychological terminology, they have attempted a precise use of words even at the occasional cost of literary effect. No word or phrase, however, has been employed which has not already received the sanction of well-known psychological writers.

J. E. CREIGHTON.
E. B. TITCHENER.

CORNELL UNIVERSITY, ITHACA, N.Y.

AUTHOR'S PREFACE TO SECOND EDITION

WHEN I was asked some years since by the publisher of this work to undertake its revision, I felt some hesitation in complying with his request. The first edition of the Lectures appeared thirty years ago; and during that time there had not only been a great advance in experimental psychology, but my own scientific knowledge and convictions had been considerably increased and modified. Thirty years ago the science was no more than a programme for the future. Except in psychophysics, where Fechner had just broken ground, everything remained still to do; and distrust and suspicion met the investigator at every turn. As for myself, I had had but little experience in the difficult work of psychological analysis, which the gradual development of the experimental methods has done so much to further; and set about my task with more zeal than discretion. So that for years before the appearance of the first edition of my *Physiological Psychology*, in which I took up the same problem with more modesty and caution, I had learned to look upon the Lectures as wild oats of my youthful days, which I would gladly have forgotten. But, unfortunately, hypotheses and views represented in them would every now and again be confused with or counted among my more mature convictions.

That I have resolved to undertake a second edition despite these somewhat discouraging facts, and in preference to the more grateful task of writing a new work of similar character, is due in the main to two circumstances. In the first place, I thought that, though the original volumes were defective both in general execution and in many points of detail, still a certain number of their chapters might stand unchanged, while I should perhaps be unable to attain again the freshness and force which characterised the first expression of my views. Secondly, every statement in the former edition about which I had modified or abandoned my original opinion seemed to lay upon me an obligation which I would fain discharge to the limit of my ability. Nevertheless, I will not omit in this place the express declaration that I no longer recognise as mine any view formulated in the earlier edition which is not admitted into the present. The elimination of everything that more recent inquiry had superseded has considerably diminished the size of the work. It has also suffered curtailment by the total exclusion of the discussions of social psychology which occupied a large portion of the second volume of the original book. It has been a matter of principle with me to restrict the contents of the Lectures to the individual psychology of man and the animals. As a matter of fact, the amount of material which social psychology has at its disposal is now so great, and the position of the science with regard to the points discussed has undergone so radical a change, that revision of the old chapters would necessarily mean rewriting. But within the prescribed limits, I have adhered to my former plan of not attempting any completeness of exposition, but rather of taking full advantage of the lecture form and confining myself to the treatment of topics which I thought especially characteristic of the spirit and trend of modern psychology. At the same time, it seemed permissible to make the work in some sense supplementary to my other writings by devoting few words to subjects which I have elsewhere discussed in detail, and giving more attention to topics which are less prominent, particularly in my *Physiological Psychology*. Thus I have based the discussion of Weber's law entirely upon the method of just noticeable differences, although this is the most imperfect of the measurement-methods and would hardly now be employed in investigations which made any claim to scientific accuracy. Again, in developing the theory of spatial localisation I have retained my previous plan of elucidating its much-misunderstood fundamental conceptions, and of the sensations attaching to movement have only dealt with muscle-sensations, although the *rôle* of certain complexes of pressure-sensations in the surrounding parts is really not less important. The reader who desires a deeper insight into psychology will, I hope, not omit to refer in such cases to my more systematic work, which is more especially devoted to the investigation of the physiological correlates of psychical processes.

The first edition of these Lectures was principally based upon Fechner's *Psychophysik* and my own *Beiträge zur Theorie der Sinneswahrnehmungen*, which appeared between 1858 and 1862. The lectures dealing with these subjects have undergone the least alteration in the second edition. I may perhaps be also allowed to state that the treatment of the problem of the causality of will in Lecture XXIX. stands for the most part precisely as it did in my previous exposition. The following lectures of the second edition present portions of the older work in revised form:—I. (I., II., of the former edition), II. (VII.), III. (VIII.), IV. (IX.), VIII. (XIV.), IX. (XV.), X., XI. (XVI., XVII.), XII. (XXI.), XIII. (XXII.), XXIX. (LV., LVI.) ; entirely rewritten are—V. (XI.), VI. (X.), VII. (XIII.), XIV. (XXX.), XXV. (XXXI.), XXVI. (LI., LII.), XXVIII. (XLII.) ; new are :—XV., XVI., XVII., XVIII., XIX., XX., XXI., XXII., XXIII., XXIV., XXVII., XXX. Very little of the lectures of the first edition not quoted here has been included in the present volume.

W. Wundt.

Leipzig, April, 1892.

CONTENTS.

LECTURE I

§ I

PSYCHOLOGY, even in our own day, shows more clearly than any other experiential science traces of the conflict of philosophical systems. We may regret this influence in the interest of psychological investigation, because it has been the chief obstacle in the way of an impartial examination of mental life. But in the light of history we see that it was inevitable. Natural science has gradually taken shape from a natural philosophy which paved the way for it, and the effects of which may still be recognised in current scientific theory. That these effects are more fundamental and more permanent in the case of psychology is intelligible when we consider the problem which is set before it. Psychology has to investigate that which we call internal experience,—*i.e.*, our own sensation and feeling, our thought and volition,—in contradistinction to the objects of external experience, which form the subject matter of natural science. Man himself, not as he appears from without, but as he is in his own immediate experience, is the real problem of psychology. Whatever else is included in the circle of psychological discussion,—the mental life of animals, the common ideas and actions of mankind which spring from similarity of mental nature, and the mental achievements of the individual or of society,—all this has reference to the one original problem, however much our understanding of mental life be widened and deepened by the consideration of it. But the questions with which psychology thus comes into contact are at the same time problems for philosophy. And philosophy

B

had made various attempts to solve them long before psychology as an experiential science had come into being.

The psychology of to-day, then, neither wishes to deny to philosophy its right to occupy itself with these matters, nor is able to dispute the close connection of philosophical and psychological problems. But in one respect it has undergone a radical change of standpoint. It refuses to regard psychological investigation as in any sense dependent upon foregone metaphysical conclusions. It would rather reverse the relation of psychology to philosophy, just as empirical natural science long ago reversed its relation to natural philosophy,—in so far, that is, as it rejected all philosophic speculations which were not based upon experience. Instead of a psychology founded upon philosophical presuppositions, we require a philosophy to whose speculations value is ascribed only so long as they pay regard at every step to the facts of psychological, as well as to those of scientific, experience.

It will, therefore, be a matter of principle for us in these lectures to stand apart from the strife of philosophic systems. But since the thought of to-day is subjected on all sides to the influence of a philosophic past' which counts its years by thousands, and since the concepts and general notions under which an undifferentiated philosophy arranged the facts of mental life have become part of the general educated consciousness, and have never ceased to hinder the unprejudiced consideration of things as they are, it is our bounden duty to characterise and justify the standpoint which we propose to adopt. We will, therefore, first of all glance for a moment at the history of philosophy before the appearance of psychology.

In the beginnings of reflective thought, the perception of the external world preponderates over the internal experience of idea and thought, of feeling and will. The earliest psychology is therefore Materialism : the mind is air, or fire, or ether,—always some form of matter, however attenuated this matter may become in the effort to dematerialise it. Plato was the first among the Greeks to separate mind from body. Mind he regarded as the ruling principle of the body. And this separation paved the way for the future one-sided dualism which considered sensible existence as the obscuring and debasing of an ideal, purely mental being. Aristotle, who combined with the gift of speculation a marvellous keenness of observation, attempted to harmonise these opposites by regarding mind as the principle which vitalises and informs matter. He saw the direct operation of mental powers in the forms of animals, in the expression of the human figure at rest and in movement, even in the processes of growth and nutrition. And he generalised all this in his conclusion that mind is the creator of all organic form, working upon matter as the sculptor works on marble. Life and mentality were for him identical terms ; even the vegetable world was on his theory endowed with mind. But, apart from this, Aristotle penetrated more deeply than any of his predecessors into the facts of mental experience. In his work upon the mind, the first in which psychology was ever treated as an independent science, he sharply separates from one another the fundamental mental activities ; and, so far as the knowledge of his time allowed, sets forth their causal connections.

The Middle Ages were wholly dominated by the Aristotelian psychology, and more especially by its basal proposition that mind is the principle of life. But with the dawn of the modern period begins in psychology, as elsewhere, the return to Platonism. Another influence combined with this to displace Aristotelianism ; namely, the development of modern natural science and the mechanical metaphysics which this development brought with it. The result of these influences was the origin of two psychological schools, which have disputed with one another down to the present day,—Spiritualism and Materialism. It is a curious fact that the thought of a single man has been of primary importance in the development of both these standpoints. Descartes, the mathematician and philosopher, had defined mind, in opposition to Aristotle, as exclusively thinking substance ; and following Plato, he ascribed to it an original existence apart from the body, whence it has received in permanent possession all those ideas which transcend the bounds of sensible experience. This mind, in itself unspatial, he connected with the body at one point in the brain, where it was affected by processes in the external world, and in its turn exercised influence upon the body.

Later Spiritualism has not extended its views far beyond these limits. It is true that Leibniz, whose doctrine of monads regarded all existence as an ascending series of mental forces, attempted to substitute for the Cartesian mind-substance a more general principle, approximating once more to the Aristotelian concept of mind. But his successor Christian Wolff returned to the Cartesian dualism. Wolff is the originator of the so-called theory of mental faculties, which has influenced psychology down to the present day. This theory, based upon a superficial classification of mental processes, was couched in terms of a number of general notions,—memory, imagination, sensibility, understanding, etc.,—which it regarded as simple and fundamental forces of mind. It was left for Herbart, one of the acutest thinkers of our century, to give a convincing proof of the utter emptiness of this ' theory.' Herbart is at the same time the last great representative of that modern Spiritualism which began with Descartes. For the works of Kant and of the other philosophers who came after him,—Fichte, Schelling, and Hegel, —belong to a different sphere. In Herbart we still find the concept of a simple mind-substance, which Descartes introduced into modern philosophy, but pushed to its extreme logical conclusion, and at the same time modified by the first principles of Leibniz' monadology. And the consistency of this final representative of speculative psychology makes it all the more plain that any attempt to derive the facts of mental life from the notion of a simple mind and its relation to other existences different from or similar to itself must be vain and fruitless. Think what lasting service Herbart might have done psychology, endowed as he was in exceptional measure with the power of analysing subjective perception, had he not expended the best part of his ingenuity in the elaboration of that wholly imaginary mechanics of ideation, to which his metaphysical presuppositions led him. Still, just because he carried the concept of a simple mind-substance to its logical conclusion, we may perhaps ascribe to his psychology, besides its positive merits, this negative value, —that it showed as clearly as could be the barrenness of Spiritualism. All that is permanent in Herbart's psychological works we owe to his capacity of accurate observation of mental fact ; all that is untenable and mistaken proceeds from his metaphysical concept of mind and the secondary hypotheses which it compelled him to set up. So that the achievements of this great Spiritualist show most plainly that the path which he travelled, apart from all the contradictions into which it led him, cannot ever be the right road for psychology. This notion of a simple mental substance was not reached by analysis of mental phenomena, but was superimposed upon them from without. To assure the pre-existence and immortality of the soul, and (secondarily) to conform in the most direct way with the logical principle that the complex presupposes the simple, it seemed necessary to posit an indestructible and therefore absolutely simple and unalterable mind-atom. It was then the business of psychological experience to reconcile itself with this idea as best it might.

§ II

When Descartes denied mind to animals, on the ground that the essence of mind consists in thought, and man is the only thinking being, he could have little imagined that this proposition would do as much as the strictly mechanical views which he represented in natural philosophy to further the doctrines which are the direct opposite of the Spiritualism which he taught,—the doctrines of modern Materialism. If animals are natural automata, and if all the phenomena which general belief refers to sensation, feeling, and will are the result of purely mechanical conditions, why should not the same explanation hold of man? This was the obvious inference which the Materialism of the seventeenth and eighteenth centuries drew from Descartes' principles.

The naïve Materialism with which philosophy began had simply ascribed some kind of corporeality to mental existence. But this modern Materialism took as its first principles physiological hypotheses; thought, sense, and idea are physiological functions of certain organs within the nervous system. Observation of the facts of consciousness is of no avail until these are derived from chemical and physical processes. Thought is simply a result of brain activity. Since this activity ceases when circulation is arrested and life departs, thought is nothing more than a function of the substances of which the brain is composed.

More particularly were the scientific investigators and physicians of the time inclined, by the character of their pursuits, to accept this explanation of mental life in terms of what seemed to them intelligible scientific facts. The Materialism of to-day has made no great advance in this or in any other direction upon the views promulgated in the last century, *e.g.* by de la Mettrie, and developed by Helvétius, Holbach, and others. But this equating of mental process and brain function, which makes psychology a department of cerebral physiology, and therefore a part of a general atomic mechanics, sins against the very first rule of scientific logic,—that only those connections of facts may be regarded as causal which obtain between generically similar phenomena. Our feelings, thoughts, and volitions cannot be made objects of sensible perception. We can hear the word which expresses the thought, we can see the man who has thought it, we can dissect the brain in which it arose ; but the word, the man, and the brain are not the thought. And the blood which circulates in the brain, the chemical changes which take place there, are wholly different from the act of thought itself.

Materialism, it is true, does not assert that these *are* the thought, but that they *form* it. As the liver secretes bile; as the muscle exerts motor force, so do blood and brain, heat and electrolysis, produce idea and thought. But surely there is no small difference between the two cases. We can prove that bile arises in the liver by chemical processes which we are able, in part at least, to follow out in detail. We can show, too, that movement is produced in muscles by definite processes, which are again the immediate result of chemical transformation. But cerebral processes give us no shadow of indication as to how our mental life comes into being. For the two series of phenomena are not comparable. We can conceive how one motion may be transformed into another, perhaps also how one sensation or feeling is transformed into a second. But no system of cosmic mechanics can make plain to us how a motion can pass over into a sensation or feeling.

At the same time modern Materialism pointed out a more

legitimate method of research. There are numerous experiences which put beyond all doubt the connection of physiological cerebral function on the one hand and of mental activity on the other. And to investigate this connection by means of experiment and observation is assuredly a task worth undertaking. But we do not find that Materialism, even in this connection, has made a single noteworthy contribution to our positive knowledge. It has been content to set up baseless hypotheses regarding the dependence of mental function upon physical process; or it has been concerned to refer the nature of mental forces to some known physical agency. No analogy has been too halting, no hypothesis too visionary, for its purpose. It was for some time a matter of dispute whether the mental force had more resemblance to light or to electricity. Only on one point was there general agreement,—that it was not ponderable. —

In our day the conflict between Materialism and Spiritualism, which was raging in the middle of the century, has almost worn itself out. It has left behind it nothing of value for science ; and that will not surprise any one who is acquainted with its details. For the clash of opinion was centred once more round the old point : in the questions concerning mind, the seat of mind, and its connection with body. Materialism had made the very same mistake which we have charged to the spiritualistic philosophy. Instead of plunging boldly into the phenomena which are presented to our observation and investigating the uniformities of their relation, it busied itself with metaphysical questions, an answer to which, if we may expect it at all, can only be based upon an absolutely impartial consideration of experience, which refuses to be bound at the outset by any metaphysical hypothesis.

§ III

We find, then, that Materialism and Spiritualism, which set out from such different postulates, converge in their final result. The most obvious reason of this is their common methodological error. The belief that it was possible to establish a science of mental experience in terms of speculation, and the thought that a chemical and physical investigation of the brain must be the

first step towards a scientific psychology led alike to mistakes
in method. The doctrine of mind must be primarily regarded
as an experiential science. Were this otherwise, we should not
be able so much as to state a psychological problem. The
standpoint of exclusive speculation is, therefore, as unjustifiable
in psychology as it is in any science. But more than this, so
soon as we take our stand upon the ground of experience, we
have to begin our science, not with the investigation of those
experiences which refer primarily only to objects more or less
closely connected with mind, but with the direct examination of
mind itself,—that is, of the phenomena from which its existence
was long ago inferred, and which formed the original incentive
to psychological study. The history of the science shows us
that mind and the principal mental functions were distinguished
before there was any idea that these functions were connected
with the brain. It was not any doubt as to the purpose of this
organ which led to the abstraction which lies at the foundation
of the doctrine of mind, but simply observation of mental
phenomena. Sense, feeling, idea, and will seemed to be related
activities ; and they appeared, further, to be bound together by
the unity of self-consciousness. The mental processes began,
therefore, to be looked upon as the actions of a single being.
But since these actions were found again to be intimately
connected with bodily functions, there necessarily arose the
question of assigning to mind a seat within the body, whether
in the heart, or the brain, or any other organ. It was reserved
for later investigation to show that the brain is the sole organ
which really stands in close connection with the mental life.

But if it be sensation, feeling, idea, and will which led in
the first instance to the assumption of mind, the only natural
method of psychological investigation will be that which begins
with just these facts. First of all we must understand their
empirical nature, and then go on to reflect upon them. For it
is experience and reflection which constitute each and every
science. Experience comes first ; it gives us our bricks : reflection is the mortar, which holds the bricks together. We cannot
build without both. Reflection apart from experience and experience without reflection are alike powerless. It is therefore
essential for scientific progress that the sphere of experience be

enlarged, and new instruments of reflection from time to time invented.

But how is it possible to extend our experience of sensations, feelings, and thoughts? Did not mankind feel and think thousands of years ago, as it feels and thinks to-day? It does, indeed, seem as though our observation of what goes on in the mind could never extend beyond the circle to which our own consciousness confines it. But appearances are deceptive. Long ago the step was taken which raised the science of psychology above the level of this its first beginning, and extended its horizon almost indefinitely. History, dealing with the experience of all times, has furnished us with a picture in the large of the character, the impulses, and the passions of mankind. More especially is it the study of language and linguistic development, of mythology and the history of religion and custom, which has approached more and more closely, as historical knowledge has increased, to the standpoint of psychological inquiry.

The belief that our observation is confined to the brief span of our individual life, with its scanty experience, was one of the greatest obstacles to psychological progress in the days of the earlier empiricism. And the opening up of the rich mines of experience to which social psychology gives us access, for the extension of our own subjective perceptions, is an event of importance and of promise for the whole circle of the mental sciences. Nor is that all. A second fact, of still greater import for the solution of the simplest and therefore, most general psychological problems, is the attempt that has been made to discover new methods of observation. One new method has been found; it is that of experiment, which, though it revolutionized the natural sciences, had not up to quite recent times found application in psychology. When the scientific investigator is inquiring into the causes of a phenomenon, he does not confine himself to the investigation of things as they are given in ordinary perception. That would never take him to his goal, though he had at his command the experiences of all time. Thunderstorms have been recorded, indeed carefully described, since the first beginnings of history; but what a storm was could not be explained until the phenomena of electricity had become familiar, until electrical machines had been constructed and experiments made with them.   Then the matter was easy.   For when once the effects of a storm had been observed and compared with the effect of an electric spark, the inference was plain that the discharge of the machine was simply a storm in miniature.   .What the observation of a thousand years had left unexplained was understood in the light of a single experiment.   Even astronomy, a science which we might think must of its very nature be confined to observation, is in its more recent development founded in a certain sense upon experiment.   So long as mere observations were taken, the general opinion that the earth was fixed, and that the sun and stars moved round it, could not be overthrown.   It is true that there were many phenomena which made against this belief; but simple observation could not furnish means for the attainment of a better explanation.   Then came Copernicus, with the thought: 'Suppose I stand upon the sun!' and henceforth it was the earth that moved, and not the sun ; the contradictions of the old theory disappeared, and the new system of the universe had come into being.   But it was an experiment that had led to this, though an experiment of thought.   Observation still tells us that the earth is fixed, and the sun moving ; and if the opposite view is to become clear, we must just repeat the Copernican experiment, and take our stand upon the sun.

It is experiment, then, that has been the source of the decided advance in natural science, and brought about such revolutions in our scientific views.   Let us now apply experiment to the science of mind.   We must remember that in every department of investigation the experimental method takes on an especial form, according to the nature of the facts investigated.   In psychology we find that only those mental phenomena which are directly accessible to physical influences can be made the subject matter of experiment.   We cannot experiment upon mind itself, but only upon its outworks, the organs of sense and movement which are functionally related to mental processes.   So that every psychological experiment is at the same time physiological, just as there are physical processes corresponding to the mental processes of sensation, idea, and will. This is, of course, no reason for denying to experiment the character of a psychological method. It is simply due to the general conditions of our mental life, one aspect of which is its constant connection with the body.

The following lectures are intended as an introduction to psychology. They do not attempt any exhaustive exposition of the methods and results of experimental psychological investigation. That would have to assume previous knowledge which cannot here be presupposed. Neither shall we include in the range of our discussion the facts of social psychology, whose contents is extensive enough to demand an independent treatise. We shall confine ourselves to the mental life of the individual ; and within those limits it will be the human mind to which we shall for the most part devote ourselves. At the same time it appears desirable, for the right understanding of individual mental development, that we should now and again institute a brief comparison with the mental life of animals.

LECTURE II

§ I

SO soon as ever the dawn of knowledge had broken upon us through the portals of the senses, we began to compare objects, to reflect upon them. The first work of thought was to set things in their places, to transform the chaos of sense-impressions into an intelligible cosmos. But after everything else has been arranged, there still remains something which has as yet no place,—ourselves : our feeling, willing, and thinking ; so that the question arises : how can our own mental life be made the subject of investigation like the objects of this external world of things about us ? And yet—can such a question be asked ? Is it not really self-contradictory ? It is as though we required that the tone should hear itself, or the ray of light be sensed by itself.

It is, indeed, true that here, as we enter upon the study of psychology, a peculiar difficulty presents itself. If we try to observe our mental activities, the observer and the observed object are one and the same. But the most important condition of a trustworthy observation is always thought to consist in the mutual independence of object and observer. Nevertheless, we should be overhasty if we disputed the possibility of psychological observation in general because of this unavoidable limitation of the science. Only so much is true : that the peculiarities of the object, in this case as in others, imply special conditions of its observation. These can be stated in two rules. First : so long as we confine ourselves to introspection, without calling in any assistance from outside, mental processes may not be observed directly while they are taking place. We must limit ourselves to analysing them, so far as possible, from the effects which they leave behind in our memory. Secondly: wherever it is possible, we must endeavour so to control our mental processes by means of objective stimulation of the external organs (particularly of the sense-organs, with the physiological functioning of which definite psychoses are regularly connected) that the disturbing influence which the condition of observation tends to exercise upon them is counteracted. This control is given by experiment. Not only does experiment, here as elsewhere, enable us to produce a phenomenon, and to regulate its conditions, at our pleasure: it possesses in psychology an especial importance, in that it alone renders self-observation possible during the course of a mental process.

Let us now seek, in accordance with the first rule which we established, to recall the general impression which any particular mental experience has left upon us. This impression will always be that of a composite process. Some parts of it, images of external objects, we designate Ideas; others, the pleasurable or painful reactions of our own mind upon these ideas, Feelings; others, again, we term Efforts, or Impulses, or Volitions. It is certainly true that these elements of mental life never occur separately, but always in connection with, always in dependence upon, one another. Nevertheless, it seems absolutely necessary, at the beginning of a psychological investigation, to follow the example of discrimination already set by language, and to separate out the most important factors of this complex inner life and subject each of them in turn to a special analysis.

Now, if these elements are all interconnected and interdependent, it is clear that, other things equal, we might begin the analysis which we contemplate with any one of them. Nevertheless, external reasons render it hardly possible to choose any other method of procedure than that of commencing with an investigation of ideas. We conceive of an idea as the image of some external object. We can, therefore, transfer to these images of external objects the abstraction which we always make in the case of the logical notions of the objects; we can consider them just as though the feelings, impulses, and volitions, which in fact invariably accompany them, did not exist. On the other hand, in the case of these feelings and impulses themselves, it is impossible to carry out an abstraction of the kind, because we are not in a position even to describe them without constant reference to the ideas with which they are associated. Granted that this results merely from the fact that all our designations took their origin from distinctions made between objects of the external world, and were only applied to our inner experiences at a comparatively late date, still it remains true that this general trend of the development of our knowledge necessarily determines the manner in which psychology sets to work to analyse those inner experiences.

By an idea, then, we shall understand that mental state or mental process which we refer to something outside of ourselves, whether this attribute of externality be thought of as directly applicable in the present, or as applied to an object which has been directly given us in the past, or even as applied to an object which is only possible, and not actual. Under ideas, therefore, we include — (1) sense-perceptions, which depend upon direct excitation of the organs of sense ; (2) memories of such sense-perceptions; and (3) images of fancy, be these what they may. The terminology adopted in many Psychologies, according to which the images of memory and fancy are alone designated 'ideas,' while the direct effects of sense-impression are termed exclusively ' perceptions,' we must judge to be unjustifiable and misleading. It lends colour to the view that there is some essential psychological difference between these two kinds of mental process, whereas such a difference is nowhere discoverable. Even the reflection upon which the distinction is based,—the thought that images of memory and fancy do not correspond to objects actually presented to us,—is not universally valid. And, in the same way, sense-perceptions may very well be themselves taken for illusions of sense. So that the characteristics, by means of which two kinds of ideas are distinguished, can never be more than secondary, while the distinction itself cannot always be satisfactorily carried through.

An idea, in the general sense in which we are here using the word, is always something composite. A visual image is made up of spatially distinguishable parts ; a sound is constituted of clangs, while it is also conceived of as coming to us in a certain direction,—*i.e.*, is associated with spatial ideas. Our first problem in analysing ideas, therefore, consists in the determination of their simplest constituent elements, and in the investigation of the psychological properties of these. We call the psychological elements of ideas Sensations. Thus we speak of the idea of a house, of a table, of the sun or moon, but of the sensations of blue, yellow, warm, cold, or of a tone of definite pitch. This use of the word 'sensation,' we must notice, like the use of 'idea' in the general sense mentioned above, has only become current in recent psychology. In the earlier treatises, and still to some extent in popular writings and *belles lettres*, we find the word 'sensation' employed with the same meaning as 'feeling.' Here, and in what follows, we shall consistently adhere to the definition just given, according to which sensations are merely the simplest and most elemental psychological constituents of the idea.

§ II

But the analysis of ideas into sensations does not conclude the task which we have set ourselves,—the analysis of those mental processes which are referable to external objects. For in every sensation, again, we distinguish two properties,—one which we name its strength or intensity, and another which we call its quality. Neither can exist in the absence of the other. Every sensation, be it of sound, heat, cold, taste, or what not, is possessed at once of a certain intensity and a certain quality. But, as a general rule, the two attributes can be varied independently of each other. We can sound a musical note, *e.g.*, at first quite softly, and then, by gradually increasing its strength, pass it through all possible degrees of intensity, while its quality remains unaltered. Or we can strike different notes one after the other, and so obtain different qualities, while we still keep, if we will, one and the same intensity of tone throughout. Here quality has changed ; intensity remains constant. This possibility of varying the two constituents of sensation independently of each other depends upon the fact that the motions in external nature, by the operation of which upon our sense-organs sensation in general was originally occasioned, present two aspects, either of which may also vary without affection of the other.

The processes of motion which, by their operation upon our senses, give rise to sensations, we commonly denominate *stimuli*, or more particularly *sense-stimuli*. Accordingly, we generally understand by stimulus the external motion-process, which, after it has acted upon the sense-organ and been conducted by sensory nerves to the brain, is accompanied by the mental process of sensation. Thus we regard the sound-waves of the air or the light-waves set up in surrounding space as stimuli, corresponding to our sensations of sound and light. In the same way, those motion-processes which are aroused, by the agency of such external stimuli, in our sense-organs and in the brain, may also be regarded as processes of stimulation or as constituents of the entire stimulation-process. For the sake of clearness, we will call these last *internal* stimuli. If we seem always to have the *external* stimuli primarily in mind when we are speaking of the relation of 'stimulus' to sensation, this is only because they are the more easily accessible to objective investigation. But wherever we can show good reason for the belief that the peculiar form taken on by a stimulus-process in the sense-organs, the sensory nerves, and the sense-centres of the brain exercises a determining influence upon a particular sensation, we shall, of course, be constrained to take into consideration the character of the internal stimuli and the transformations which occur in the conversion of an external into an internal stimulus.

Now, in whichever of these two senses we employ the notion of 'stimulus,' we are able to vary both the intensity and the form of any stimulation-process. But the intensity of stimulus corresponds to the intensity of sensation, the form of stimulus to its quality. (Thus, in the case of sound and light, the intensity of the sensations corresponds to the extent or amplitude of vibration, their quality to its rapidity. The quality of tone we call pitch ; the quality of light, colour.) Although, therefore, intensity and quality of sensation do not exist independently of each other, yet psychological analysis is able to distinguish them for its own purposes. In doing this, it is only completing an abstraction which was begun when ideas were separated out from the totality of mental life, and continued a step farther in the subdivision of ideas into elementary sensations.

§ III

We begin, then, with an investigation of the *intensity* of sensations. And we leave for the present out of account everything which has reference to their qualitative aspect.

If we compare with each other two different sensations of the same modality, we are undoubtedly able to pass judgment regarding their intensities. Our judgment runs either : The sensations are of equal intensity, or: They are not of equal intensity. The midday sun we assert to be brighter than the moon, the roar of a cannon louder than the crack of a pistol, a hundredweight heavier than a pound. These comparative .judgments are taken directly from sensation. We really state in them merely this : that the sensations which the sunshine, the cannon, and the hundredweight arouse in us are more intensive than the sensations which we have from the moon, a pistol-shot, or a pound-weight. There is therefore possible a quantitative comparison of sensations. We can say of two sensations that they are of equal intensity, or that this one is of a greater or less intensity than the other. There our measurement of sensation ordinarily rests. We are not able to say how much stronger or how much weaker one is than another. We cannot estimate in the least whether the sun is a hundred or a thousand times brighter than the moon, the cannon a hundred or a thousand times louder than the pistol. Our ordinary measurement of sensation tells us only of ' equality,' of a ' more,' or of a ' less,' never of a 'so much more ' or 'less.' And this natural measurement is, therefore, as good as none at all when an exact determination of intensity is required. Although, perhaps, we may be able to observe that, as a general rule, intensity of sensation increases and diminishes with intensity of stimulus, yet we have not the remotest idea whether the two vary in the same ratio, or whether one increases more slowly or more quickly than the other. In a word, we know nothing of the law of the dependence of sensation upon stimulus. If we are to discover this, we must necessarily begin, by finding a more exact measurement for sensation. We must be able to say : a stimulus of the intensity 1 occasions a sensation of the intensity 1, a stimulus of the intensity 2 a sensation of the intensity 2 or 3 or 4, and so on. But, to do this, we must know what it means to say that 'this sensation is twice,' or 'three times,' or 'four times as great as that.'

Now, we have said above that it is possible to strike a note first of all very gently, at an intensity at which it can only just be heard, and then gradually to increase this intensity, until we reach a point at which the note is as loud as it can be made. Between these upper and lower limits the tone-sensation has passed, not by leaps and bounds, but smoothly and uniformly, through all its possible intensities. And the same is true of other sense-impressions. From every sensation-quality we can construct a one-dimensional series of sensation-intensities, which pass over into one another without break or gap. In such a series we may, first of all, quantitatively distinguish every member from every other member ; we say that the one of two compared sensations is the stronger, the other the weaker. But more than that. We find no difficulty in stating, after successive comparisons, that the difference of intensity in one case was greater than it was in another.

Now, as the result of these very obvious considerations, there arise for psychological investigation two separate questions. The first is : what is the basis of this natural measurement of sensation-intensities, which enables us directly, without knowing anything about the external affection of our senses, quantitatively to compare different sensations? And the second,—which, as soon as stated, becomes a problem in experimental psychology,—runs : may not this crude and inaccurate natural measurement be transformed into an exact one; so that, *e.g.*, we might be able to state how much stronger or weaker a given sensation was than another with which we compared it? We will try to answer this second question first.

§ IV

At first sight the attempt to measure the intensity of sensations may appear overbold. How can we hope to reach any result when no definite measure is contained in the sensation itself? But if we take a little time to consider how it is that the measurement of magnitude in general is carried out, matters will begin to look more hopeful.

For all measurement there is required a standard. And this standard can never be the measured object itself. Thus we may measure the time of an occurrence by a clock; and what the clock shows us is a uniform motion. Or we measure longer periods of time by days, months, and years; and these correspond to uniformly repeated changes in external nature. That is, we measure time by space. But to measure space, on the other hand, we employ time. The length of the road over which we have travelled we estimate by the time that the journey has taken. And when we mark the successive divisions upon a scale, we must do it in a time order. So that the original measurement-units of space and time always coincide: an hour is just as much an hour of space-experience as an hour of time-experience. Space gives us our only means of measuring time, and time our best means of measuring space. Nevertheless, there is a noteworthy difference in the way in which each of these two measures depends upon the other. For space-measurement it is only necessary that time should be already existent; it is not requisite that we should possess an exact measure of time. When we are constructing a scale, we must mark in one unit after another; but, that once done, we do not need in every particular measurement to compute the number of units which the scale embraces. We measure directly with the whole scale; that is, we take all at once, simultaneously, what was constructed gradually. To carry out the most exact spatial measurement we need have no more than the general notions 'earlier,' 'later,' 'simultaneous.' Then, when space has been measured, we come back to time, in order to divide it up by the help of our spatial measurements.

All exact measurement is, therefore, spatial measurement. Times, forces, everything that can be considered as magnitude, we measure by a spatial standard. Now, when we talk of comparing the intensities of sensations, we imply that sensations are magnitudes. And although a direct comparison of sensation-intensities does not enable us to do more than pronounce them 'less' or 'greater' or 'equal,' that is in itself no obstacle in the way of obtaining an exact measurement. For at first we possessed only the vague ideas of 'earlier,' 'later,' and 'simultaneous' in the case of time ; and yet we are now able to measure with very great accuracy temporal differences, the mere cognition of which would have far transcended our original powers.   Indeed, it is just the same with sensation as with time, and with all the other magnitudes which, like these two, are primarily mental magnitudes.   Temporal and spatial magnitudes are alike distinguished in the first place only as 'equal,' 'greater,' or 'less.'   We quickly arrive at an exact determination of the latter, since we are able to measure each new space-magnitude by magnitudes already known.   But the measurement of mental magnitudes is apparently attended with greater difficulties.   In this sphere it was until recently only the *movement* of thought, time, which had been subjected to an exact measurement, by the substitution, for movement of ideas in us, of movements of objects without us, and especially those movements with which the impression of uniform regularity was invariably connected.

An exact means of measuring time cannot, then, be obtained from time alone ; we must call in the aid of movement in space. In the same way, we shall never be able to discover a means of measuring sensation in sensation itself, but must take into consideration the relation of its magnitude to other measurable magnitudes.   And there is no magnitude which presents itself more obviously for this purpose than that of the stimulus, from which the sensation arises.   Indeed, the stimulus furnishes us not merely with the most obvious, but with our only possible, means of measuring sensation.   There is no other magnitude which stands in any such direct relation to the magnitude of sensation.

The only assistance which sensation itself renders us in this measurement is that of the ordinary distinction of sensation as of 'greater,' 'less,' and 'equal' intensity.   Everything else must be derived from the measurement of stimulus.   If two sensations are of equal intensity, our first thought is that the external stimuli are also of the same intensity in the two cases.   But measurement of them shows not seldom that this surmise is wrong ; that stimuli of different intensity may occasion sensations of equal intensity.   A weak eye finds ordinary daylight so intense that it involuntarily closes ; but the normal eye displays no such tendency, except when looking directly at the sun. If we fall into a swoon, or into deep sleep, we do not sense the prick of a needle which, in the waking state, would cause us acute pain. Indeed, facts of that kind have been observed from the beginning of time. This greater or less receptivity of the organism, in face of external stimuli, we call *sensibility* or *excitability*. We say that a weak eye is more excitable than a strong one ; that we are more sensitive awake than asleep. But we do not ordinarily think of measuring this excitability. And yet the measure is given at once, if we only ascertain the intensities of the stimuli which, on different occasions, give rise to a sensation of equal intensity. If the stimuli are of equal intensity in both cases, the excitability is the same ; if the stimulus in the first case were twice or three times as strong as it is in the second, the excitability in the former experiment was half or a third as great as it is in the latter. In short, excitability is inversely proportional to the intensity of the stimuli employed for the production of equally intensive sensations.

Already, then, we have gained one result, which is not unimportant for our proposed measurement. We have discovered a method of eliminating the differences of excitability which may be found to exist in different individuals or in the same individual at different times. And we are thereby in a condition to propose and define a unit of excitability, such as has been universally accepted for time,—supposing, of course, that its proposition shall prove to possess any real significance.

A further basis of measurement is given with the increase and decrease of sensation-intensity. What we all know with regard to this is only that the intensity of sensation increases and decreases with the intensity of stimulus. If the 'sound in our ear' increases, we know that the external sound has become louder, always provided that we have no reason to assume a change of sensibility in our sense-organs. Originally this conclusion regarding increase of the external stimulus was merely an inference from increased intensity of sensation. Not until we have made those physical processes which constitute the stimulus the object of separate investigation can we attain to the definite conviction that this conclusion was correct. But in pursuing such an investigation we come to make stimulus inde-

pendent of sensation, and so are on the road towards the discovery of a valid measurement of stimulus.

Now, if our entire knowledge were confined to this fact, that
sensation increases and decreases with stimulus, we should not
have gained very much. But there are facts of direct and un-
assisted observation which tell us something, even if in the most
general terms, of the law which governs the intensive relations
of stimulus and sensation.

Every one knows that in the stillness of night we hear things
which are unperceived in the noise of day. The gentle ticking
of the clock, the distant bustle of the streets, the creaking of the
chairs in the room, impress themselves upon our ear. And
every one knows that amid the confused hubbub of the market-
place, or the roar of a railway-train, we may lose what our
neighbour is saying to us, or even fail to hear our own voice.
The stars which shine so brightly at night are invisible by day ;
and although we can see the moon in the day-time, she is far
paler than at night. Every one who has had to do with weights
knows that if to a gramme in the hand we add a second gramme,
the difference is clearly noticed ; but if we add it to a kilogramme, there is no knowledge of the increase.

All these experiences are so common that we think them
self-evident. Really, that is by no means the case. There
cannot be the least doubt that the clock ticks just as loudly by
day as by night. In the clamour of the street or amid the
noise of the railway we speak, if anything, more loudly than is
usual. Moon and stars do not vary in the intensity of their
light. And no one will deny that a gramme weighs the same
whether it is added to one gramme or to a thousand.

The sound of the clock, the light of the stars, the pressure of
the gramme weight,—all these are sensation-stimuli, and stimuli
whose intensity always remains the same. What, then, do
these experiences teach us ? Evidently nothing else than this :
that one and the same stimulus will be sensed as stronger or
weaker, or not sensed at all, according to the circumstances
under which it operates. But what kinds of change in the circumstances are there, which can produce this alteration in sensation ? On considering the matter closely, we discover that
the change is everywhere of one kind. The tick of the clock is

a weak stimulus for our auditory nerves, which we hear plainly when it is given by itself, but not when it is added to a strong stimulus of rattling wheels and all the other turmoil. The light of the stars is a stimulus for the eye ; but if its stimulation is added to the strong stimulus of daylight, we do not notice it, although we sense it clearly when it is joined to the weak stimulus of twilight. The gramme weight is a stimulus for our skin which we sense when it is united to a present stimulus of equal strength, but which vanishes when it is combined with a stimulus of a thousand times its own intensity.

We can, therefore, lay it down as a general rule that a stimulus, in order to be noticed, may be so much the smaller if the stimulus already present is weak, but must be so much the larger the stronger this pre-existing stimulation is. From this alone we can see, in a general way, how our apprehension of a stimulus depends upon the intensity of it. It is plain that this dependence is not quite so simple as might have been expected beforehand. The simplest relation would evidently be that we should estimate increase of sensation in direct proportion to increase of stimulus-intensity. So that if the sensation 1 should correspond to a stimulus of the intensity 1, sensation 2 would correspond to intensity 2, and sensation 3 to intensity 3, and so on. But if this simplest of all relations prevailed, a stimulus added to a present strong stimulus would occasion as great an increase in sensation as if it were added to a present weak stimulus ; the light of the stars would make as large an addition to the daylight as to the night. This we know not to be the case ; the stars are invisible by day. The increase which they occasion in our sensation is not noticeable, whereas this increase is very considerable indeed in the twilight. So that this much is made out as regards our comparative measurement of sensation-intensities, that they do not increase proportionally to the increase of stimulus, but more slowly. But when we attempt to decide what the relation which obtains actually is, everyday experiences do not suffice. We have need of exact and special measurements.

However, before we apply ourselves to the task of making these measurements, it is necessary that we should be quite clear as to the meaning of the questions which are before us

and the importance of the answers which we may expect to find
to them.   If we increase two stimuli of different intensities,—
e.g., a gramme and a kilogramme,—by the same unit,—*e.g.*, by
the pressure of a gramme,—we come upon the fact that the addition to the smaller weight is quite plainly perceived, whereas
the addition to the larger one is almost or altogether imperceptible.   This fact may be interpreted *a priori* in two ways.
(1) It may be that the addition made to the stronger stimulus
produces absolutely a smaller increase in sensation than the
same addition made to the weaker.   (2) Or it may be that the
sensation-increase is the same in both cases, but that the
stronger stimulus requires a greater increase in sensation than
the weaker, if the differences are to be equally clear in consciousness.   If the first hypothesis is correct, the measurements which
we are to make will have direct reference to the relation between stimulus-increase and the corresponding sensation-increase ; if the second, then the law of which we are in search
will refer only to our apprehension and comparative estimation
of sensations, and not to these themselves.   Now, without these
activities of apprehension and comparison, it is impossible for
us to formulate any judgment whatsoever concerning sensation-
intensities, from which it follows that the results of our measurement of sensation must, in the first instance, be interpreted on
the alternative hypothesis : that all that we can get at directly
is the relation between alteration of stimulus and our apprehension of this alteration.   It was with this in mind that I was
careful to say above, not that a given stimulus-increase produces
a smaller sensation-increase when added to a strong, than when
added to a weak, stimulus, but that in our estimation this increase is smaller.   If the absolute sensation-increase is smaller,
that can only be due to the working of another law,—that of
the parallelism of our estimation of a sensation-increase and its
actual magnitude.   Now, obviously, an answer to the question
of the validity of such a hypothesis as that can only be looked
for at the conclusion of a detailed investigation of the relation
existing between the intensities of stimulus and sensation.   This
is the investigation upon which we are now to embark.   You
will, perhaps, allow me, for the sake of brevity, to speak in what
follows simply of 'sensation,' when I should more correctly say

'apprehension' or 'estimation of sensation.' But I shall do so with the repeated caution that this mode of expression is only provisional, and with the assurance that I shall not fail in a later lecture to enter fully upon the question whether the implicit assumption that our apprehension of alterations in sensation-intensity runs parallel with the alterations themselves is correct, or whether it must ultimately give place to some other.

This being understood, then, the problem immediately before us takes the following shape. We are to determine what increase of sensation corresponds to equal increases of stimulus, or, in other words, to discover what stimulus-increase corresponds to equal increases in sensation.

How to execute these measurements is something which our everyday experiences suggest. A direct measurement of sensation-intensities we saw to be impossible. It is only sensation-differences which we can take account of. Experience showed us what very unequal sensation-differences might correspond to equal differences of stimulus. In most cases we find that the same stimulus-difference would be sensed or not sensed according to circumstances ; that, *e.g.*, a gramme is sensed when added to another gramme, but not when added to a kilogramme. We should think very much less of the statement that a gramme added to a gramme produced a considerable difference, added to a kilogramme a slight difference, in sensation. And the reason is not far to seek. It is difficult to say whether one sensation-difference is just smaller or just larger than another ; but we have generally no hesitation in calling two sensations equal. We are quite sure that the stars are invisible by day ; but we might be in doubt as to whether the full moon is brighter by night than in the day-time. Our inquiry will, therefore, lead to results most quickly, if we start out with some arbitrary stimulus-intensity, observe what sensation it arouses, and then see how long we can increase the stimulus without having the sensation seem to change. If we carry out such observations with stimuli of varying magnitude, we shall certainly be obliged to vary the stimulus-increase which is just capable of producing a difference in sensation. A light, to be just visible in the twilight, need not be nearly so bright as starlight ; it must be far brighter to be just perceptible by day. If now we institute these observations for all possible stimulus-intensities, and note for each intensity the magnitude of the stimulus-increase necessary to produce a just perceptible increase of sensation, we shall get a series of numerical values, in which is definitely and immediately expressed the law according to which sensation alters as stimulus is increased.

Experiments by this method are especially easy to carry out upon the sensations of light, sound, and pressure. We will consider the last of these first, since they are the most simple. The experimenter lays his hand comfortably upon a table. The chosen weight is placed upon it. Then a very small weight is added to this, and the question put whether the observer, who, of course, must not look at his hand during the experiment, notices any difference. If the answer is negative, a somewhat larger weight is taken, and the same procedure is continued until the increment of weight is found, which is just large enough to be sensed clearly. When an experiment has been concluded with one standard weight, a second and third are taken, and so on, until the magnitude of the just necessary increment of weight has been determined for a sufficient number of standards.

We find a surprisingly simple result. The addition to the original weight, which is just enough to produce a noticeable difference in sensation, always stands in the same proportion to it. Suppose, *e.g.*, that we had found that the necessary addition to a gramme was a quarter of a gramme. Then if, instead of grammes, we took pennyweights or ounces or pounds, we should have to add a quarter of a pennyweight to the pennyweight, a quarter of an ounce to the ounce, a quarter of a pound to the pound, in order to obtain a just noticeable difference. Or, if we confine ourselves to grammes, we must add two and a half to ten, twenty-five to a hundred, two hundred and fifty to a kilogramme.

These figures explain the familiar fact that the difference between heavy weights, to be cognisable, must be larger than the difference between light ones. But they also give us the exact formulation of the law which governs the relation of sensation of pressure to force of pressure exerted. You can hold this law in mind by remembering a single number, the number expressing the proportion of the added weight to the standard. Experimental results show that this proportion is, on the average and approximately, that of $1 : 3$. Whatever magnitude of pressure may be exerted upon the skin, we sense its increase or decrease so soon as the amount added to or subtracted from it is one-third of the original.

Experiments of the same kind, but in greater number and with greater accuracy, have been made with lifted weights. Here, of course, the conditions are not so simple. When we lift a weight, we have not only a pressure-sensation in the hand which holds it, but also a sensation in the muscles of the arm which raise hand and weight together. This second sensibility is much finer than that of pressure proper. Indeed, it has been experimentally shown that if lifting is allowed, an addition of merely $\frac{6}{100}$ to the original weight produces a difference in sensation. Our sensibility to weight with lifting is, therefore, some five times as great as our sensibility to weight which simply exerts pressure. And the law of the dependence of sensation upon stimulus may be similarly expressed in terms of the sensation of lifting, the fraction $\frac{1}{8}$ being replaced by $\frac{6}{100}$ or $\frac{1}{17}$. This proportion holds whether the weight is large or small, whether we are speaking of ounces, pounds, or grammes. It tells us that there must be added to a hundred grammes six, to a thousand grammes sixty, to every standard weight $\frac{6}{100}$ of its own amount, if a difference in sensation is to be apprehended.

To determine the objective magnitude of weights, we employ the balance ; to measure accurately the objective intensity of light, we use a *photometer*, or light-measurer. This is in principle an instrument by means of which the brightness of a given light is measured by reference to, and expressed in units of another light of constant brightness. A very simple form of the photometer is that schematically outlined in Fig. I. A vertical rod, s, is fixed in front of a white screen, w. Behind the rod is placed the light n, the intensity of which is regarded as the unit of measurement. Beside n is set the light l, whose

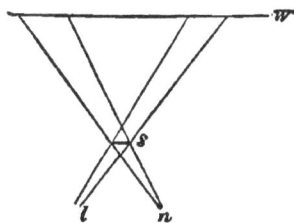

FIG. I.

intensity is to be measured. Both lights throw a shadow on the white screen. Neither shadow is as dark as it would be if there were present simply the one light which produces it ; each is illuminated by the other light, and the greater the intensity of this other light, the brighter will the shadow appear. Suppose that both shadows are equally bright ; that would mean that the intensities of the two lights are equal. But suppose, again, that the shadow cast by the normal light, the brightness unit, is darker than the other ; this means that the intensity of the light which is to be measured is less than unity. We can readily determine how much less by moving the normal light somewhat farther back, for it is a law of optics that the intensity of a light is inversely proportional to the square of the distance of the luminous body. If the light, which was standing at the distance of one metre from the white screen, is moved in a straight line to a distance of ten metres, the intensity of the light falling upon the screen is reduced from 100 to 1 ; at the distance of ten metres it is a hundred times less than it was at the distance of one metre. We can easily institute in this way a quantitative comparison of a light of unknown intensity with a given normal light. We have only to shift the two lights to such distances that the darkness of the shadows cast upon the screen appears to be precisely the same ; then we measure the distance of each light from the screen, and the inverse ratio of the squares of the two distances gives us the relation of the intensities of the lights.

We can turn this same method to good account for the measurement of the dependence of light-sensations upon intensity of light-stimulus. The strong illumination of the shadowless part of the screen and the weak illumination of the shadows both give rise to light-sensations, which are of course the more different the darker the shadows. If we set out with two lights of equal intensities, situated at the same distance behind the rod (say, two exactly similar stearine candles), the two shadows will be of the same intensity ; *i.e.*, they are equally different from the bright background upon which they are thrown. If now we move one candle farther and farther away, the shadow cast by it becomes weaker, and its difference from the illuminated background less, till finally a point is reached at which this difference vanishes altogether. By measuring first the distance of the stationary candle from the screen, and secondly the distance of the candle which has been moved so far back that its shadow has just disappeared, we obtain the data necessary for the formulation of the law of the increase of light-sensation with increasing intensity of light-stimulus. So long as only the stationary candle was there, the total illumination of the screen was due to it. When the other candle is moved up from a distance its light adds something to the whole amount of illumination present. But this increase is at first unnoticeable ; the point where it becomes noticeable is fixed by the appearance of a second shadow of the rod. The place which this shadow comes to occupy is, of course, illuminated by the nearer candle, and not by the more distant one ; and as soon as the latter has approached near enough to produce a noticeable increase in the total illumination the shadow must appear ; it is an index, so to speak, pointing to an increase of illumination. And we now possess, in the inverse ratio of the squares of the distances of the two candles from the screen, the relation of those light-intensities which condition a just noticeable difference of light-sensation. Suppose, *e.g.*, that the first candle was placed at a distance of one metre, and the second (which casts a just noticeable shadow) at a distance of ten metres, then the light-intensities stand to one another as 100 : 1 ; or, in other words, the intensity of the first candle must be increased by one-hundredth, if its increase is to effect an increase of sensation. We have here pursued exactly the same method as in our experiments with weights. There we added to a heavy weight a lighter one, which just noticeably increased the sensation of pressure ; here we add to a strong illumination a weaker one, which just noticeably increases the light-sensation. It only remains to extend these observations to different stimulus-intensities, as was done in the experiments with weights. Just as we varied our normal weights, so must we vary the luminosity of the standard candle by known amounts. That is very easily done. It is only necessary to move the candle backwards or forwards, and to calculate its luminosity from the distance at which it stands from the illuminated screen. Experiments made in this way soon convince us that the distances of the two candles always bear the same relation to one another. If the second candle had to be placed at a distance of ten metres when the first stood at one metre, it must be placed at a distance of ten feet when the latter stands at one foot, at twenty metres or twenty feet when the distance in the other case is two metres or two feet, from which it follows that light-intensities which condition a just noticeable difference of sensation always preserve the same relation to one another. They stand to each other as 1 : 100, as 2 : 200, etc. But this is the law which we discovered in our experiments with weights, and the law can just as well here be expressed by the number defining the relation of the just noticeable increase of illumination to the original illumination. This number is approximately $\frac{1}{100}$; that is, every light-stimulus must be increased by $\frac{1}{100}$, if its increase is to be sensed.

It is not hard to institute similar experiments in the sphere of sound. The intensity of the sound produced by the fall of a body upon some underlying surface increases with the magnitude of its weight and the height of its fall. If we always employ the same body, we can vary the intensity of the sound at will by varying the height of fall. Intensity and height of fall are directly proportional to one another. A fall from twice or three times the standard height produces a sound twice or three times as loud as the normal sound. A good way of turning this principle to account for the investigation of sound-intensities which do not differ very greatly from one another is indicated in the schematic representation of the *sound-pendulum*

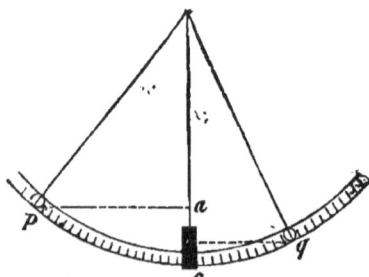

FIG. 2.

given in Fig. 2. We take two ivory balls, p and q, of exactly the same size, and suspended by cords of equal length. Between the balls is placed a block of hard wood, c. If one of the two balls is let fall from any chosen height against the block, the resulting sound is directly proportional to the height of its fall, which can be measured by the angle through which the ball was raised from the position of rest. The angle is read off from a graduated circular scale placed behind the block. The height of fall of the ball *p*, *e.g.*, is the distance *ac*; for the ball *q*, the distance *bc*. That is, the balls strike the block with the velocity which they would have possessed had they fallen vertically from the points *a* and *b*. If *ac* and *bc* are made equal by moving both balls through the same angle, the two sounds are naturally of equal intensity ; but if they are different, the sounds are also of different loudnesses. As we pass by slow degrees from equality to larger and larger differences of height of fall, dropping the balls in quick succession, so that the sounds may be accurately compared, we find that for some time there is no noticeable difference of sound, despite the difference in height of fall. Not until this difference has reached a certain magnitude does the difference of sound begin to be noticeable. At that point the height of fall is measured for both balls. The difference, of course, gives us the amount by which a standard sound-intensity, measured by the total height of fall, must be increased if we are to obtain a just noticeable difference of sensation. Suppose, *e.g.*, that the first ball had fallen through ten centimetres and the second through eleven. That would mean that the standard sound-intensity must be increased by $\frac{1}{10}$ before a difference could be sensed. By making similar measurements over a very large number of heights of fall, we shall learn whether this relation is constant when the sound-intensity is increased or diminished. Just the same is found to hold here as in the case of weights and light-intensities : the relation of stimulus-increment to stimulus-intensity always remains the same. Every sound must be increased by about one-third for the production of a clear increase of sensation.

We have found, then, that all the senses, whose stimuli we can subject to exact measurement, obey a uniform law. However unequal may be the delicacy of their apprehension of sensation-differences, this law is valid for all : that the increase of stimulus necessary to produce an equally noticeable difference of sensation bears a constant ratio to the total stimulus-intensity. The figures which express this ratio in the several sense departments may be shown in tabular form as follows :—

Light-sensation . . . $\frac{1}{100}$
Muscle „ . . . $\frac{1}{17}$
Pressure „ ⎫
Sound „ ⎭ . . . $\frac{1}{3}$

These figures are far from giving as exact a measure as might be desired. But they are at least adapted to convey a general notion of the relative sensibility of the different senses. First of all stands the eye. Next comes muscle ; the muscular sensation affords an accurate measure of the differences of lifted weights. Last, and on an approximate equality, stand the ear and the skin.

This important law, which gives in so simple a form the relation of our apprehension of sensation to the stimulus which occasions it, was discovered by the physiologist Ernst Heinrich Weber, and has been called after him Weber's law. He, however, examined its validity only in special cases. That the law holds for all departments of sense was proved by Gustav Theodor Fechner. Psychology owes to him the first comprehensive investigation of sense, the foundation of an exact theory of sensation.

LECTURE III

§ I

THE question might, with some show of reason, be raised as to whether the law which we have discovered is valid for our quantitative estimation of sensation-magnitudes in general, or whether it possesses only a more limited importance. For all that we have directly ascertained is this : in what proportion the just noticeable sensation-difference stands to the stimulus-increment which conditions it. But, as a matter of fact, it will be easily seen that the determination of this proportion is simply a special case in the determination of a more general relation of dependency.

No one will doubt that it is possible to pass gradually by very small sensation-differences to very large ones. Suppose that we take a sensation which has increased by a just noticeable magnitude, and that we allow this second sensation to increase again by a just noticeable difference ; the difference between the first and third will be clearer than that between the first and the second. And if we proceed in this way, always increasing by a just noticeable increment, we shall finally arrive at a sensation-intensity which is very much greater indeed than that of the sensation from which we set out. And we shall have correspondingly reached a very considerable difference of stimulus-intensity. Had we passed directly from the weak stimulus to the strong, and therefore from the weak to the strong sensation, we should never have been able to gain any exact information as to the dependency of sensation upon stimulus. Taking steps of such length from sensation to sensation, we should not have been able to decide whether the sensation had increased in the same proportion as the stimulus. A result which we could only have attained to with difficulty, if we had tried to alternate between large sensation-differences, comes out of itself if we gradually increase the stimuli in such a way as to pass invariably from one just noticeable sensation-difference to another. By how much any one sensation exceeds any other is just as difficult to determine from their immediate comparison as it would be to say how many more grains of wheat there are in one heap than in a second. If we want to know that, we must just set to work and count every single grain. And, similarly, if we wish to learn how much more intense a second sensation is than a first, our best method will be to analyse the sensations into those elements which are the equivalents of just noticeable differences.

It is true that in following this method we can never compare more than one sensation with another. But if we have once established a sensation-unit, we can easily determine by comparison with it the magnitude of any other sensation whatever. Let us assume that we have adopted, as the unit of cutaneous pressure-sensibility, the sensation occasioned by the pressure of 1 gramme. We have found that the relation in which sensation increases with increase of stimulus is expressed in the case of pressure-sensations by the fraction $\frac{1}{3}$; *i.e.*, the external pressure must increase by $\frac{1}{3}$ of its intensity, if it is to produce a just noticeable increase of the pressure-sensation. We can, therefore, just distinguish $1\frac{1}{3}$ grammes from 1 gramme; while we can only distinguish $2\frac{2}{3}$ from 2, or $3\frac{3}{3}$,—*i.e.*, 4 grammes from 3, etc. Now if we regard all equally noticeable sensation-increments as equal magnitudes, then obviously the magnitude of the just noticeable sensation-increase occasioned by the pressure of 1 gramme is equal to the just noticeable increase of the sensation occasioned, *e.g.*, by a pressure of 10 grammes. So that we may think of any increase of a sensation of whatever intensity as being entirely made up of a number more or less of just noticeable sensation-increments. We may assume that these begin at the point where the external stimulus just suffices to excite a sensation. Now, then, we are in a position to give quantitative expression to sensation-intensity, however great or small this may be. One sensation is twice, three times, or four times as intensive as another, when it is made up of twice, three times or four times as great a number of equal sensation-increments. This system of measurement presupposes that we follow up sensation in its gradual increase. But that is the case, strictly speaking, in all measurement. All the measures which we possess consist of a series of measurement-units. The unit which we have chosen for sensation is the just noticeable increment. If a sensation is made up of four times as many units as another, then it is four times as great as that other ; just as a scale on which four inches are marked is four times as long as one which measures only one inch. If we merely *estimated* the relation of the two scales as regards length, our comparison would perhaps not be very accurate. An exact judgment is only possible by the application to each of the same measurement-unit. And it is precisely similar with sensation.

The method of measuring sensations of various intensities by the addition of just noticeable differences would, however, be very cumbrous. We can plainly reach our end very much more quickly so soon as we have learned the law according to which sensation-increase is correlated with increase of stimulus. Having formulated such a law, we could predict that exactly so great an increase of stimulus would condition so great an increase of sensation.

As a matter of fact, we possess a law of this kind. Weber's law tells us that a stimulus must always increase in a like ratio, if the corresponding increase of sensation is to be equally noticeable. So that, for practical purposes, any question of sensation-measurement may now be put in the form : by how many units, or by how many equally noticeable magnitudes will, on Weber's law, a given sensation be increased, if we increase the stimulus by a definite number of its units ? Or conversely : how great must a given stimulus be made, in order that the sensation may increase by a definite number of sensation units ? Let us take pressure-sensations once more, for purposes of illustration. You will remember that the sensation occasioned by 1 gramme must be intensified by $\frac{1}{3}$ gramme for it to increase by 1 unit. Suppose now that we wished to learn how much the pressure must be intensified for the sensation to

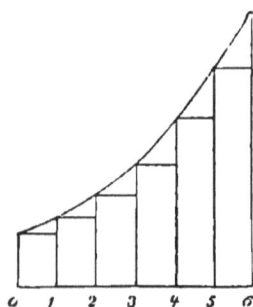

increase by 6 such units. We imagine the sensation-units arranged upon a scale. At the zero-point of this scale, which we will place for the moment arbitrarily at a stimulus of 1 gramme, we draw a perpendicular of any length to represent the gramme. In order now to represent the magnitude of pressure for a sensation increased by 1 unit, we must lengthen the perpendicular at 1 by $\frac{1}{3}$ of the perpendicular at o.

FIG. 3.

Similarly at 2, we must lengthen the perpendicular 1 by $\frac{1}{3}$; at 3, the perpendicular 2 by $\frac{1}{3}$, etc. Since the perpendiculars constantly increase, these incremental parts will also of course become larger; we have to draw upon our scale lines of continually increasing length. And it is plain that the magnitude of each of these lines stands to that of the perpendicular drawn at zero in the same relation in which the weight, occasioning the sensation-increase marked upon the scale, stands to the initial weight of one gramme. The question being, what weight has to be applied to produce a sensation-difference equal to 6 sensation-units, we have only now to measure how much longer the perpendicular at 6 is than the perpendicular at o.

If we connect the upper ends of the perpendiculars drawn upon our sensation-scale to represent stimulus-magnitudes, we obtain a curved line ascending more steeply as we approach the higher values of the scale. This curve obviously shows the dependence of our measurement of sensation-intensities upon the corresponding stimuli, not only for the points 1, 2, 3, etc., but also for all points situated between these, *e.g.*, for $1\frac{1}{4}$, $1\frac{1}{2}$. If we wish to discover what intensity of stimulus corresponds to some particular point lying between two unit values, we need only connect the point in question by a perpendicular with the curve representing the alteration of stimulus. The magnitude of the required stimulus is represented by the length of this perpendicular. The sensation-difference which corresponds to a point on the scale lying in this way between two unit values is, of course, not perceptible by us; but it would be quite wrong to infer from this that it has no existence whatsoever. For we can only reach perceptible differences by heaping up, as it were, a great number of imperceptible differences. It is mere chance that the just noticeable sensation-differences in our illustration fall exactly at the points 1, 2, 3. If we were to take as our initial weight ½ or ¾ gramme instead of 1 gramme, the whole scale would be shifted to the left, and the points where the numerals now stand would then fall between two numerals of this second scale. But the law of the variation of sensation-with stimulus-intensity would remain precisely as before. Our measurement on any scale is discrete, but the scale itself is continuous. We cannot, you see, proceed from one weight to another so as to pass through all possible intermediate weights ; but we interpolate between 2 grammes $\frac{1}{10}$, $\frac{1}{100}$, $\frac{1}{1000}$, or perhaps even $\frac{1}{10000}$ of a gramme, if we wish to be exceedingly accurate in weighing. But no one would maintain that a weight of less than $\frac{1}{10000}$ of a gramme is no weight at all. And just as there are differences of weight, which no balances can detect, so there are differences of sensation, which we are unable to cognise.

Now there can be no doubt that the scale which we have been using to measure sensations, is not one particularly suited to its purpose. We started out from the simplest possible stimulus-magnitude, from the pressure of 1 gramme, our unit of weight. We made the zero-point of our scale correspond to this point, and proceeded to fill in our sensation-units to the right of it. But when we have done this, we have not put ourselves in a position to determine anything more than by how much we must increase the weight of a gramme in order to obtain a definite increase of sensation-units ; or how many sensation-units have been added to the pressure sensation of one gramme, when we are being stimulated by a weight of definitely greater magnitude. We do not know in the least how great the sensation is which is occasioned by 1 gramme ; *i.e.*, how many sensation-units are to be reckoned to the left of the zero point on one scale. The way to determine this is obviously to set out, not from a definite stimulus-unit, but from the unit of sensation ; and to measure onwards in terms of this, from the point where sensation begins. If, then, we wish our scale to be a natural one, we shall take the point at which sensation begins for our zero-point. But this is not at the same time the zero-point of stimulus. Some stimuli are so weak that they are not sensed at all. In order to occasion a sensation, the stimulus must have attained a definite magnitude, which in each case is determined by the character of the sense organ. The case here is similar to that of sensation-differences. These are only perceived if the stimulus-differences are of a certain intensity. In the same way sensations in general are only perceived when the stimulus has attained a certain magnitude. It might, perhaps, be supposed that the two cases are not only similar, but identical; that the intensity of the stimulus necessary to produce a sensation at all is equal to the intensity of stimulus-difference which gives rise to a just noticeable difference of sensation. But it may be easily seen that this is impossible. The intensity of a stimulus-difference is always directly dependent on the total stimulus-intensity, and decreases with decrease of the latter. So that if the stimulus becomes infinitely small, we should be forced to assume that the stimulus-difference must also become infinitely small. That however is contradicted by experience, which shows us that every stimulus must have attained a definite measurable magnitude, if it is to produce a sensation.

If, therefore, we follow our former method, and erect perpendiculars to express the stimuli which correspond to the series of sensations, we must draw at the zero-point a line whose length represents the magnitude of the stimulus which occasions a just noticeable sensation. If we keep to our sensations of pressure, and find that $\frac{1}{50}$ of a gramme is the magnitude of weight sufficient to excite a just noticeable pressure-sensation, we shall represent this weight by a perpendicular at the zero point. At 1, which is removed from o by a just noticeable difference, the vertical representing the stimulus will, in accordance with the dependency of sensation upon stimulus, be $\frac{1}{3}$ longer; *i.e.*, the stimulus whose original magnitude was $\frac{1}{50}$ or $\frac{3}{150}$ will here be $\frac{4}{150}$, etc. In short, we obtain the same relative increase of stimulus and sensation that we had upon our former scale (Fig. 3), the only difference being, that the new vertical at o now stands for $\frac{1}{50}$ of a gramme, and not for 1 gramme.

To answer all the questions that come up in any sense-department, then, two measurements are in general sufficient; first, the measurement of the constant relation in which sensation-intensity varies with variations in the intensity of the stimulus; and secondly, the measurement of the just noticeable sensation. The first measurement enables us to divide up the sensation-scale; by calling in the aid of stimuli we can mark it off into equal parts. The second measurement gives us its zero-point, and thus renders the scale ready for practical use. If we have found in the sphere of pressure-sensations that the constant ratio is $\frac{1}{3}$, and that the just noticeable sensation is produced by $\frac{1}{50}$ gramme, we can dispense with all further measurement, and solve any problem presented to us. Suppose that we wish to know the intensity of a sensation excited by the pressure of 1 gramme. We take our scale, and begin with the zero-point. The pressure at 0 is $\frac{1}{50}$ gramme; the pressure at 1 is $\frac{1}{3}$ greater; the pressure at 2 is $\frac{1}{3}$ greater than it was at 1, etc. We proceed in this way till we come to a pressure of 1 gramme, and then count up how many units of our sensation scale have been employed up to that point. We shall find that we have used nearly 14 units; so that if we press upon the skin first with $\frac{1}{50}$, and then with 1 gramme, we have passed over 14 just noticeable differences. And the nearer we come to 1 gramme, the greater are the pressure-differences to which the just noticeable differences correspond. The first unit corresponds to $\frac{1}{3}$ of the original stimulus, or $\frac{1}{150}$ gramme. If the sensation increased directly as the stimulus, our 14 units would correspond to an increase of $\frac{14}{50}$ or not quite $\frac{1}{3}$ gramme; while, as a matter of fact, they require an increase of pressure of $\frac{49}{50}$, or almost a whole gramme.

§ II

This method of determining the intensity of sensation by proceeding gradually from weak to strong stimuli through just noticeable 'differences would, however, be exceedingly tedious in practice. Direct observation would possess over it the advantage of greater brevity. The question, therefore, suggests itself, whether we cannot discover some shorter method, which would permit us to pass at one step from $\frac{1}{50}$ to 1 gramme, instead of using, as we did above, no less than 14 intermediate stages. This question may be answered in the affirmative, as a somewhat closer consideration of the dependency existing between sensation and stimulus will convince us.

Sensations and stimuli are interdependent magnitudes. Both

are capable of numerical expression. The numerical values which stand for sensations increase with the increase of the numerical values of stimulus. The simplest relation in such a case would plainly be this : that corresponding to the stimuli expressible by the numbers 1, 2, 3, etc., there existed sensations which were also expressible by those numbers. We should then say that sensation-intensity is directly proportional to intensity of stimulus. This simple relation, however, does not hold ; stimuli increase far more rapidly than sensations. Now there are, of course, countless forms of the relations of dependency existing between numerical values, where one numerical series increases faster than the other. If, for instance, we multiply every number by itself, we obtain from the series, 1, 2, 3, 4 . . . another series, 1, 4, 9, 16. . . . The first numbers are known as the square roots of the second ; the latter are called the squares, or second powers, of the first. So that if these two series expressed the relation of stimulus and sensation, we should say the sensation is equal to the square root of the stimulus. A similar numerical series, differing from this only by its more rapid increase, can be obtained by multiplying each number by itself twice or three times, and so obtaining its third or fourth power. If either of these series expressed the rate of stimulus increase, we should say that the sensation is equal to the third or fourth root of the stimulus. But sensation-intensity increases neither as the square root, nor the cube root, nor as any other root of the stimulus-intensity. This is plain from the fact that the stimulus-increments which condition definite increases of sensation-intensity stand in a constant ratio to the total stimulus-magnitude. Since, therefore, the relative stimulus-increments always remain equal, the relative numerical increments in the series of numbers representing the stimuli must also be constant. This is not the case in the series cited. In the series, 1, 4, 9, 16 . . . *e. g.*, the numerical increments are successively 3, 5, 7, and the numbers to which these increments are referable, 1, 4, 9 ; but the ratios $\frac{3}{1}, \frac{5}{4}, \frac{7}{9}$, are not equal. If this case actually corresponded to the sensation-law, we must have obtained the fractions, $\frac{3}{1}, \frac{6}{2}, \frac{12}{4}$, etc., or others which gave a constant result when the division was made. But neither the second nor the third nor any other powers give such a series.

On the other hand, there is another numerical relation of very general application which exactly corresponds to the relation between stimulus and sensation.

If we cast a glance at an ordinary table of logarithms, we notice that the numbers in it are entered in two columns; one contains the ordinary numbers, the other the logarithmic numbers. We see at once that these latter increase more slowly than do the ordinary numbers; just as magnitudes of sensation increase more slowly than magnitudes of stimulus. If the number 1, *e.g.*, stands on the one side, we find 0 on the other, as its logarithm. The logarithm of 10 is 1, of 100 is 2, etc. Here also, then, in the case of numbers and their logarithms, we have two series which increase in very different ways. And if we look more closely, we find that this similarity is more than merely external. The logarithms of 1, 10, 100, 1,000, are 0, 1, 2, 3. What is the relation of the increase of those numbers to their magnitude? When 1 is increased to 10, 9 is added; when 10 is increased to 100, 90; when 100 to 1,000, 900. The ratios of this increase are, therefore, $\frac{9}{1}$, $\frac{90}{10}$, $\frac{900}{100}$. But these ratios are all equal, *i.e.*, all equal to 9. Now this is an expression of the law which regulates the increase of sensation. Sensations increase by equal magnitudes, when the increase of stimuli is such that each increment stands in a constant relation to the particular total stimulus-magnitude; and the logarithms increase by equal magnitudes, when the increase of their numbers is such that each increment stands always in the same ratio to the corresponding numerical magnitude. So that we can say that sensations increase as logarithms when stimuli increase as their numbers; or, still more shortly—since we may express any stimulus-magnitude by some definite number—*sensation increases as the logarithm of stimulus.*

Logarithmic tables were naturally in use long before psychology felt the necessity of them. Indeed, the expression of the dependency of sensation upon stimulus is merely that of a very simple relation, of frequent occurrence in the expression of the dependency of magnitudes in general. The logarithms 0, 1, 2, 3, *e.g.*, differ each from its neighbour by the same amount, 1; while the corresponding numbers 1, 10, 100, 1,000, differ from one another by the same multiple: *i.e.*, by ten times their value in each instance. But if this were the only rule we possessed for finding logarithms, the process would be exceedingly tedious. The matter is happily very much simpler. If we raise a number to all its possible powers, we get from it, of course, other numbers. Thus $10^1 = 10$; $10^2 = 100$; $10^3 = 1,000$. It is clear that by thus raising the powers of a single number we can obtain any number whatsoever. For if we take the $1\frac{1}{4}$, $1\frac{1}{3}$, $1\frac{1}{2}$ powers of 10, they give us numbers lying between 10 and 100; the powers $2\frac{1}{4}$, $2\frac{1}{3}$, $2\frac{1}{2}$, give numbers between 100 and 1,000. And if we take all the possible fractional powers, we shall obtain all the possible numbers between 10 and 100, between 100 and 1,000, etc. In order to obtain also the numbers which are smaller than 10, we must not multiply the number 10, but divide it so many times by itself. We must raise it, as the mathematicians say, to negative powers. Thus $10^{-1} = \frac{1}{10}$; $10^{-2} = \frac{1}{100}$, etc. But between 10^1 and 10^{-1} stands 10^0 or 10^{1-1}: *i.e.*, 1. If we take as well the intermediate fractions of these negative powers, there result all the possible fractional numbers; while between the powers 0 and 1 come all the numbers between 1 and 10. We have, therefore, obtained every possible number simply by raising the single number 10 to all its powers. Now, if we compare the powers 0, 1, 2, 3, with the corresponding numbers 1, 10, 100, 1,000, we see that the latter stand to one another in the same ratio as the logarithms to their numbers. The former increase by equal increments, when the numbers resulting from the involution increase by equal multiples. The indices of the powers are therefore nothing but the logarithms of the numbers which we obtain by the process of involution. And we can now formulate the sensation-law as follows : sensations stand to their stimuli as the indices to the numbers arising from involution.

§ III

But now a certain doubt may arise with regard to this paralleling of indices and logarithms with sensations. There are negative indices, as we have seen ; and, consequently, negative logarithms. If we divide the number 10 by itself once, twice, three times, and four times, we obtain the powers 0, -1, -2, -3, or the logarithms 0, -1, -2, -3. The number of these negative logarithms is just as unlimited as the number of the positive. This will be perfectly intelligible when we remember that the negative powers and logarithms signify fractions. If we continue the series 10^{-1}, 10^{-2}, 10^{-3}, or $\frac{1}{10}$, $\frac{1}{100}$, $\frac{1}{1000}$, we reach successively smaller and smaller fractions. Just as the series of whole numbers only terminates at infinity, so with the series of fractional numbers. If, then, we wish to reach zero by the method which we have described, it will be necessary to divide 10 by itself an infinite number of times. Thus the logarithm corresponding to zero is negative, and infinitely large. But is all this applicable to sensations? Are sensations ever negative? And can there be sensations which, besides being negative, are also infinite? .

When we speak of negative sensations, we ordinarily understand by the term sensations which are opposite in direction to other sensations which we call positive. Cold, *e.g.*, is a negative sensation as opposed to hot. But it would be equally correct to call cold positive, and thus to make hot a negative sensation. The terms 'positive' and 'negative' are, here as elsewhere, the expression of an opposition. The negative is by no means nothing : it is just as much a real magnitude as the positive ; and the terms we apply are in themselves arbitrary. A shopkeeper reckoning up his effects, counts everything which he has in the till, or that others owe him, as positive ; his own debts he regards as negative. If, on the other hand, he is estimating his debts, he considers them as positive, and the contents of the till and his loans as negative. The result is the same in both cases. Or if a geometrician wishes to distinguish directions in space, he names that direction negative which he does not name positive ; which becomes which is quite immaterial. Just in the same way we characterise the logarithms of fractions as negative because we have already used the positive denomination for the logarithms of whole numbers. We must guard ourselves against supposing that we have here anything more than a mere convention, even though this convention is the most natural and obvious.

The question arises then whether we may not speak of negative sensations, using the word in the above sense of simple opposition. No one will hesitate to answer this question in the affirmative, if it can be once shown that such an opposition exists among sensations. It is of course unnecessary to say that oppositions like that of hot and cold do not concern us in the present instance. Hot and cold are differences of sensation-quality, about the nature of which we have here as little to inquire as about the differences between agreeable and disagreeable, pleasant and unpleasant. It is true that these attributes are predicated of sensations of opposite character. And if we were subjecting these to a special investigation, we might not only justifiably, but very naturally, express the antitheses of hot and cold, pleasurable and painful, by positive and negative magnitudes. But our business in this first instance is only with the intensity of sensation ; and all other sensation-properties are, therefore, excluded from our consideration.

We found the natural zero-point of our scale to be the point where sensation begins, where we first sense at all. Can there be sensations which are not sensed ; or does the putting of that question involve a contradiction of terms ?

There certainly is a contradiction. But it is only an apparent one, due to an equivocal use of the word 'sense.' We have already seen that there exist sensation-differences which are not sensed (p. 22). It is obvious that two different meanings have been given to the word. In its first signification the sensation is simply something which depends upon an alteration of stimulus, no matter whether we detect this alteration or not. But, secondly, it is our discovery of such alteration, which is denoted by sensation. And this is equally true for sensations taken absolutely. In speaking of sensations which are too weak to be sensed we are regarding them as something independent of our apprehension of them ; we are considering them merely as conditioned by external stimuli. We can put the matter in this way. A sensation-difference is not at all identical with a sensed difference ; the latter implies a definite intensity of the former. And a sensation may exist long before it can be sensed. We only sense it when it reaches a definite intensity. But though in this statement we recognise the equivocation, we have not done away with it. The equivocation is explained by the fact that when the word first appeared in language the naïve consciousness which produced it knew only those sensations and sensation-differences which it was itself able to recognise as such. Not till scientific reflection had arisen was the human mind forced to the conclusion that there must be sensations and sensation-differences which it was inadequate to recognise for the reason that sensations neither arise nor alter abruptly, but only through continuous gradations.

So that there is nothing left for us but to use the word ' sensation ' here and in what follows to express all those sensations and sensation-differences which we do not perceive, but whose existence we must assume to explain those which we do perceive, as well as sensations in the narrower sense of processes which we are able clearly to apprehend. Where it becomes necessary to make a distinction we will call sensations and sensation-differences of the latter class ' noticeable,' and of the former ' unnoticeable.' Now, since we observe that a sensation must have attained a certain magnitude if it is to become noticeable, and that, other things being equal, it gains in intensity the greater its magnitude becomes, we are surely justified in taking as the zero-point of our sensation-scale the point where sensation becomes just noticeable. That settled, we shall naturally call the noticeable sensations, to the right of that point, positive ; the unnoticeable sensations, to the left of it, negative. For noticeable and unnoticeable denote a direct antithesis, as valid as that of cold and hot, or of opposing directions in space.

We conclude, therefore, that our comparison of the relation in which sensation stands to stimulus with the relation of logarithms to their numbers holds with regard to this further point of the opposition between positive and negative. And we can now produce our scale beyond the zero-point in a negative direction until the stimulus vanishes, as has been done in Fig. 4. And now at length we have our sensation-law in its most general form. How many units must we enter on the negative side to the left of o before we reach the zero-point of the stimulus ? The stimulus zero-point in this connection is not, of course, the external process of movement

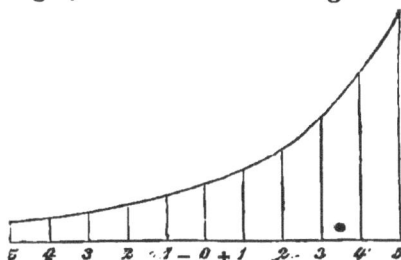

FIG. 4.

affecting our sense-organs, and which has just attained the lower limit of efficiency, but the internal stimulus in the brain resulting from the former, and paralleled as physical process with the mental process of sensation. For it may be assumed that there are external stimuli too weak to reach the brain, whether because of their inability to affect the organ of sense, or because they cannot be conducted from it to the brain. This assumed, where will the line which expresses the increase of stimulus with increase of sensation cut the sensation-scale? We can obviously extend our negative sensation-units to infinity without arriving at that point; for if we suppose, *e.g.*, that the stimulus decreases by $\frac{1}{3}$ of its magnitude at each division of the scale, it yet decreases more and more slowly; and though at last it becomes exceedingly small, it does not disappear so long as the negative sensation-units which we are positing are expressible in numbers. Only when these numbers become infinite may we assume that the corresponding stimulus-magnitudes are also infinitely small, *i.e.*, so small that we may without hesitation regard them as zero. Once more, then, we have the same relation as that of logarithms to their numbers. If we extend further and further the fractional series $\frac{1}{10}$, $\frac{1}{100}$, $\frac{1}{1000}$, we do not come upon any fraction, however small, which is not greater than o. We should only reach o at infinity; and, therefore, the negative logarithm corresponding to it is infinitely large. In the same way, we may conceive of a stimulus as divided and subdivided as long as we please, and nevertheless the smallest particle of it would still be a stimulus. The stimulus only becomes equal to zero at infinity, and the negative sensation corresponding to a stimulus equal to zero must, therefore, be infinitely great; and since a negative sensation means the same thing as an unnoticeable sensation, an infinitely great negative sensation will simply be that sensation which is less noticeable than any other, just as it may be asserted of o and ∞ that the first is smaller and the latter larger than any other number.

Our analogy between the logarithmic law and the law of sensation is now incomplete in one point only. We saw that all possible numbers can be obtained by raising a single number to all its possible powers. The positive powers give us the whole numbers; the negative, the fractions; and the zero power gives us unity. All these facts we have found to possess a definite significance in the case of sensation. But we have left one point still undetermined; that is the number whose involution gives us all the other numbers that are possible. In the instance which we took, we raised the number 10 to the powers 0, 1, 2, 3, and obtained the series 1, 10, 100, 1,000. Had we taken some other number than 10 and raised it to those powers, we should have obtained a different series. It is important, therefore, to know what number it is which has been chosen as the base by whose involution the other numbers are expressed.

It is obvious that this must also be an important question for the sensation-law, since sensations stand to stimuli as their indices to the numbers obtained by involution; and it is evident that we can only say what stimulus-magnitudes correspond to the sensations 1, 2, 3, if we know what definite number was taken as the base in this case of involution. Our choice of that number is entirely arbitrary. For our sensation-scale it is immaterial; it conditions only the divisions of the scale. We shall plainly have the most convenient division if it is so carried out that magnitudes of sensation may be calculated directly from magnitudes of stimulus, and *vice versâ*. But this is possible only when sensation is the simple logarithm of stimulus, and not some multiple or fraction of this logarithm; and this depends entirely on the absolute magnitude of our unit of stimulus and our unit of sensation. Both of these magnitudes may be arbitrarily chosen when we have once made it clear to ourselves what they mean. We have already seen that the stimulus must be taken as equal to 1 where the sensation is equal to 0— *i.e.*, is just noticeable, for $1°$, $10°$, $100°$, are all equal to 1; or, in other words, the logarithm of 1 is always 0. That determines once for all the magnitude of the stimulus-unit. Now, if the sensation 1 is also to come at the point where its stimulus is the number corresponding to the logarithm 1, we must mark it (10 being, *e.g.*, the base employed) at the point where the stimulus has attained the magnitude 10. Had 100 been the basal number, we must have placed 1 where the stimulus had the magnitude 100, and so on. For $10^1 = 10$, $100^1 = 100$, and every number raised to the first power is equal to itself. Further, if we mark in more of our sensation-units, the divisions 2, 3, 4, take their necessary places where the stimulus-magnitudes are 100, 1,000, 10,000, etc. For $10^2 = 100$; $10^3 = 1,000$; $10^4 = 10,000$. This is all required by our law, as we have seen, if the stimulus 10 corresponds to the sensation 1 ; so that now we have also determined our sensation-unit. It is equal to the number which we have chosen as base. Under these conditions, when the stimulus is represented by the number obtained by involution, the sensation corresponds to the index ; or the sensation is equal to the logarithm of the stimulus.

In our ordinary logarithmic tables 10 is the base by whose involution all the numbers are obtained. So that, if we wish to calculate sensations from stimuli, we have only to call that sensation 1 which is occasioned by a stimulus-magnitude ten times as great as that which lies at the limit of noticeability. Having done this, it is only necessary, when a particular stimulus-intensity is given, to look up in the logarithmic tables the number which expresses that intensity ; the logarithm in the next column gives at once the magnitude of sensation. To return to our previous example,—if a weight of $\frac{1}{50}$ gramme produces a just noticeable sensation, we call $\frac{1}{50}$ gramme stimulus 1. Pressure by ten times this stimulus,—*i.e.*, by $\frac{1}{5}$ gramme,—we call sensation 1. Now it is easy to determine at what weight · the sensation is any whole or fractional number of times greater, or by how much the weight must be increased in order to condition a particular increase of sensation. If we wish to get a sensation $2\frac{1}{2}$ times as intense as sensation 1, we refer to our table, and find for the logarithm 2·5 the number 316. That means 316 stimulus-units, or $\frac{316}{50} = 6\cdot3$ grammes. Or if the problem is to determine how great the sensation is which is occasioned by a stimulus of 5,000 units (100 grammes), we look up the number 5,000, and find its logarithm, 3·698. That is, a pressure of 100 grammes produces a sensation which is 3·698 times as great in intensity as the sensation arising from the pressure of $\frac{1}{5}$ of a gramme.

We have now completely answered the question which was before us. Not only have we discovered the law of the dependency of sensation upon stimulus, but we have indicated the method by which the intensity either of sensation or of stimulus can be calculated when the intensity of its correlate is given. This method is simplicity itself, for it presupposes no more knowledge than that of the multiplication-table and no more apparatus than a book of logarithms.

LECTURE IV

§ I

FOR the solution of all the problems which may arise in any definite sense-department, there are required, as we have seen, two kinds of measurement. First, we must know the constant relation in which alteration of sensation-intensity stands to alteration of the intensity of stimulus ; and, secondly, the magnitude of the just noticeable sensation must be determined. The first of these measurements we have carried out ; the second now remains to be performed.

Pressure-sensations afford us the simplest conditions for our investigations. We lay upon that portion of the skin whose sensibility is to be tested small weights, preferably of cork or pith, and seek to ascertain what magnitude of weight is necessary for the production of a just noticeable sensation. Observations made in this way have shown that the sensibility of the skin at different parts of its surface is very far from being uniform. The most sensitive portions are the forehead, temples, eyelids, the outer surface of the fore-arm, and the back of the hand. We can usually sense on these parts weights of only $\frac{1}{500}$ gramme. Less sensitive are the inside of the fore-arm, the cheeks, and the nose, and very much less sensitive than these the palm of the hand, the abdomen, and the thigh. Here the sensibility sinks to about $\frac{1}{20}$ gramme. On some specially protected parts,—e.g., the nails and the heel,—the just noticeable weight rises as high as a whole gramme.

Far more adequate for the apprehension of weak stimuli is our organ of hearing. A mere touch of the external auditory meatus or any contact with the tympanic membrane excites, as we all know, a fairly strong sound-sensation. And even a distant sound must be very weak indeed to be imperceptible. In making observations for the determination of the limit of auditory sensibility, we must, of course, never forget to take into account all the conditions upon which the intensity is dependent. If, *e.g.*, we measure the sensibility of an ear by the sound produced by a falling weight, we must know, not only the magnitude and material of the weight, but also the material of the body upon which it falls. And we must, further, determine the rapidity of its fall and the distance of our ear from the place where the sound is produced. It has been discovered that a normally sensitive ear can just sense the sound made by a cork pellet, weighing 1 milligramme, in falling through a height of 1 millimetre, at a distance of 91 millimetres. That we may expect to find considerable difference in different individuals is a matter of course, justified by our everyday experience. Diseases of the sense-organ affect our hearing ; and, in addition to this, as old age draws on, the acuteness of this sense usually declines, passing through the most various stages from hardness of hearing to complete deafness—one of the commonest defects of sense.

If we are to use the sound-magnitude which we have just determined as a unit of stimulus, we must be able to compare with it the intensity of all other sounds which are employed as stimuli. The comparison is not difficult. Given the sound whose intensity is to be measured, we need only to remove it to the distance at which it just disappears. It is then precisely as great as the sound made by a cork weighing 1 milligramme, falling through a height of 1 millimetre upon a sheet of glass, at a distance of 91 millimetres from the ear. That distance tells us at once how many times greater the given sound at the place of its production is than the just noticeable sound-intensity. An ordinary musket-shot is just audible at a distance of 7,000 metres. This distance is rather more than 70,000 times as great as the distance 91 millimetres. Since the intensity of the sound decreases as the square of the distance, it follows that the sound-intensity of the musket-shot is more than 4,900,000,000 times that of the cork pellet which we adopted as our unit.   A similar comparison with the unit may be carried out for any other sound.   We could easily determine, *e.g.*, how many sensation-units are comprehended in a definite stroke of the sound-pendulum (p. 30).   And since we can easily take the next step, and compare the various sound-intensities with one another, it is perfectly possible to express intensities of sound by means of a single scale like that which we used for weights.   There is only one condition which we must be careful not to disregard : we must never make an observation while other noises are affecting the ear, or while the movement of the air renders the propagation of sound irregular. The quiet of night is therefore especially suitable for the experimental measurement of sound-sensations.

The conditions are different when we are dealing with the sense of sight.   It is obvious that we can only attempt to determine the just noticeable sensation, if there is possible for the sense-organ a state of absolute inactivity, during which there is no sensation whatever.   This condition is realisable for the ear.   We clearly distinguish noise from silence, as a state of things where auditory sensations are wanting.   The corresponding distinction for the eye would be that between dark and bright.   But visual darkness is something quite different from auditory silence.   By greatly diminishing the intensity of light we may obtain darkness without there necessarily being any actual disappearance of the external light.   Or if we close our eyes we are also in darkness ; but it does not follow that we are completely destitute of light-sensations.   In nearly every case, a certain amount of external light penetrates to the closed eye.   And not only that, but the closing of the eye is the cause of a light-sensation, the pressure on the eyeball serving as a retinal stimulus.   You may easily convince yourselves of this by making the pressure somewhat stronger ; the weak shimmer which you still see, though you have closed your eyes, is thereby intensified, till finally the whole of the darkened field of vision is flooded by a sea of light.

But even in the absence of this mechanical stimulus, and even in the darkest night, our eyes are never free from light-stimulation.   With a little attention we can see that the darkness deepens and lifts, gives place here and there to a brighter twilight, which is in its turn followed by a still denser blackness. We can even persuade ourselves at times that we recognise the blurred outlines of external objects ; now and again a brilliant flash of lightning seems to irradiate the shadows. So that the eye is always active, however complete the darkness, and we may easily find ourselves doubting whether it is due to a light from our eye or from the night itself that we are able to see. But we may readily convince ourselves that it is no external light which we have to thank for these phenomena of light and darkness. If we move, they accompany us ; they correspond to no external object ; they persist though we have assured ourselves with all possible care of our complete isolation from external light. But more, not only this changing shimmer which we observe in the dark, but even the deepest black that we can see, is always a light-sensation. When we close our eyes, our darkened field of vision possesses the same form as the bright field of the open eye. All that lies within the limit of this field we see black ; whatever lies outside of it we see not black, but not at all. When it is daylight, and our eyes are open, we do not say that the objects behind our back appear black to us. So that the blackest black which we can see is our weakest sensation of light. To sense this is not the same as to have no sensation whatever. And it follows that there are degrees of darkness, as of light ; that there are differences of blackness, that we may pass gradually from the deepest black to a brighter, from that to grey, and so finally to white.

We see then that the view of the ancients, that the eye is itself the source of light, is not without a certain foundation. Only we can never see and recognise external objects by means of this light. The light-sensation which we have in the dark is caused by a stimulus within the eye. But if we are to see objects, the light-stimulus must proceed from them. That there should be a continual excitation of a sense-organ is certainly a peculiar state of things, probably not occurring elsewhere ; but it becomes intelligible when we remember that the eye is by far the most sensitive of the sense-organs. A stimulus which is not nearly strong enough to occasion a sensation of hearing or of pressure is considerably more than just noticeable for the eye. In this latter case the normal physiological conditions of the organ may very probably furnish the occasion for a sensation ; the chemical processes which constitute nutrition may possibly serve to stimulate the neural epithelium of the eye. Less constant stimulation is caused by the pressure exerted upon the eyeball by the muscles which move it. This stimulus will always be operative even in rest, since the muscles are never entirely relaxed ; but it increases in intensity during movement. We can observe the same phenomenon in the light-sensations which we have in the dark. They, too, become more intensive when the eye is moved.

It is now self-evident that the conditions of vision prevent us from measuring the magnitude of the stimulus which corresponds to a just noticeable light-sensation. The eye always has a sensation which is more than just noticeable, and all stimuli which affect us can, therefore, simply increase the intra-ocular light-sensation which is inevitably present. It only remains for us in this case to determine the least light-intensity, which is in absolute darkness just noticeably brighter than the black of the field of vision. We can most easily obtain very weak light-intensities of this kind by passing a constant current through a metal wire. As we increase the intensity of the current, the wire becomes hotter and hotter, till at a definite temperature it begins to be luminous. And since we can graduate the strength of a galvanic current at our pleasure, the intensity at which the luminosity of the wire becomes just noticeable can be readily determined. We have then only to compare its objective value with that of other known light-intensities. It has been found in this way that the just noticeable intensity of light is approximately $\frac{1}{300}$ of the light of the full moon reflected from white paper.

The investigations to which we have referred furnish us approximately with our units of sensation and stimulus for pressure, sound, and light, though in the latter case with the limitation rendered necessary by the existence of the intra-ocular light. No successful attempt has yet been made to determine these units for the other sense-impressions,—for taste, smell, and temperature. This is partly due to the fact that we are not able to control the operation of stimuli in these departments with sufficient accuracy; and it is in part caused by the general impossibility of putting the organ into a condition of total freedom from stimulation, a condition,—that is, which would correspond to the zero-point of our stimulus-scale.

Now that we have determined in this way the just noticeable stimulus-difference and the just noticeable stimulus-magnitude, the two magnitudes upon which our measurement of sensation depends, there arises a further question : do these two magnitudes stand in any definite relation to each other ? If our sensibility to stimulus shows a certain variability, will not also our sensibility to stimulus-difference be variable ? We saw that this latter is expressible by certain constant fractions ; that, *e.g.*, our sensibility to differences of pressure is $\frac{1}{3}$, to differences of light is $\frac{1}{100}$: in other words, that a pressure must be increased by $\frac{1}{3}$ of its magnitude, a light by $\frac{1}{100}$ of its intensity, if the difference is to become noticeable. Are these relations really constant, as we have asserted ; or is it not rather highly probable that they vary with variations in sensibility?

Obvious as it may appear to answer the latter question in the affirmative, more careful reflection will at once convince us that the opposite is to be expected if the general law of the dependence of sensation upon stimulus holds. This law informs us, you remember, that a stimulus, whether great or small, must always increase in the same ratio, in order to condition a definite sensation-difference. Suppose, therefore, that the sensibility of some sense were, in an exceptional case, reduced by one-half. It would then, of course, be necessary to take twice as great a stimulus as before in order to occasion a noticeable sensation ; and if we wished to increase this sensation again by a noticeable magnitude, the larger stimulus would naturally need, as the law says, a relatively greater increase. But there is not the least reason for supposing that this increase must be greater than the proportion originally required.

This hypothesis is completely confirmed at every point by observation. If sensibility has changed, every stimulus is sensed more or less intensely than before ; but if two stimuli are compared, their difference is just as great in sensation as it was previous to the change. If the sensation 1 is doubled, the sensation 2 is also doubled. If a stimulus 1 had to be increased by $\frac{1}{3}$ to alter sensation, then when, on account of the decrease of sensibility, the stimulus 2 must be substituted for it in order to produce the same sensation, this latter stimulus must be increased by $\frac{2}{3}$, if sensation is to be altered, etc.   In short, sensibility to stimulation does not affect in any way the law of dependency of sensation upon stimulus.

§ II

We may now return to the point from which we set out.   Our object was to investigate the dependency of sensation upon stimulus.   Stimulus, as the physical process directly parallel to the sensation, means here of course the internal stimulus, operative in some sensory centre of the brain.   But, to make our problem easier, we began by investigating the dependency of sensation upon external stimulus.   The time has now come for raising the question whether it is at all probable that the translation of external into internal stimulus has in any way influenced the connections which we have found.   We have, in fact, already seen that stimulation-processes are only set up in the sense-organs and nerves when the external stimulus has attained to a certain intensity ; and since it is not until this process reaches the brain that it is immediately accompanied by sensation, a stimulus which is weaker than this is naturally the same to us as no stimulus at all.   On the other hand, it is equally conceivable that the *internal* stimulation-process must have reached a definite intensity before it gave rise to a noticeable sensation.

As a matter of fact, there can be no doubt that both these conditions are realised.   It is a necessary consequence of the more or less protected position of the sensory nerves and their peripheral end-organs that exceedingly weak stimuli cannot affect them.   And it is just as certain that the stimulation-process in the brain is only perceived by us at a certain intensity.   This is sufficiently obvious if we consider the causes which condition change of sensibility.   If we direct our attention to the impressions of any sense-organ, we can apprehend much weaker stimuli than is the case when our attention is first aroused through the force of the impressions themselves.   But it is not probable that the conditions of conduction to the brain have altered in the two cases. We are always subject to a large number of external impressions, but only a few of them are perceived by us. Nevertheless it is true that these impressions, acting jointly, are capable, unless they are very weak, not only of exciting the sensory nerves, but of passing along them to the central organ.

/Now, just as there is a lower limit, below which the external stimulus is too weak to occasion an internal stimulation, may there not also be an upper limit, above which it is impossible to arouse any stronger neural excitation ?) If this is so, we shall expect to find that the law which is valid for moderate stimuli does not hold in the case of the strongest.

As a matter of fact, it may easily be proved that neural excitations can never be increased beyond a certain point. The preservation of the nerves and their end-organs renders this necessary. If we stimulate the eye with stronger and stronger light, we shall at last injure the power of vision, or, indeed, entirely destroy it. The processes in the sensory nerves depend upon the constant renewal of the substances provided by the blood. The more intensive the sensory processes, the more energetically must the renewal be attended to. And since this cannot go on indefinitely, it is evident that the intensity of the neural processes has also its limit of increase. We do not as a rule reach this limiting point suddenly in the process of stimulation, but rather approach it gradually. At first the neural process increases in intensity in direct proportion to the external stimulus ; later this increase becomes somewhat slower ; finally it ceases altogether, however much we may continue to increase the intensity of the stimulus. We must, therefore, necessarily expect that the relation of the just noticeable sensation-difference to the total magnitude of stimulus is in reality not altogether constant, but slowly varies with the gradual increase of stimulus. If, *e.g.*, a moderate pressure upon the skin must always be increased by $\frac{1}{3}$, a very intense pressure will require a somewhat greater increase ; and finally there will be a certain sensation of pressure an increase of which is absolutely impossible, however heavy the weights we place upon the stimulated part.

Many phenomena of our everyday life are to be explained on this principle. It is well known that extreme pain admits of no degrees or distinctions; that a very intense light blinds us; an excessively loud sound deafens us. But the possibility of sensation-increase does not cease abruptly, but by degrees. If we compare the shadow thrown by an object in moonlight with the shadow cast by the same object in sunlight, it will be at once seen that the former appears much darker than the latter. In a landscape seen by moonlight, this stronger contrast of light and shade makes the illumination far brighter, although it is absolutely much less intense. And from this fact we can distinguish at the first glance whether a picture represents a moonlight or a day-light scene. It is not in the power of the artist to mark this difference by an absolute difference of light-intensity. Both his paintings are equally bright; but he makes the difference between light and shadow greater in the first picture than in the second, and by this single device enables us to distinguish in a moment the night scene from the day scene. This device would be impossible if it were exactly and invariably true, as our law puts it, that an equal sensation-difference always corresponds to an equal difference-relation of light-intensity. For our two land-scapes are a case in point. The moonlight shadows differ from moonlight by a quantity of light which is relatively to the intensity of the moonlight just as great as that separating the sun-light shadows from the sunlight; so that the light-intensity of the moonlight shadows stands to that of the sunlight shadows as moonlight to sunlight. Nevertheless the light of the moon appears much brighter in relation to its shadow; *i.e.*, the sensation-difference is greater here, where the light-stimulus is less, than it is in the case of sunlight, where the light-stimulus is more intense.

All the various influences which condition divergences from the simple law of the dependency of sensation upon stimulus have proved to be due to the intermediary processes of neural excitation. And it is a justifiable assumption that the law is literally valid as between internal stimulus and sensation. So that if we were able to measure directly the intensity of the stimulation-process in the brain, instead of the external stimulus, we should find the law holding without exception. In investigating the relations of sensation to stimulus, we have, as a matter of fact, been observing the effect of two laws: the law of the dependency of internal upon external stimulus, and the law of the dependency of sensation upon the former class of stimuli. If we suppose that the intensity of the internal stimulation remains within certain limits proportional to that of the external stimulus which occasions it, but that, as the external stimulus continues to increase, it increases more and more slowly, we have a simple explanation of the deviations from the law of the logarithmic relation between stimulus and sensation. We have been unable to investigate this relation without the constant intrusion of the nervous excitation,—an intermediary which has unfortunately remained hitherto inaccessible to the method of exact physiological examination. '

And now we approach the final question which is suggested by a consideration of the law of the dependency of sensation-intensities : the question of the psychological interpretation of the facts which we have hitherto ascertained.

§ III

The discovery of a law only becomes of cardinal importance when we have learned to know its connections. The relation between sensation and stimulus is of importance because the knowledge of it allows us for the first time in the history of psychology to apply principles of exact measurement to mental magnitudes. But this measurement will not have its proper value until we have learned in what peculiarities of sensation, or of the organ which transforms the stimulus into a sensation, the law has its basis. Is the relation physically conditioned by processes in the nervous system ? Or is it psychically conditioned by the nature of the mind ? Or, finally, does it express the interconnection of the world without and the world within, which is conditioned by both these factors ? Is it, in a word, to be explained in terms of psychophysics ?

It has been often assumed that our law possesses only a physiological significance. As the stimulus, even in acting upon the external sense-organ, must have reached a certain intensity, if it is to cause an excitation in them, this process of excitation will perhaps meet with increasing obstacles in the sensory nerves, but especially in the central organ. May not now, it is said, these central obstacles increase with increasing stimulation just so that finally, in the sensory centre, where alone the processes of stimulation are sensed, the magnitude of the stimulation is only proportional to the logarithm of the external stimulus?

As yet this conjecture can neither be proven nor contradicted. For we know almost nothing regarding the law of the transmission of stimulation-processes in the brain. Considered more closely, however, it falls into two hypotheses. According to the first of these, it should result from a comparison of the internal stimulation-processes in the sensory centre with the external stimuli, if such a comparison were possible, that the former does not increase in direct proportion to the latter, but more slowly ; and in such a way that the logarithmic law of dependency arises. This view seeks support in the general postulate of the parallelism of mental and physical processes, according to which the bare fact of the logarithmic increase of sensation-intensities requires that the physical processes in the brain, corresponding to the sensations, shall also behave as these latter behave. According to the second hypothesis, this behaviour is the result of the gradual decrease of the intensity of the stimulation-process during its propagation through the central organ. We shall hardly find any facts that can be adduced in favour of this latter assumption. At any rate, what little we do know of the propagation of stimulation in the central organ (*e.g.*, the laws of reflex movement, when the intensity of the stimulation is increased) speaks rather against than for it. Moreover, we cannot regard the hypothesis as probable, even if we grant the applicability of the law of psychophysical parallelism to the present case. To imagine that the processes of sensation, of the apprehension of sensation, and of its comparative measurement depend on their physical side on a simple transmission from a sense-organ to a definite part of the brain is to have a very crude and inadequate idea of that principle. It is surely plain that the different degrees in the clearness and relative noticeability of sensations are secondary sensation-characteristics, which must occupy the most important place in any explanation of Weber's law. And there must be certain physical processes running parallel to these characteristics, if that law is to be physiologically interpreted. But mental processes of so complex a character would necessarily be paralleled by complex physical concomitants,—by the complicated interaction of various central areas. Only by their means could justice be done to the fact, expressed in the law, of the relative decrease of stimulation with absolute increase of stimulus-intensity. However, be this as it may, no unverifiable hypotheses of this kind should prevent us from raising the question whether it is not also possible to discover a psychological interpretation. This would have to take its place alongside of the physiological, just as the physical and mental processes themselves, as we have seen, are parallel and not mutually dependent.

This latter consideration tells equally against the third of the interpretations of Weber's law which we mentioned above,—the psychophysical. According to this view, neither physiological nor psychological conditions suffice for the establishment of the law. It is rather to be regarded as a specific principle of the interaction of physical and psychical, a fundamental law which as such is not capable of any further explanation.

We must urge, in the first place, that it is extremely difficult to conceive of the existence of a law of this kind. It belongs neither to the one territory nor to the other, but only to their borderland ; and it disappears when we leave that on the one side or the other. It seems as though such a hypothesis must inevitably take us back to the spiritualistic doctrines which proved so unfruitful for the explanation of the facts of mental life,—those which regarded bodily and mental existence as two generically distinct modes of being, whose interconnection is merely external. We shall, therefore, decline to adopt this theory of mutual influence, which explains nothing, unless we find that a physiological or a psychological explanation is impossible. But we must remember that it was through the observation of psychological facts, of sensations, that the law was discovered. It is, therefore, only natural that we should ask for a psychological interpretation of it. The physiological interpretation must remain at present a general postulate, because a relation of external to internal stimulus, like that expressed by the law, is as yet only a matter of hypothesis based on the principle of psychophysical parallelism, and can by no means be proven. As a matter of fact, we shall see as we proceed that the view which makes our law a psychical uniformity is supported by many other phenomena of our mental life. A psychological value is assured to it by its actual universality. But more than that, we are also able to show its dependence on definite psychological conditions which are everywhere valid.

What the law tells us, first of all, is simply this : that our sensation furnishes no measure of absolute, but only of relative magnitudes ; or, in other words, that we can only estimate magnitudes by comparison. If an excitation of a pressure-nerve is increased from the intensity 1 to $1\frac{1}{3}$, this is the same as if an excitation of the intensity 2 were increased to $2\frac{2}{3}$. The two differences are equal, if we compare them in ignorance of the absolute intensity of the two excitation-processes. We possess no mental measure whatsoever of absolute mental magnitudes. We are no more able to conceive of an absolute magnitude of sensation than we can have an idea of absolute time-magnitude, or of any other magnitude of a mental nature. It is well known that we continually make mistakes in estimating absolute distances with the unaided eye, without the assistance of instruments of measurement,—whereas the eye is an exceedingly accurate instrument for estimating differences of distance. And the same holds in every case where we are restricted to the means with which nature has provided us : we can only measure relatively ; we can only compare the magnitudes which are directly given us.

Measurement of sensation in general is rendered possible simply and solely by our reference of all qualitatively similar sensations as regards their intensity to an arbitrary sensation-unit. We cannot compare all possible intensities at once with one another, and in this way refer them to the unit which we have chosen. That method of procedure is excluded by the very notion of comparison. In the first place, we can only compare individual things. We can, therefore, never unite more than two units of sensation in a single comparison. We represent to ourselves first one and then the other of these intensities, and so determine which is the stronger sensation. We can only pass to a third sensation, and estimate its intensity by comparing it with one of the two sensations which have been already compared. In this way it becomes possible for us to bring a large number of sensations into one continuous series. We can only do this by proceeding successively from sensation to sensation, from comparison to comparison. But if we can never compare simultaneously more than two, never three or more, magnitudes, the consequence is obvious that our measure of sensation is relative, *i.e.*, is always limited to a determination of the ratio in which two sensations stand to each other. It is no argument against this relativity to urge that we can always proceed to new comparisons, and so measure all possible intensities ; for the series thus obtained is still only composed of individual comparisons. In fine, then, this law of the logarithmic relation of sensation to stimulus is a mathematical expression for a psychological process of universal validity.

But in saying this we have already answered the question which we left undecided when we entered upon our consideration of the measurement of the intensity of sensation. The fact that the more intensive stimulus requires a greater addition for the production of an equally noticeable sensation-increase admits, as we saw, of a twofold interpretation. Either the more intensive sensation demands the operation of a more intensive stimulus for its increase by an equal sensation-magnitude, or the more intensive sensation demands a more intensive sensation-increase, if this latter is to appear equally noticeable. Our reference of Weber's law to the principle of the relativity of sensations favours the second of these interpretations. Our comparison is always relative. In order that a more intensive sensation-magnitude may increase by as much as a lesser sensation, the sensation-increase must be correspondingly greater; and two sensation-increases which lie at different parts of the sensation-scale will be equally noticeable when they stand in equal relations to the stimulation-intensities to which they are added.

§ I

THE intensity of a sensation is only one side of it. Not only has every sensation its own intensity, but also a definite quality which renders it distinguishable from other sensations.

The most extreme instances of qualitative difference are furnished by the sensations of the different sense-organs,—sensations of eye, ear, and skin. A colour and a tone, a sensation of pressure and one of warmth, are simply incomparable with each other. Hence they are denominated disparate sensations. But qualitative differences are also to be observed within the sphere of one and the same sense. Thus red, green, blue, and yellow are entirely different sensations, although they are all sensations of sight. The one thing that proves a closer interconnection among these sensations of a single modality is the possibility of a continuous passage between any two of them, the one passing by slow degrees into the other. Thus red may pass into green, or a low into a high tone. So that the relation between two different sensations within one sense is analogous to that between two points which lie within one and the same spatial continuum ; while two disparate sensations may be compared to points which belong in entirely different spaces, and whose position with regard to each other is unknown.

There is no sense-organ within whose sphere qualitative differences do not occur in greater or less number. Sometimes those differences are very few, as in the case of temperature, where cold and hot appear as the only two sensible qualities. Sometimes the differences are of such a nature that they will not submit to any definite method of classification. The sensations of pressure, *e.g.*, obviously present marked qualitative differences; an ordinary pressure upon the skin is a very different sensation from the prick of a needle-point or the scratch from a rough surface. But plain as these differences are, it is impossible for us to state them in terms of any definite reciprocal relation.

Nor are we in a much better position with regard to the sensations of smell and taste. It is true that certain groups of odoriferous substances, which are for the most part chemically related, give rise to similar scents; *e.g.*, many of the ethereal oils, the volatile fatty acids, the metals, etc. But we are entirely in ignorance of the relation in which these various scents stand to one another.

With taste-sensations we can go one step farther. Here there can be no doubt that the number of existent sensations is more limited, and therefore their investigation easier. If we exclude everything which does not belong to the sense of taste itself, there remain, it seems, only six sharply differentiated sensations : sweet, sour, alkaline, metallic, bitter, and salt. In saying this, we do not mean to assert that these six are the only taste-sensations possible at all. It is clear that, *e.g.*, by combining sweet and bitter, we can produce a taste which is neither sweet nor bitter, although it has something of both qualities. The result is a mixed sensation, not a qualitatively simple one. In ordinary life we are apt to think that we possess a far greater number of taste-sensations. But this is only because we do not commonly distinguish taste from smell. When we are tasting, we are smelling at the same time, and so there arises a combination of smell and taste resulting in a mixed sensation, which is referred solely to taste simply because our attention is principally directed to that sense. How much really depends upon the sense of smell can be readily seen by· recalling a bad cold. In that state we discover with amazement. that many things have absolutely no taste at all. Or, again, the influence of smell can be still more certainly eliminated by filling both nostrils with water. In this experiment we are entirely confined to the sensations of taste proper. And we find that our tongue cognises no more than those six definitely characterised sensations.

This examination of the sensations of taste serves to show us the method to be pursued in a more exact investigation of the quality of sensation in general. In every case our first question must be, whether there are not discoverable certain sensation-qualities which are incomparable with one another, and which are therefore to be regarded as pure and simple. When we have found these, and definitely established their number for some particular modality of sense, we have to ask further : what are the compound or mixed sensations which arise from the simultaneous occurrence of two or more of the simple ones ? That is to say, in investigating any given sensation, we adopt a similar mode of procedure to that which the chemist employs in the investigation of a given body. We must first determine the elements of which the sensation is composed, and then go on to show what relations these elements bear to one another in the combination. Here, just as in the case of the measurement of intensity, we have to set out from definite units of measurement. But then, of course, these were units of quantity ; now we shall be dealing with units of quality. These units are comparable to atoms, from which the sensation is built up. But, as you all know, the term ' atom ' means two different things. For the physicist it is a unit of quantity, for the chemist a unit of quality. So that in splitting up our sensations into quantitative and qualitative units we are analysing these mental states in a way which recalls the two chief directions of analysis of the material world without us.

In the case of the sensations which we have had under consideration up to the present time, this analysis into qualitative units has been either not carried out at all, or only very imperfectly. It is quite otherwise with the two senses whose high degree of functional development has gained for them the title of the 'higher senses': those of sight and hearing.

§ 11

The quality of auditory sensations is given, first of all, with their *pitch*. With this is always connected the *clang*, a peculiar colouring of tone-sensation. *Noise* we distinguish from both, as a sound-impression in which pitch can be perceived either only uncertainly or not at all.

The simplest of these three forms of auditory sensation is the pitch, although in reality this can never be separated from the clang, since it is only in a clang that a particular pitch is perceived. That need not, however, prevent us from disregarding for the time being everything that gives to a tone its peculiar clang-character, and attending simply to that property of it which we call its pitch. Indeed, a psychological analysis of sensations demands that such an abstraction be made, for its business is to continue analysing every sense-impression until it reaches the ultimate elements, which cannot be any further divided. Now the pitch is easily separable from the other elements of a musical clang. It may remain unchanged, while the clang-character of the impression varies. This happens, *e.g.*, when we strike the same tone upon a number of different musical instruments. On the other hand, the pitch may vary to a certain extent, while the clang-character of the impression does not change. This happens when we sound neighbouring tones on the same instrument. When, however, the tonal pitch of the two impressions is very different, the clang-quality generally changes with it, as is easily seen by comparing, *e.g.*, two widely distant tones upon a piano.

It was known to the ancients that tones consist objectively of vibrations of the sounding bodies and of the air carrying the sounds. Indeed, in the case of the very deepest tones, these vibrations are actually perceptible by the eye. In the same way the vibrations of sounding strings can be easily perceived by the eye. The best means of showing the origin of tones from vibrations is afforded by the siren, a physical instrument especially constructed for this purpose. It consists of a disc provided with a series of circular holes, and moving across a current of air in such a way that within any given time the current is

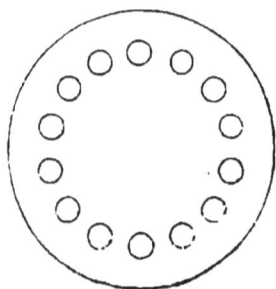

FIG. 5.

interrupted as often as unperforated portions of the disc alternate with perforated. By regulating the velocity of the rotation of the disc we can produce at will high or low tones. The slowest rate of air-vibrations which can give rise to the perception of a tone is about 16 in 1″, though under favourable conditions it may sink to 8. The best means of producing these very deepest tones is given by large tuning-forks or vibrating steel rods. As we approach the limen of perception, however, the tone becomes so faint that, however extensive the vibrations, it can only be heard at a quite short distance. The deepest tones of the musical scale lie between 32 and 100 vibrations in 1″. As the number of vibrations increases the pitch steadily rises. When the vibration-rate has increased to about 40,000, tone ceases altogether, and we hear only a hissing noise.

It is only in the case of the very deepest tones, which cannot be employed for musical purposes, that we are able to distinguish the air-beats corresponding to their vibrations. Our knowledge of the increase of vibration-rate in the case of the higher tones does not, therefore, depend upon the immediate perception of the vibrations, but upon another observation, which is closely related to it. As early as the days of the Pythagoreans it was a familiar fact that a string shortened to half its length vibrates twice as rapidly as the whole string; that one reduced to a third of its original length vibrates three times as rapidly, one reduced to a quarter four times, and so on. Now, the tone of the half-string is the octave of the tone of the whole string; the tone of the third part, the fifth of this octave; that of the fourth part, the double octave. So that this law of the uniform relation of the length of a string to its vibration-rate contains in it another important law, that those relations of tones which are apprehended as harmonious correspond to simple ratios of their vibration-rates.

Harmonious relations of tones were originally distinguished from inharmonious only by the more pleasing character of the impression which they made in a tone-series. Singing and playing in unison came long before part-singing and harmony. But as soon as the custom arose of employing several voices of different register in the rendering of a melody, there came to light other phenomena, connected with the simultaneous sounding of tones, and with the consequent simultaneity of air-vibrations of different velocities. Not only, that is, can we distinguish a single tone from a clang compound of several tones, but we can easily hear out of such a clang, supposing that it is harmonious, the separate single tones which compose it. It is a matter of direct perception, *e.g.*, that the common chord of *c* major is composed of the three tones *c e g*. Whenever a compound clang is harmonious, the separate simultaneous vibrations unite to produce a common movement of the air, itself consisting of very brief and uniformly recurring periods. Fig. 6 shows this for the three compound clangs of a tone plus its octave, its fifth, and its major third. The points at which a new period begins are indicated in each case by dotted vertical lines. In the octave the two vibration-rates which unite to form the compound clang have in each period the ratio 1 : 2 ; in the fifth, the ratio 2 : 3 ;

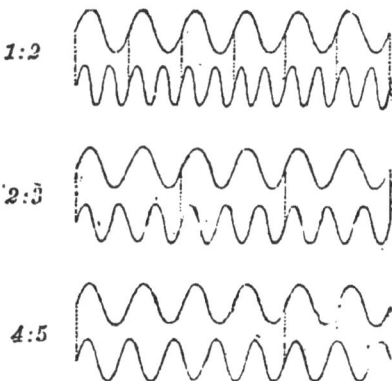

FIG. 6.

in the major third 4 : 5. Similar simple periods are found to recur in the other harmonious two-clangs ; the ratios of the constituent vibration-rates are in the case of the fourth 3 : 4, in the minor third 5 : 6, in the sixth 3 : 5. Since all these periods of compound vibration-rates are repeated just as regularly as the periods of simple vibration-rates, we can understand how it is that a harmonious compound clang produces upon us as uniform an impression as a single tone. It is true that we distinguish in it the presence of more tones than one ; but these unite to form a total sensation, which runs its course with perfect evenness.

But it is quite different when two tones are sounded together whose vibration-rates do not stand in any simple and harmonious ratio, but bear a more complex relation to each other. In such a case there can be no production of the uniform periods, recurring at very brief intervals of time, which we have found in the harmonious compound clangs. As a result of this, the interaction of the vibrations causes a disturbance in the uniform course of sensation. Wherever two movements in the same direction coincide, as at *a* and *b* (Fig. 7), they strengthen each other; and where two coincident movements have an opposite direction, as at *m*, they weaken each other. It

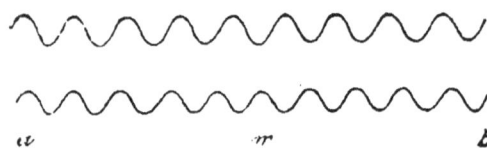

FIG. 7.

depends, of course, on the difference of vibration-rate, how often these pendular to-and-fro movements of the air particles agree or disagree with each other. If one tone makes exactly one vibration more in 1″ than another, there will occur in each second one such increase and decrease. For if at the beginning of the second, at *a*, both vibrations start at the same stage, there will meet in the middle of the second, at *m*, a forward movement of one wave and a backward movement of the other, so that the two cancel one another; while again at the end of the second, at *b*, they will be travelling in the same direction, and will, therefore, assist each other. It is clear that just the same will happen if the difference of the two tones is one of a greater number of vibrations; there will be as many increases and decreases, as many *beats*, as there are more vibrations in the one case than in the other. If the difference is very small, amounting, *e.g.*, to one vibration in the course of several seconds, it will be scarcely noticed, the diminuendo and crescendo of the tone occurring continuously and gradually. If the change is spread over a sufficiently long space of time, it will not be perceived at all. But if one or more beats occur in 1″, they are clearly noticeable; and if their number increases to 10 or more, their quick succession will be sensed as a very unpleasant whirring.

The limit of rapidity at which the beats of dissonant tones may be perceived cannot be determined with any degree of certainty. For, in the first place, the beats, as they follow one another faster and faster, give rise to a general impression of harshness, more or less comparable to that which a rough surface produces in the sense of touch; and when the rapidity becomes still greater, while the tones are not heard as a harmony, the beats and even the roughness of the clang also disappear. The extreme limit at which this harshness can still be distinguished appears to be reached in the neighbourhood of 60 beats in 1″.

Now these observations upon inharmonious compound clangs seem to imply a contradiction between the perception of the beats and the laws which we formulated above in terms of the vibration-rates of tones. For it is found that tones can still give rise to clearly perceptible beats when the difference between their vibration-rates amounts to considerably more than 60 in 1″. If we take, *e.g.*, the two neighbouring tones *c* and *d* from the lower or middle region of the scale of pure temperament, and strike the clangs together, we shall obtain loud beats. This is perfectly intelligible from what has been said above. For if the tone *c* makes 128 vibrations in 1″, *d*, which is higher by the interval of a second, will make 9/8 × 128, or 144 vibrations The two tones must, therefore, give 16 beats in 1″. But if we strike with *c*, not the *d*, but the octave *d′*, we are giving a tone of 2 × 144, or 288 vibrations. Its difference from *c* amounts to 160 vibrations. Yet although it is quite impossible to hear beats which follow one another as quickly as that, the compound clang is not merely inharmonious, but is also clearly accompanied by beats similar to, if not quite so strong as, those arising from the striking of two notes which are a single whole tone apart. What reason is there for the fact that the higher *d′* makes beats with *c*, while the tones *c′* or *g*, octave or fifth of the purely tempered scale, the differences of whose vibration-rates from that of *c* are smaller, give no noticeable beats at all? The reason may be discovered from the following simple experiment.

When we strike a piano or guitar-string that is stretched over the sounding-board, the result is, of course, a tone. If a bridge is placed in the middle of the string, so that only half of it can vibrate, the resulting tone rises, as we have said, an octave higher. By striking first the ground-tone, and then the octave, we come to see that the latter was really contained in the former ; that it sounded, though weakly, along with the ground-tone. The case is the same if first the whole string, and then one-fourth of it, is struck. Here the double octave is seen to be sounded, though very weakly, with the ground-tone, and so on. If we have trained our ear in the comparison of clangs, we are able to hear out these higher tones, the *over-tones*, from the ground-tone. It is found that every tone of our musical instruments and of the human voice contains a large number of over-tones ; so that strictly speaking, we never have the sensation of a simple tone, but always that of several simultaneously sounding tones, one of which, however, the ground-tone, is so much stronger than the rest that we usually fail to hear them. The phenomenon of over-tones finds its physical explanation in the fact that, in most forms of tonal stimulation, the wave-movement set up in the air is a compound one. When the string is struck, *e.g.*, not only does it vibrate in its entire length, and so transmit the ground-tone to the air, but either half of it vibrates also, though not so violently, on its own account, and so produces the octave. In the same way each third and each fourth of the string vibrate ; thus giving rise to the third of the higher octave, the double octave, and so forth in decreasing series. These separate tones persist just as independently of one another as if several instruments were sounding at the same time. The only difference consists in the greater weakness of the over-tones.

And now we are able to explain the very curious fact that the tone *c* beats not merely with the neighbouring *d*, but also with the *d'* of the higher octave. Simultaneously with the ground-tone *c* there is given the octave *c'*, and this beats with the *d'*, which stands next to it. The beats are certainly not so pronounced as if the *c'* had been directly sounded, partly because the over-tone is weaker, partly because the beats follow one another more quickly ; but they are clear enough to be heard.

This simultaneity of ground-tone and over-tones is of importance not merely as throwing light on the consonance and dissonance of the tones in a compound impression, but also as influencing our apprehension of separate tones. The tones of a musical instrument and of the human voice are characterised not only by pitch, but also by a definite clang-character. If all tones depended simply upon the vibration-rate which determines their pitch, then, apart from the noises that may chance to accompany them, every tone of the same pitch would possess the same character, however it might be produced. This, of course, is not the case. The same tone sounds quite differently when given by flute, violin, clarionette, organ, etc. The vibration-rates must have still other properties, varying for one and the same tone with the source of sound. As a matter of fact, we have found the over-tones regularly accompanying tones, and presenting differences which depend upon the mode of origin of clangs. There are tones in which scarcely any over-tones are noticeable. Those of the flute-pipes of an organ come very near to absolute purity, and those of a tuning-fork standing upon its resonance-box nearer still. If the resonance-space is exactly fitted to the primary tone of the tuning-fork, all the secondary tones are so weak in comparison that they are not heard as the tone rings off. On the other hand, wind and string instruments and the human voice always allow a large number of over-tones to be heard beside the ground-tone. As a general rule the intensity of the over-tones decreases with their height. The octave can be heard more clearly than the double octave, this than the third, etc. But there are considerable differences in the different instruments. Sometimes the higher octaves sound most strongly, as on the piano ; sometimes the higher fifths and thirds, as in the clarionette ; sometimes the first over-tones are heard at a comparatively uniform intensity, as on the harmonium ; and sometimes single very high over-tones are preferred, as in the trumpet and trombone.

We have now inquired into all the conditions of the peculiar colouring of the different kinds of clangs. This depends partly upon the intensity of the over-tones in general, partly upon the character of the strongest among them.

§ III

If the laws of concurrent vibrations which we have been discussing are true, no tone is ever entirely free from over-tones. Even though it should be so objectively, there would probably still attach to it subjectively some kind of clang-colour, due to the presence of very weak over-tones caused by the concurrent vibrations of certain parts of the auditory organ, attuned to the particular tone.

The tone-sensation which, in virtue of its attendant partial tones, is possessed of a definite clang-quality, we call a *clang*; the particular clang-quality produced by the over-tones, *clang-colour*. Every clang consists, therefore, of tone-pitch and clang-colour, the latter component being in its turn made up of a number of weaker tone-sensations accompanying the primary tone. The clang, that is, is a compound sensation, and since all tones are in reality clangs, our tone-sensations are never given in any other than a compound form. We can separate out the individual simple pitches only by either subjectively abstracting from the attendant secondary tones in the clang, or strengthening the primary tone to such an extent that they disappear, as happens when a tuning-fork vibrates on its resonance-box. But even if a tone is comparatively rich in secondary tones, we apprehend it in idea as perfectly unitary and relatively simple, as is shown by the fact that we ascribe to it only a single pitch. Over-tones, on the other hand, even though they are strong enough to be clearly perceived, are not apprehended as separate pitches, but appear merely as a peculiar modification of the principal tone. It is plain that this cannot be explained simply in terms of the lesser intensity of the secondary tones. But it becomes intelligible when we consider that wherever definite simple sensations are given in constant connections these connections blend to form unitary ideas; and that when it is a tone that is connected with harmonious secondary tones this blending process must be very materially furthered by the possibility of the co-existence, without mutual disturbance, of harmonious vibration-rates. Regarded from this point of view, the clang-idea presents to us a simple and typical example of a psychological process which we shall frequently meet with, for the most part in a more complex form,—the process of sensation-*fusion*. All the elementary constituents in this fusion-process have lost the character which they possessed in their isolated condition; in the stable connections into which they have entered they are determined by the character of the other elements present. Thus the octave c' of a tone c, when it appears in the first over-tone of the latter, is something entirely different from what it is when sensed alone. In the latter case it would be an independent tone; in the former it is perceived directly in its relation to the simultaneously given principal tone, and, since this is much the stronger, appears as a mere modification of its clang-character.

The compound clang is distinguished from the simple clang only by the number and relative intensity of the tones which enter into it. If we strike the chord *c e g*, we are sounding three tones which form part of the series of over-tones belonging to a lower *C*. The major third *c e* corresponds to the proportion of vibration-rates 4 : 5 ; the fifth *c g*, to 2 : 3 or 4 : 6. That is, the three tones occupy the fourth, fifth, and sixth places in the complete tonal series of a simple clang: 1, 2, 3, 4, 5, 6, . . . But while in a single clang these tones appear only as secondary tones of a lower ground-tone, whose clang-colour they determine, in the second they constitute the chief elements in the whole impression, and are of equal intensity. In the compound clang, therefore, we sense at once a plurality of tones. Since compound clangs are only harmonious if the ratio of their vibration-rates is one of simple whole numbers, they may be regarded in every case as intensifications of the separate members of the tonal series of a single clang.

But there is still another element in compound clangs. It consists in the appearance of lower tones, which are in harmony with the principal tones, and play a part in the determination of the character of a compound clang similar to that of the over-tones in the clang-colour of a single clang.

Whenever there are set up simultaneously two harmonious sound-waves, whose vibrations strengthen and weaken one another alternately, at short and uniformly recurring intervals, there arises from this interaction a new tone, the vibration-rate of which corresponds to the difference between the vibration-rates of the two original tones. Look for a moment at the second pair of curves in Fig. 6, representing the fifth, *c g*. During two vibrations of the first and three of the second tone, there is one coincidence of hill and valley and one each of hill with hill and valley with valley. So there is set up a third wave-movement, which makes one vibration for every two of the first or three of the second tone. Such tones, which may be called either *under-tones* from their relation to the over-tones, or *difference-tones*, from the relation of their vibration-rates to those of the original tones, may be intensified in complex chords by the fact of several of them falling upon the same note. In the chord *c e g*, for example, in which the vibration-rates of the tones stand in the ratio 4 : 5 : 6, both *c* and *e*, and *e* and *g*, produce the same under-tone 1,—a *c* lying two octaves below the lowest tone of the chord,—while *c* and *g* give a tone 2, *i.e.*, a *c* only one octave below it. [1]

To these harmonious under-tones, which we have found accompanying the compound clang, must always be added the over-tones of the single clangs. They may also strengthen one another in certain cases, where different clangs have similar terms in their series of partial tones. So that every chord, even the comparatively simple, is made up of a very large number of sensation-elements, some of which, the more intensive primary tones, stand out as clearly distinguishable qualities, while the others merely serve to determine the clang-character of the chord. The triple clang *c e g*, for instance, gives us the following tones :—

Under-tones.	Principal tones.	Over-tones.							
C_1 C	*c* *e* *g*	c^1	e^1	$\underline{g^1}$	b^1	$\underline{c^2}$	d^2	$\underline{e^2}$	$\underline{g^2}$
1 2	4 5 6	8	10	12	15	16	18	20	24

The first over-tones are usually the strongest ; only these are entered in the schema. Under-tones which appear as difference-tones of more than one two-clang, and over-tones which belong to more than one single clang, are underlined. The difference-

[1] When harmonious tones are simultaneously sounded, we have formed not only difference-tones, but a second kind of resultant tonal wave, depending upon the fact that the hills and valleys of the primary waves are not perfectly coincident. The vibration-rate of these new tone-waves is the sum of the vibration-rates of the original tones. The tones themselves are, therefore, of higher pitch than the principal tones in the chord, and are termed, from their mode of origin, *summation-tones*. Thus the fifth, 2 : 3, has a summation-tone of the vibration-rate $2 + 3 = 5$. Difference- and summation-tones together are sometimes called *combination-tones*. However, the interpretation of the summation-tones is not beyond doubt, many psychologists regarding them as high over-tones of the principal clangs. In any event, they are so weak as to exert no influence upon the clang-character of the chord unless they coincide with over-tones. We may, therefore, leave them here out of account.

tones which the over-tones form with one another, or with the principal tones, are not set down. In most cases they are so weak as to be sensed with difficulty, or not at all. You see that, even in a completely harmonious chord, the over-tones of the second octave stand so near together as to produce very considerable dissonance. Indeed, the most perfectly attuned chords of an instrument whòse clangs are rich in over-tones (organ, harmonium) allow the beats of these over-tones to be clearly perceived. They combine with the quality of the under and over-tones to determine the general character of the different chords.

§ IV

The compound clang arises from the single clang by the strengthening of secondary tones to principal tones. The compound clang in its turn may pass over into the third general sound-quality, into *noise*, as soon as the dissonant elements, which we have found to be not altogether absent even in harmonious chords, multiply to such an extent that harmonious tonal ratios cannot be any longer perceived. You may easily convince yourselves of the origin of noise from the compound clang by striking simultaneously upon some instrument of wide range, such as piano or harmonium, a large number of inharmonious tones. The separate tones make such strong beats with one another that the resultant sensation tends to lose its clang-character altogether.

But when we seek to determine the point at which the clang ends and the noise begins, we find that there is no sharp line of division discoverable. In most noises we can distinguish one or more deep tones, but these are accompanied by a crowd of indistinguishable secondary tones, strong or weak, and of the most different pitch. That is, the difference between clang and noise is only one of degree. Noise and clang alike depend upon the simultaneity of several tone-sensations. Even in the clang, it is impossible to distinguish and identify the greater· part of these tone-sensations ; they merely serve to colour the principal tone in a particular way, and it requires a sensitive ear and close attention, or special experimental aids, to refer the effect to its true cause. And the fact that the clang-colour depends upon the occurrence of over-tones is further obscured by the presence of the principal tone itself. With noise the conditions are directly contrary : it is the mixture of tones which plays the principal part, and the separate tone, in consequence, tends entirely to disappear.

It is, however, probable that this, though the customary, is not the only, mode of origin of noise. There is another, which sometimes co-operates with the first, sometimes appears alone. A vibration-rate the rapidity of which is high enough to transcend the upper limit of tonal sensation is perceived as a hissing noise, while very slow vibration-rates, which do not reach the lower limit of tone, give rise to a roaring noise. It is supposed that these sensations are caused not by the excitation of the cochlear apparatus which is attuned to tone, but by the vibrations of more simple organs, connected with the fibres of the auditory nerve, and situated in the vestibule of the labyrinth of the ear. Since the vestibule belongs to a much earlier stage of development than the cochlea, we might interpret these simple, absolutely toneless sensations as more primitive than clang-sensations, and as constituting the whole series of sound-sensation for most of the lower animals. When once the tone-sensation has been developed, however, these toneless sensations are completely overshadowed, even in perceptions of noise, by the dissonant clang-constituents which enter, as we have seen, into the majority of noises.

§ V

If we abstract from these elementary noise-sensations, which are probably of high importance for the development of auditory sensation in the animal kingdom, but which play so small a part when once audition has been perfected, we may say that all kinds of auditory sensation,—clangs, compound clangs, and noises,—are combinations of simple tone-sensations. The simple tone-sensation itself, however, cannot be analysed into still more simple constituents : like every simple sensation, it possesses only the two attributes of intensity and quality, *i.e.*, intensity and pitch. Pitch, like intensity, can vary only in two opposite directions, up and down ; we can pass from a given tone either to a higher or to a lower one, just as from any given point on a straight line we can proceed only in two directions and keep within the line. That is to say, our whole system of tone-sensations may be considered as a plurality in one dimension, or as a linear plurality.

The analogy between the quality of tone-sensations and the intensity of any particular pitch holds in yet another connection. Intensity varies step for step with variations in the strength of the external stimulus. In the same way, tone-quality follows step for step variations in vibration-rate ; and we are as little able to appreciate any the least alteration in vibration-rate as a change in sensation-quality as we are to notice minute alterations in strength of stimulus as changes of sensation-intensity. In both cases there is a lower limit of discrimination. It is, of course, not possible to determine this limit on ordinary musical instruments with fixed note-values, like the piano, because the tones are separated by far more than just noticeable intervals. But if we take two similarly tuned strings or tuning-forks and gradually alter the pitch of one of them, we have no difficulty in discovering the point at which one tone sounds just noticeably deeper than the other. It is necessary, in making this experiment, to strike the strings or forks successively, and to check the vibrations of one before the other is sounded, since otherwise beats would be produced, and the qualitative difference of the tones known from them, and not from the differences in sensation. On the other hand, when the point at which a sensation-difference becomes noticeable is once found, we may employ the beats produced by the simultaneous sounding of the tones to tell us the objective difference of vibration-rate, corresponding to the bare possibility of their qualitative discrimination in sensation. The number of beats, you will remember, corresponds exactly to the difference between the vibration-rates of the two sounding bodies. For instance, if two similar tuning-forks have been so far removed from unison, in the manner described above, that their successive tones are just distinguishable as different pitches, and if we find that they produce two beats in 10″ when struck simultaneously, we may conclude that at their particular pitch the just noticeable sensation-difference is represented by a difference of 0·2 vibration in 1″. Careful experimentation on these lines has shown that the differences of pitch, which are just noticeable for successive stimuli, remain absolutely constant over a large part of the musical scale. Between the limits of 200 and 1,000 vibrations in the 1″, we can sense a tone-difference represented by 0·2 vibration. For lower tones the fraction is somewhat smaller, for higher tones correspondingly larger; but within the sphere of musical applicability there is no considerable deviation from this average value. Where the tones are high or low enough to approach the limit of sensibility, discrimination becomes, as we should expect, far less certain. You may convince yourselves of this by striking successive notes at either end of the key-board of a piano: differences of a whole half-tone are hardly perceptible.

Let us apply the general conclusions which we arrived at in considering the question of the measurement of sensation-intensity to this particular case. The result of our observations may be summed up in a single sentence. Within wide limits, we have found, *equal differences of tonal quality correspond to equal differences of vibration-rate*; in other words, the sensation of pitch varies in direct proportion to the objective variation of tonal vibration. And there is another path which will lead us to the same point. We possess an especial capacity for the quantitative comparison of more than just noticeable differences of tone-sensation. Suppose that we strike first the two tones *c* and *d* in succession, and then the *d* and *a* of the same octave. Even the most unmusical person, who has no notion of the intervals between the tones in the technical sense, is perfectly certain that *a* and *d* are further apart than *c* and *d*. Here, then, is another experimental method. Two tones at any distance from one another on the musical scale may be given in succession, and the observer required to estimate their exact sensation-mean, the pitch which is just as far removed from the first as it is from the second. It has been found that the tone selected as the mean is always approximately the tone whose vibration-rate lies midway between the vibration-rates of the two extremes.

But there are facts, recurring in the musical experience of all times, which stand in apparent contradiction to these experimental observations. These facts are expressed in the tonal relations of the musical scale. We have seen that the vibration-rate of the octave is invariably twice that of the ground-tone, that of the fifth 3/2 of it, that of the major third 5/4, and so on. The octave of the tone of 32 vibrations in 1″ makes 64 ; the octave of this, 128, etc. That is to say, the higher the pitch, the greater becomes the difference between the vibration-rates constituting any particular interval. Nevertheless, the sensation-difference between a tone and its octave appears to remain the same from whatever region of the scale the interval is taken : the difference in pitch seems unchanged whether we compare the tone of 32 vibrations with that of 64, or the tone of 64 with that of 128.

The law regulating musical intervals, therefore, lays it down not that pitch varies in direct proportion to variation of stimulus, but that it varies more slowly than stimulus. And this law of slower variation is, again, a very simple one. To increase tonal qualities by equal increments, we must increase the rapidity of their vibration-rates by a magnitude which always bears the same relation to the rate to which it is added. To obtain the octave of a given tone, its vibration-rate must always be multiplied by two ; to obtain the fifth, third, and fourth, its original rate multiplied respectively by 3/2, 5/4, 4/3. And this result is precisely parallel to that which we obtained when we were discussing the pressure of weights, the strength of sound and light, in short, the intensity of sensation in general. In all departments of quantitative comparison we found that for the sensation to increase by absolutely equal magnitudes the external stimulus must be increased by relatively equal magnitudes. We have, then, only to write 'tonal quality' for 'sensation-intensity,' and we have our law,—the same law which held for the general relation of stimulus and sensation in the sphere of intensity. If pitch is to increase by absolutely equal magnitudes, vibration-rate must be increased by relatively equal magnitudes ; or, more briefly, *pitch increases in direct proportion to the logarithm of vibration-rate.*

Here, then, we are met by a very curious contradiction. According to the law which regulates the musical scale, the sensation of pitch is dependent upon stimulus in just the same way as sensation-intensity is dependent upon it. . But so soon as we apply the methods which we used to measure intensity to the measurement of quality in this particular case, we find, within certain limits, a direct proportionality between variation of sensation and alteration of vibration-rate.

The contradiction is only apparent. The most obvious way to remove it, perhaps, would be to point out that sensation-intensity and tone-pitch are different things. If Weber's law is proved to hold of one of them, there is not the least reason to assume that it will hold of the other. How were the musical intervals established in the first place? Not, at any rate, with conscious reference to the fact that the same interval presents the same difference for sensation from whatever region of the tone-scale it is taken. We must look for other conditions, conditions of our tonal sensibility in general, which give a definite character to each tone-interval, quite apart from its position upon the whole musical scale. They are not far to seek; think of the over-tones which accompany every simple clang. When a pitch changes by the amount of some particular musical interval, the character given to the clang by its over-tones must change in the same way. Suppose that the change is that of a fifth. The vibration-rates of the principal tones bear the ratio 2 : 3. The clang-character of the lower one is determined by the over-tone series : 4, 6, 8, 10, 12 . . . ; that of the higher by the series : 6, 9, 12, 15. . . . The relations of these two series remain the same whatever the absolute pitch of the principal tones may be.

At the same time, the explanation is not satisfactory. Granted that this constancy of the relations of the secondary tones in every interval gives a reason for the dependency of the musical intervals upon constant relations of their constituent vibration-rates, still the problem is only pushed one step farther back, not solved. If we are to know that an interval is the same when it is given, say, first in the upper and then in the lower region of the scale, we must apprehend in sensation the likeness of the interrelations of all the partial tones in the two cases. But what is true of secondary tones will be true of their primaries. As a matter of fact, we are able to cognise the harmonic intervals of pure tones, which are practically free from over-tones, with almost as much accuracy and certainty as the intervals of clangs whose over-tones are numerous and intensive. That is to say, though our apprehension of harmonic intervals may be furthered by the compound character of the single clang, there must be operative some more ultimate influence than this, which we have not yet found. To which must be added that the contradiction resulting from the application of the customary methods of sensation-measurement to the intensity of sensation on the one hand, and to the pitch of tones on the other, is not in the least degree removed by the adoption of the proposed explanation. The answer that intensity and quality are two different things simply gives up any attempt at an adequate explanation of the incongruity of the two sets of results. As a matter of fact, there is a complete parallelism between the continuous gradations of stimulus-intensity and vibration-rate, of sensation-intensity and tonal pitch.

But there is one way of escape still open to us. Recall to mind the psychological interpretation which we offered of Weber's law. We explained the law by assuming that in estimating differences of sensation it is the relative, and not the absolute magnitude of the compared sensations which we regard. But there is always the possibility of an absolute estimation by the side of this relative one. And we shall expect to find the possibility realised in all cases where a sensation, for some reason or another, is apprehended by itself, in isolation from the remaining terms of the sensation-series to which it belongs. That, again, will occur, and only occur, when the effect of the sensations upon consciousness is not such as to necessitate a reference to other sensations of the same kind. Now this reference is inevitable in every apprehension of an intensity. A loud sound makes greater demands upon consciousness, so to speak, than a weak one. For a sensation-increment to possess the same magnitude in the two cases, therefore, the increase of the stronger stimulus must be greater in proportion as the stimulus itself and, consequently, its effect upon consciousness is greater. But with tone-pitch the case is quite different. A high and a low tone may well be upon a perfect equality as regards the intensity of their effect upon consciousness. That is, the criterion of our discrimination of two tones whose qualitative difference is just noticeable can only be their absolute difference in sensation, which is paralleled by the absolute difference between their vibration-rates. And a comparison of this kind will be possible for tones whose difference is more than just noticeable; so that in subdividing such a total difference into two equal smaller differences we shall always have in mind an absolute, and not a relative, standard of measurement. It is, of course, still another case, if we propose to find two tones at one part of the scale, the relation of whose qualities is similar to that of the qualities of two given tones from another part. In this case the difference is stated in the formulation of the problem to be a relative one ; and the interval is chosen with a view to this relativity. Coincidence of over-tones will certainly help us in this case to cognise the likeness of the two intervals compared ; all that we insist on is, that it is not the sole determinant of our estimation. It is true that the repetition of the fifth, *cg*, in a higher octave allows of a readier cognition of the particular interval than the giving of the tones *da* or *fc*. But, nevertheless, the likeness of these two intervals to the first does not remain doubtful for a single moment.

Our views as to the significance of Weber's law are, then, partly confirmed and partly supplemented by the facts which we have learned as to our apprehension of differences of tonal quality and of tone-intervals. Partly confirmed : for we have found still further reason for our supposition that Weber's law is to be interpreted as a law of the *relative* estimation of sensation-magnitude. Tone-intervals furnish us with particularly convincing evidence of the truth of this law of relativity. Partly supplemented ; for we have found that where the conditions of our apprehension 'of different sensations suggest an absolute rather than a relative comparison, a simple proportionality takes the place of the logarithmic relation between stimulus and sensation. This fact serves at the same time to disprove once and for all the psychophysical theory of Weber's law, which saw in it an expression of the universally valid relations obtaining between the psychical and the physical. Such a hypothesis could only be maintained, if the sensation itself, apart from the psychological processes involved in its comparative apprehension, were subject to the logarithmic law. Nor is the physiological theory,—that is, in its customary form,—less clearly disproved. It supposes that the conduction of the sense-excitation in the brain meets with obstacles, which increase as the magni-

tude of the stimulus increases, so that the excitation in the
central organ itself increases more slowly than the external
sense-stimulus, the exact amount of its inhibition being expressed in the logarithmic formula. The fact that, under conditions which exclude the influence of comparison in estimation,
the course of sensation and stimulus is within limits a perfectly
parallel one, makes against this supposition. It must rather be
the case that within these limits there is a direct proportionality between central and peripheral excitation. So that if we
are led by the principle of psychophysical parallelism to look
for a physiological basis, as well as for a psychological explanation of Weber's law, our field of search must be the relations of
the stimulation-processes in some sensory centre of a higher
order, where are aroused the physical excitations which underlie
a relative apprehension of sensations.

Our general conclusion, therefore, will be this : wherever we
are able to bring about a continuous alteration in the intensity
or quality of stimulus and sensation, we shall find certain limits
within which the alteration of sensation runs directly parallel to
the alteration of stimulus. On the other hand, when we are
comparing different sensations with one another, we shall expect
to find our estimation of their magnitude determined as absolute
or relative by the special conditions of the investigation. A
just noticeable difference in the intensity of sensation is always
apprehended in relative terms ; for the amount of increase which
is equally noticeable in different cases depends upon the demands made upon consciousness by the particular sensation.
The greater the intensity of the sensation, of course, the greater
its effect in and upon consciousness. Our estimation of tone-
intervals is also relative : it is the relation of the terms, not
their absolute value, upon which we have to direct our attention.
Nevertheless, it is not difficult to perceive that the same interval
represents a greater absolute sensation-difference in the upper
region of the tonal scale than in the lower, unless the tones
which compose it are so very low or high that our discrimination
fails us. If we strike first the tone *c* followed by *g*, and then
g followed by *d'*,—if we give, *i.e.*, two opposed fifths in succession, and concentrate the attention exclusively upon the absolute
sensation-difference,—we have no hesitation in declaring the

distance *gd'* to be greater than *cg*. And this helps to explain the converse fact : that when we are halving more than noticeable tonal differences simply in terms of their absolute value in sensation, and without regard to them as musical or unmusical intervals, or when we are determining just noticeable differences of tonal quality, we estimate absolutely, and not relatively.

So that if we are to sum up the result of this whole discussion in one general proposition, that will run somewhat as follows :— *Unless a sensation approaches the upper or lower limit of sensibility, alteration in it is directly proportional to the absolute magnitude of alteration of its stimulus. But our apprehension of alteration in sensation remains only relative so long as its absolute perception is not made possible by the express introduction of especial conditions.*

LECTURE VI

§ I

VISUAL sensations have two qualities,—*colour* and *brightness*. The latter term includes black, white, and all the various shades of grey. The number of distinguishable colours is enormous: it has never been determined. But it is a fact of immediate perception that the variety of colour in nature is not altogether heterogeneous : there are very many intermediate tints between this colour and that. And if we try to divide up the whole multifarious colour field, and to separate out the colours which are clearly and definitely unlike the rest, we can reduce the list of 'pure' colours to very small dimensions. Red, yellow, green, blue, together with black and white, prove to be the simple and ultimate qualities which we are able, so to say, to abstract from the innumerable specifically different presentations of nature. All other distinguishable colour-tones are intermediates,—a fact which is very frequently expressed in the names given to them (purplish red, orange-yellow, yellowish green, violet-blue, etc.). But these six simple qualities, again, are not co-ordinates ; they evince different degrees of variety or resemblance. We are inclined to regard green as being nearer blue than yellow is, and to look on red and yellow as pretty closely related colours, even when the intermediates blue-green and orange are not present to suggest the comparison. It might be thought that this notion of colour relationship was due to our knowledge of colour sequence in the rainbow. But children who have never observed a rainbow with any degree of attention will usually connect blue with green, and red with yellow, when required to arrange the four qualities in the order of likeness.

The hypothesis that there is a limited number of simple colours, from which we may conceive all the other qualities of light to be compounded, is thus suggested by the subjective relationship obtaining between certain colour-tones. It is further confirmed by the familiar results of the mixture of pigments. The painter has long known that green can be obtained by mixing yellow and blue, violet and purple by mixing blue and red. It was an obvious corollary that every colour which could be produced in this way by the intermixture of other colours was in itself a complex, not a simple, sensation-quality. No distinction was drawn, you see, between objective light and subjective light-sensation ; if the external light is a complex, it was thought, the sensation corresponding to it must be a complex also. And even to-day the painter is wont to look upon red, blue, and yellow, together with black and white, as the simple qualities by whose intermixture all other colours are produced.

The science of colour went still farther. Colours usually differ not only in tone, but in brightness ; red appears darker than yellow, etc. So it was thought possible to arrange all the colours in a series, the terminal members of which should be constituted by the two extremes of brightness,—black and white. Aristotle, for instance, taught that black and white are the two fundamental qualities of light, and that every colour can be obtained from their intermixture in varying amounts.

From the point of view of direct perception, the simplicity and universality of this hypothesis are very tempting. When we have once convinced ourselves that the great majority of colours in nature result from the intermixture of a small number of simple qualities, and that these themselves are sensibly related to one another, our mind will not be at rest till it has reduced all the phenomena to two polar opposites. And these can be nothing else than black and white. For all the true colours stand somewhere between these two in brightness, approaching white if their brightness is increased and tending towards black if it is diminished. If all colour is to be derived from two opposites, those opposites must be black and white.

The Aristotelian view of the origin of colour prevailed, then, until modern times. Goethe defended it, and many of his admirers are its enthusiastic champions. But it has been banished from science these two hundred years, thanks to Newton's discoveries. Newton said to himself: If there really are simple kinds of light or simple colours, which intermix in various ways, we must be able both to isolate and to recombine the simple constituents of any given compound colour. That meant that the whole question was referred to the tribunal of *experiment*, where alone it could be definitely answered. For direct perception is deceptive. Can the chemist 'see' of what elements a body is composed? Of course not. We know that bodies of very different chemical composition appear just alike. May not the same hold of light? May not similar kinds of light give rise to different mixtures, and different kinds to similar mixtures? So Newton looked round him for a means of analysing compound light, and by a happy accident found what he wanted in the refraction of light by the prism.

If we allow a ray of light coming from *a* to pass through a prism, *p*, of glass or some other transparent substance, its course is not the straight line that it would be if no prism were in its way, but

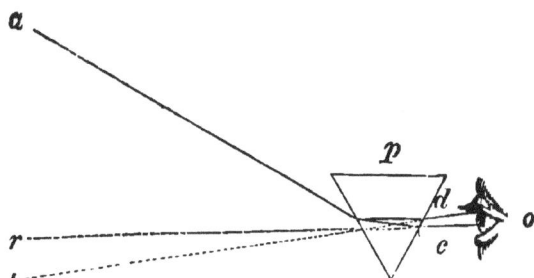

FIG. 8.

it is turned aside, refracted, as we say : so that an eye, *o*, behind the prism receives it as if coming from *b* or thereabouts ; the source of light is transposed from *a* to *b*. Moreover, the point from which the ray appears to come in the direction *bd* does not always remain the same. It varies with the quality of the light. If, *e.g.*, *a* is light of a blue colour, and the ray is seen as though it came from *b*, a red *a* will emit a ray which appears to travel in the direction *rc*, *r* lying higher than *b* and nearer to *a*. It follows that different kinds of light are not refracted in the same degree by the same prism and under similar experimental conditions. Red light is less strongly refracted than blue ; *r* is nearer *a* than *b* is. On comparing the different colours with one another, we find that they fall into a definite series in regard to refrangibility. Red is least, violet most, refracted ; and the series runs—red, yellow, green, blue, violet. Tones interpolated between two neighbouring colours possess an intermediate degree of refrangibility. Orange lies between red and yellow, greenish yellow between yellow and green, indigo-blue between blue and violet.

How is it with regard to white? White is, of course, the most widely diffused quality of light; it is that of sunlight. It is the light which we ordinarily see, if its character is not modified by the peculiar colour of an object. A ray of white light sent through a prism is affected in this way : the eye that receives it after its passage finds it not white at all, but distributed into a whole number of colours ; so that if *a* is a point of white light, the ray proceeding from it is not refracted simply like a ray of mono-chromatic light, and its source transposed from *a* to *r* or *b*, but it seems to issue from a series of sources arranged in a vertical line, each showing a different colour. Violet stands at the bottom ; and then follow blue, green, yellow, and red. White sunlight is, therefore, not simple, but capable of analysis into a large number of simpler light-qualities. These, on the other hand, are not further decomposable. However often we pass pure red or pure yellow through a prism, it retains its character unaltered. You notice that the colour-series obtained by the refraction of white sunlight, whether experimentally or naturally —the rainbow is caused by refraction in the particles of water suspended in the atmosphere,—contains all the colours which occur in nature. By mixing its tones in the right proportions, we can produce any colour that we wish. This is really self-evident, since all the light that the earth receives is derived from the sun. So that, whether a natural body reflects or absorbs light, no effect can arise whose cause is not contained in the constitution of sunlight. As the intensity of white light decreases, we gradually arrive at darkness, or black. Black, that is, is not a colour, but the minimal degree of brightness of white light.

The facts obtained by this exact analysis of light were, how-ever, difficult to reconcile with the results of the mixture of colours, which had also been reached by way of observation. The spectrum produced by the analysis of white sunlight has, you see, at least five colours,—and still more if we count in the intermediate tints. But painters had long since noticed that all possible varieties of colour could be produced from *three* simple tones. It is true that the resulting mixtures are not so saturated as the spectral colours ; but still they are as saturated as most of the colours occurring in nature. The three colours,— *fundamental colours*, they were called,—which could be so inter-mixed as to give rise to any other colour-variety, were generally given, as we have said above, as red, yellow, and blue. But it is better to take red, green, and violet ; and it is better, instead of mixing pigments, to mix directly colours that have been separated out from sunlight by the prism, or to allow colour-impressions to follow one another so quickly that they fuse for sensation. The colours to be mixed can be painted upon the sectors of a round disc, and the disc rapidly rotated on a top or by clock-work. This gives us a perfectly uniform im-pression. Red, green, and violet employed in their right amounts produce white ; and every distinguishable colour-tone corresponds to some particular mixture of the three funda-mental colours with one another and with white. White can also be obtained by the mixture of *two* colours situated at the right distance from each other in the spectral series.

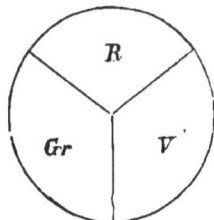

FIG. 9.

The constituents of such pairs as, taken together, give rise to white, are called *complementary* colours. Green-blue, *e.g.*, is complementary to red, blue to orange, indigo-blue to yellow. Green is the only pure spectral colour that has no comple-mentary. To produce white it must be mixed with purplish red, a combination of red and violet. That, of course, is equivalent to a mixture of the three fundamental colours.

How is this contradiction between the analysis and syn-thesis of light to be got over? It is generally left unchallenged, as it was by Newton himself. He said : There are combined in white light particles of red, yellow, green, blue, and violet light, and the prism isolates each separate ray ; but when we are putting together particles of different kinds of light, three of them,—red, yellow, and blue,—are enough to produce all the phenomena. Analysis, you see, had come into conflict with synthesis ; and physical science was not sufficiently advanced to set them at one again.

The first step towards reconciliation was taken when it was discovered that Newton's theory of light was incorrect. He held that the particles of light were themselves coloured, and that light was a substance continually emitted by the sun, and containing in it a multitude of particles of the most diverse colours. This view had often had objections urged against it; but it was reserved for the French physicist Fresnel to adduce a direct disproof of it by experiment. Fresnel showed that when light meets light it is by no·manner of means necessary that an increase of intensity should result. Were light a substance, that must be the case. But, as a matter of fact, decrease of intensity is as common a phenomenon as increase. These observations of what is known as the 'interference' of rays of light prove indisputably that light is not a substance, but a *movement*. Two intercrossing movements may result either in an increase or decrease of intensity : nothing else can. If two balls travelling in opposite directions with equal force meet each other, their movement is annihilated ; if they are travelling in the same direction, it is accelerated. If two waves of water meet, there is increase of the wave where crest and crest come together, decrease or annihilation where crest meets valley. The phenomena of interference show that there are wave-crests and wave-valleys concerned when rays of light meet one another ; that light-intensities will increase at one point of junction, while they decrease at another ; in other words, that we have to regard light as a movement analogous to the movement of a wave in water. If you throw a stone into water, you start a wave, which extends in every direction. The shock of the stone gives rise to a vibration which is transmitted from one particle of the liquid to another. Light consists of vibrations of this kind, except that the substance in which they are set up is infinitely less dense than water. ·It is a form of matter which interpenetrates all physical bodies, solids and liquids as well as gases, besides filling the space between them. The particles of this 'luminous ether' are set in vibration in the fiery atmosphere of the sun, and the movement is transmitted from particle to particle at the enormous velocity of 42,100 miles in the second. What our eye senses as an impression of light, therefore, is not a substance penetrating to it from the remote depths of space, but a movement which, to excite our sense-organ, must be continuous through all the vast distance which separates us from its place of origin. It is one and the same form of matter which occasions all the multifarious sensations of light and colour; so that difference in sensation can only mean difference in the movement of the luminous ether. Accurate measurements of the effects of interference have enabled the physicist to determine this difference in different cases, and it has been found that colour-differences depend upon differences in the velocity of the oscillations of the particles of the luminous ether. In red light, the number of oscillations is something between 400 and 500 billions in 1″; in violet, it approaches 800 billions. All the other colours lie between these extremes. Orange has 500, green 600, blue 650, and indigo-blue 700 billions in 1″; so that the spectral colours constitute a progressive series within whose limits the velocity of vibration increases by nearly 400 billions. It is worth noticing that sunlight contains, besides these coloured rays, other vibrations which are invisible, not sensed as light. There are rays less strongly refracted than the red, and rays more strongly refracted than the violet; vibration-rates, *i.e.*, both greater and less than those which the eye senses as light or colour. The invisible rays beyond the red end of the spectrum manifest themselves as heat; those beyond the violet, in certain forms of chemical action.

It is, you see, only a comparatively narrow section of the ether-vibrations which has the power of exciting a retinal sensation. The whole variety of colour-tone is included within these narrow limits, and a very slight change in vibration-rate suffices to produce a noticeable difference of colour-sensation.

These brief remarks upon the physical nature of light will serve to show you that light and colour have no objective reality, —*i.e.*, do not exist as light and colour outside of and around us,—but that all the properties by which we discriminate light as such, and the various separate colours from one another, are *within* us, originating in our colour- and light-sensations. What we call light and colour is just our own sensation of light and colour. Outside of us there is no system of sensations, but only vibrations in the ether. And the proof that light and colour are subjective phenomena marks an important step in psychological as well as in physical knowledge. We now know that a complete explanation of the phenomena of light and colour cannot be based solely on a physical examination of light, but must also take account of the conditions under which we sense it. What we sense, once more, is not the ether-vibration, but the particular reaction of our eye and mind upon that vibration. Movements which are too fast or too slow to be perceived by the eye are evidently removed from the list of visual stimuli simply by reason of their velocity. But objectively they may be light, just as much as any other movements.

So that, if we are attempting an explanation of the phenomena of light and colour, the result even of a purely physical investigation will be to refer us to the *seeing subject.* Now you remember that we had come across a contradiction. The analysis of light by the prism told us one thing ; the recomposition of the various light-qualities told us another. We can separate out from sunlight at least *five* simple colours, without counting intermediate tints, while we can produce every colour that occurs in nature by the intermixture in appropriate amounts of *three* colours only, best chosen as red, green, and violet. How is the contradiction to be met ?

It is evident from what has been said above that the fact that white and all possible colours can be obtained from three fundamental colours does not mean that objective light is compounded of those three fundamental colours. Nor does it mean, as many physiologists would still have us believe it does, that all our subjective light-sensations are derived from three ultimate sensations corresponding to the three fundamental colours. All that the results of experiments upon colour-mixture prove is this: that three objectively simple modes of vibration are sufficient, when mixed in different amounts, to set up all those stimulation-processes in the organ of vision which can be occasioned by the colours of the solar spectrum and their intermixtures.

The three fundamental colours could only possess any significance in *physics* if objective light were analysable into three modes of vibration, corresponding to them, and no more. We have seen that that is not the case. In one way the fundamental colours do occupy exceptional positions in the infinite gradation of light-vibrations. Red and violet stand at either end of the series of visible vibration-rates, and green is situated at its centre. But this, though a fact of some importance for an examination of the conditions under which the eye is stimulable by light, has no reference whatever to objective light as such, the light whose vibration-rates extend far beyond the limits within which the ether rays are visible.

The fundamental colours would only possess significance in *psychology* if we could subjectively, in immediate sensation, analyse all our light-sensations into these three particular constituents. We can, certainly, say that orange is a sort of mixture of red and yellow, violet a mixture of blue and red, and so on. But even in these cases the phrase 'intermediate between' would be more correct than 'mixture of.' It seems to me, at any rate, that my sensation of orange and violet makes of them impressions as simple as it does of red, blue, or yellow. And so much is indisputable: that no one can say he senses red and green in yellow, or red, green, and violet in white. Subjectively white is just as simple as any simple colour. And black we shall all incline to regard not only as the minimal degree of intensity of white, but at the same time as its qualitative opposite.

There is only one possibility left. If the existence of three fundamental colours is incapable either of physical or of psychological explanation, it must depend simply and solely upon *physiological* conditions. If we accept the principle that to every difference in our subjective sensations there corresponds a difference in the physiological stimulation-processes within the sense-organ, we must suppose that the three objective light-qualities in the red, green, and violet portions of the spectrum, when mixed in the right proportions, can set up as many physiological excitatory processes as there are subjectively discriminable sensations. How many of these stimulation-processes are possible we cannot directly determine: but they must be estimated from the number of distinguishable sensations, and not from the number of objective light-stimuli by which the sensations are occasioned.

§ II

Simple as these considerations are, they have not as yet been able to command anything like universal assent. In the current theories of the nature of light and colour-sensation we find all too often a confusion between the physical and the physiological stimulus, and again between this latter and the sensation correlated with it ; or, if not that, the opposite error of a derivation of the objective conditions of light-excitation from the subjective differences in sensation,—a hypothesis arbitrary in form and contradictory of experience in content.

For instance, Thomas Young, an English physician and physicist of the beginning of the present century, maintained that all our sensations of light and colour are compounded of the primary sensations red, green, and violet. There exist in the eye, he said, three kinds of nerve-fibres, sensitive respectively to the red, green, and violet rays. We can represent the laws of colour - mixture by drawing a triangle, as in Fig. 10, the three angles of which are constituted by the three fundamental colours, while the intermediate spectral tints are placed along its sides,—purple,

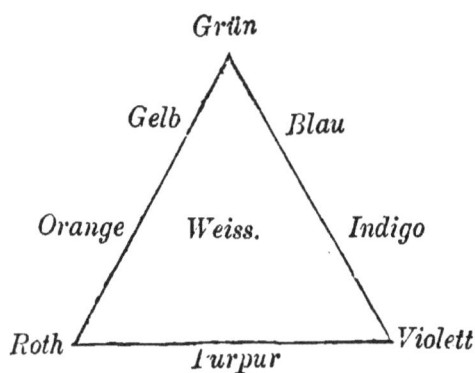

Grün

Gelb   *Blau*

Orange   *Weiss.*   *Indigo*

Roth   *Purpur*   *Violett*

FIG. 10.

e.g., between the red and violet of which it is compounded,—and white occupies the centre of its area. Such a triangle, according to Young, would express equally well the conditions of visual sensation and visual stimulation. Orange and yellow, for example, would excite the fibres sensitive to red and green, the red predominating in orange, and the green in yellow, while the sensation of white would result from the excitation of all three fibres in approximately equal intensity. The sensation of white, that is, is simply a mixture of the three sensations red, green, and violet. The impression of a whitish colour occurs, on the other hand, if one or two of the fundamental colours predominate in the mixture. These whitish colour-tones may therefore be written in upon that portion of the area of the triangle which lies between the centre (white) and the sides.

Thomas Young regarded his own assumption of three fundamental sensations as no more than a provisional hypothesis, especially useful for the explanation of the phenomena of colour-mixture. But many physiologists and physicists of recent times have imagined this supposed structure of the organ of vision to be a necessary corollary from the facts of *colour-blindness*, and so have made the hypothesis a certainty. Colour-blindness does not, as a rule, signify blindness to colour in general, but only insensibility to certain particular colours. If it is connate, the colour-blind person need not be conscious of his defect, which simply consists in the confusion of special colours,—red and green, *e.g.*,—clearly distinguishable by the normal eye. Experience shows that in the great majority of cases colour-blindness is red-blindness, though green-blindness also occurs. But red and green are fundamental colours, so that we seem to have in the phenomena a confirmation of Young's theory. The condition of the eye, you see, is quite easily explicable in terms of it. We have simply to regard one of the three sets of nerve-fibres or terminal organs of the normal eye as absent or functionless : in red-blindness those sensitive to the red rays ; in green-blindness those sensitive to green.

Nevertheless, the proof is not so unexceptionable as has often been thought. If we grant that the only forms of colour-blindness are the 'red' and 'green' types, we may, perhaps, find it to be a necessary inference that there are normally particular portions of the retina especially sensitive to red and green light, and that these, for some unknown reason, are in the defective organ either absent or insensible ; but we shall not find the least ground for supposing that the sensation of yellow, *e.g.*, is a mixture of the sensations of red and green, or that stimulation by yellow light simply means a stimulation of the elements sensitive to red and green. The first proposition is negatived by the character of the sensation ; yellow is qualitatively different both from red and green, and shows no trace of a mixture of the two. The second we can hardly regard as probable, unless we are willing entirely to give up the principle which has stood us in such good stead heretofore,—the principle of parallelism of differences in physiological stimulation with differences in sensation. But, besides this, the progress of our knowledge of the phenomena of colour-blindness has brought facts to light which are irreconcilable with Young's hypothesis. First of all, it has been found that 'red' and 'green'-blindness, though the commonest, are not the sole types of abnormal colour-sensibility. Cases are known of insensibility or diminished sensibility to other rays in the spectrum, especially the yellow and blue. Secondly, there are extant observations of unilateral or monocular colour-blindness. Now, on Young's hypothesis, white must here be composed of different primary sensations in the two eyes; in unilateral red-blindness, *e.g.*, it would be a mixture of red, green, and violet in the normal eye, of green and violet only in the other. The same white light would therefore be differently sensed by the two eyes; to the normal it would appear white, to the other, which has no sensation of red, greenish. As a matter of fact, this is not the case; the same white appears precisely the same to both eyes. Thirdly and lastly, we have instances of total colour-blindness. This generally results from disease of the eye, and is frequently confined to one retina, or even to portions of it. Black and white, with all the intermediate greys, can still be sensed, but there is no sign of a colour-sensation of any kind. A picture looks like a drawing; light and shade are clearly distinguished, but there is absolutely no perception of colour. It is obvious that such a state of things could never come about if every light-sensation were the resultant of a mixture of the three fundamental colours. Total colour-blindness would be impossible unless the sensation of brightness and that of colour were correlated with different, and under certain circumstances separable, excitatory processes in the visual organ. And here we have a fresh proof of the validity of our principle that difference of physiological stimulation-process runs parallel to difference in sensation. For this independence of white from the various colours might have been inferred before from the independence of its quality in sensation.

§ III

These facts have, of course, told against Young's theory ; and an attempt has been made in quite recent times to replace it by another. To this end the views of Leonardo da Vinci—a name as conspicuous in the history of science as in that of art—have been revived. Leonardo regarded four colours, which he called *principal* colours, as of prime importance, and looked upon all the others as intermediate or mixed colours. These principal colours were red, yellow, green, and blue. To them must be added black and white. From the six fundamental qualities so obtained and from their intermixture in various amounts all our sensations of light and colour could, he urged, be derived. Orange, *e.g.*, is given in immediate sensation as a mixture of red and yellow, violet as a mixture of red and blue.

Leonardo's view is entirely based upon the subjective character of our sensations. And it might perhaps have held its own, being not without interest as an expression of the psychological side of the matter, if the attempt had not been made to graft further hypotheses upon it for the explanation of the objective laws of light-stimulation and colour-mixture. But it was assumed, *e.g.*, that between the members of each pair of principal colours there existed an ' antagonism ' analogous to that between white and black. Antagonistic colours were defined as those which cancelled each other when mixed, leaving only the sensation of brightness, which accompanies every colour-sensation of whatever quality. Red and green, blue and yellow, were regarded as antagonistic in this sense. To make the dominant idea of the theory still more definite, it was supposed that there are intermingled in the retina three different kinds of sensitive substance, in each of which two antagonistic processes may be set up, corresponding in a manner to the processes of anabolism, or assimilation, and catabolism, or dissimilation, which are found together throughout organic nature.

For the sake of brevity, we will term these processes a and d, and call the three substances, after the sensations which they mediate, the black-white, the red-green, and the blue-yellow. The assumption then is, that the sensation of black is due to an a-process, white to a d-process, in the black-white substance ; in the red-green, red is the a and green the d-process, or *vice versâ*; and similarly with the blue-yellow. But every stimulation of a colour-substance involves the excitation of the black-white; and so it comes about that if the a and d-processes of one or both colour-substances cancel each other, we still sense brightness.

It cannot be denied that this theory has done good service. For the first time since the banishment from science of the older colour-theory of Aristotle and Goethe, it called attention to the fact that white light is just as simple in sensation as any monochromatic light, and that black and white are not only to be regarded as different intensities of a single quality, but also, and indeed predominantly, as qualitative opposites. In all other respects, however, it is simply an arbitrary combination of arbitrary assumptions. Even the analogy between the 'antagonistic' colours and the black-white pair cannot be carried through. When we mix black and white we get grey. And grey is directly sensed as a quality intermediate between the two extremes. But when we mix red and green or blue and yellow, there is no mixture, but only mutual disturbance in sensation; the only thing that is left is white, and white was present from the beginning, except that the colours were too strong for it. Again, the fundamental colours, indicated by the laws of mixture, have to be accommodated to Leonardo's principal colours. That is not altogether easy. We are obliged to change the names of the colours to suit our theory. For the antagonistic colours are not what we generally call pure red and pure green, pure yellow and pure blue; if we are to obtain complementariness, the red must be tinged with purple, and the blue with indigo-blue,—in other words, both these principal colours be mixed with a considerable amount of violet. And, lastly, it is a necessary consequence of the theory that there should be no partial colour-blindness without annihilation or diminution of the sensibility to the two members of some pair of antagonistic colours. For it can only explain the affection as the result of some defect in one of the two colour-substances. We ought, that is, to find red-green and blue-yellow blindness, but no other form. As a matter of fact, there cannot be any doubt at all that red-blindness may occur without green, and green without red-blindness.

But not only does the assumption of antagonistic colour pairs come at every point into conflict with facts : its psychological foundation is exceedingly questionable. The four principal colours,—red, yellow, green, and blue,—are, you see, the only simple sensations ; all other colours are immediately given as compound in sensation. What support is there for that assertion ? Plainly this, first of all : that when we have once recognised these four as fundamental sensations the others fall into place readily enough as subjective intermediates ; and secondly this : that the names of these four are by far the oldest colour-names, the others bearing the evidence of modernity upon their face. Now the first of these facts is only of importance in the light of the second. If language had originally contained special names, say, for orange and violet instead of for red and yellow, it is quite probable that we should be inclined to look upon these latter as intermediate tones. So that everything depends upon the answer to be returned to the question : why is it that these particular four ' principal ' colours were the first to receive definite colour-names ?

From the point of view of the theory which is under consideration, it is a matter of course that the reason can only be looked for among the immediate facts of sensation ; that red and yellow are just given as simple, while orange is sensed as a compound impression. Language, that is, preferred at the outset only to name what was simple in sensation. Now this view obviously proceeds from a wrong postulate as regards the origin of word-symbols in language. In the first place, it is not true that a separate word must exist for every qualitatively simple sensation. More than one philologist has called attention to the fact that a sharply distinguished term for 'red' appears earlier than one for .'blue.' In the ancient literatures,—*e.g.*, in Homer,—the expressions for the blue of the sky are such as could be used for any dark or grey object. Now and again the conclusion has actually been drawn from this that the Greeks of Homer's time had not yet sensed blue ; that is, that the colour-sense in man has within this comparatively short period undergone a very considerable development. We shall hardly feel tempted to assent to that conclusion. Language does not distinguish everything that sensation distinguishes ; it contents itself with creating special terms for those impressions whose discrimination is necessary for the expression of thought and its communication to others. Are we to suppose that it is only since Newton's day that mankind can distinguish orange from yellow, or indigo-blue from sky-blue? Surely not. These new names of colour-tones simply came into use when they were needed for optical or technical purposes. To make assurance doubly sure, it has been quite recently proved that the scale of colour-sensations in various savage races presents no differences from our own.

At the same time the four principal colours do in one sense constitute a special case. Whenever different colour-names occur, they are ultimately reducible to these four. So that there is some show of reason for assuming their original sensational preference. Nevertheless the law holds even here that language does not name sensations because of any subjective peculiarities that they may possess, but merely with reference to their objective significance. So that whenever we can follow a colour-name back to its original meaning we find it indicating some external object by which the colour-sensation is occasioned. Orange, indigo-blue, and violet are named from the colour of the fruit, the dye-stuff, and the flower. Now what are the colours which mankind would have named first, on this principle? Surely either those which excited in them the strongest feelings, or those which were commonest among the natural objects of their environment: the red of blood, perhaps, the green of vegetation, and the blue of the sky, against which the bare earth and the light of sun, moon, and stars looked yellow instead of white. So that there is no need for us, in our search for the origin of the four principal colours, to descend into the bottom-less abyss of theory, and postulate a sensation that is independent of any impression. And, of course, these colour-impressions, which frequent repetition or some other reason has brought more prominently than the others before consciousness, must obtain an advantage not only as regards linguistic expression, but also in sensation itself, in that all other sensations are arranged with reference to them. If once red and yellow are given, orange can only be looked upon as an intermediate tint. In the same way purple and violet fall into place between blue and red. And since there is a continuous transition from colour to colour, while the range of sensation is strictly limited, the four principal colours were amply sufficient to allow of a permanent arrangement of all the possible colour grades. Had the dominant impressions, which determined at once colour-arrangement and colour nomenclature, been orange, yellowish green, greenish blue, and violet, instead of red, yellow, green, and blue, we should undoubtedly have sensed red as an intermediate between violet and orange, and green as a tint lying between yellowish green and greenish blue. Isolate any one of these colours for yourselves from the spectrum, and ask yourselves whether the impression it makes is not that of absolute simplicity, if once you abstract from the relations into which it has been brought with other colours by the customary arrangement of the colour-system.

§ IV

The principal colours owe their prominence, then, not to anything in the quality of sensation itself, but to external circumstances which have nothing at all to do with sensation-quality. Subjectively every colour-impression is a simple quality, resembling its nearest neighbours, but only resembling them because of the continuity of tints in the colour series. And the mention of this fact of continuity leads us to another point, which the two colour theories discussed above leave entirely out of account,—the relation of the two terminal colours of the spectrum to each other. Red and violet are not the most different of all colour-sensations, as their spatial positions on the colour scale might lead us to expect ; on the contrary, they are as much alike as any two colours can be. Here again colour-sensation and brightness-sensation are diametrical opposites. The maximal differences of vibration-rate produce in the one case similar subjective effects ; while in the other the extremes of luminous intensity correspond to opposite qualities of sensation, black and white, between which the whole series of brightness sensations is arranged in continuous progression. Now, whatever hypothesis we adopt, we must seek to do justice to *all* these peculiarities of sensation, besides taking account of the laws of colour-mixture. Bearing that in mind, we arrive at something like the following theory.

We may suppose that every retinal light-stimulation is compounded of two separable constituents,—a colour-excitation and a brightness-excitation. The brightness or 'achromatic' excitation may occur by itself. When that is the case, we sense black, white, or grey. The colour or 'chromatic' excitation always implies the presence of the achromatic. When it occurs, we sense saturated or whitish colour, according to the intensity of the concomitant achromatic process. Differences in this latter we may regard as ultimately conditioned by objective differences in the intensity of light. It consists always of two qualitatively opposite part-processes, one attending stimulation by light and corresponding to the sensation of white, and the other accompanying the recuperation of the retina after stimulation and corresponding to the sensation of black. This part-process of recuperation, you observe, is not only present when the visual

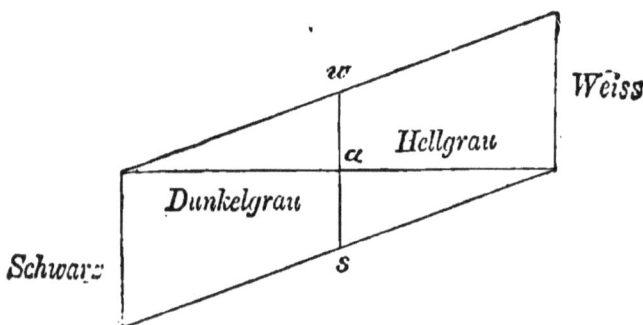

FIG. 11.

organ is entirely free from stimulation, but accompanies the more moderate degrees of stimulation, as a reaction of the stimulated substance in the direction of restitution of the decomposed chemical compounds. In the event of very weak excitation, its intensity may even exceed that of the other part-process,—that of stimulation. We can therefore represent the system of brightness-sensations by a straight line, terminating in black and white and having the various shades of dark grey, grey, and bright grey arranged along its length in all their manifold but perfectly continuous gradations. Such a line is drawn in Fig. 11. The opposed processes are represented by vertical lines falling upon it, the intensity of the excitation-process being shown by the length of the ascending verticals, and the intensity of the recuperation-process by that of the descending ones. At absolute black the excitation-value must be regarded as zero, while recuperation is at its maximum ; at the brightest white recuperation is at its vanishing point, and excitation at a maximum. Every intermediate brightness-sensation presupposes a mixture of both processes. The total physiological process corresponding to a mean grey, *e.g.*, is composed of a stimulation of the magnitude $a\,w$ and of a recuperation of the extent $a\,s$. These two part-processes do not cancel one another, but intermix, so that in sensation grey is a quality intermediate between those of black and white, and equally related to both.

If we attempt to construct a similar geometrical diagram for colour-sensations on the basis of their subjective peculiarities, we must substitute for the straight line a curved one ; its two ends must approximate, to indicate the subjective similarity of red and violet. We may choose a circle, as the simplest line of the description required. Then, as in Fig. 12, all the saturated colours may be arranged round its periphery. But as the colours of the solar spectrum leave a gap between red and violet, we must fill the series out by introducing purple. Purple is obtained by mixing red and violet, and all its various shades lie between those two qualities. The process of chromatic stimulation, on the principle of parallelism of sensation and physiological stimulation, must be regarded as recurrent or periodic ; the processes set up in

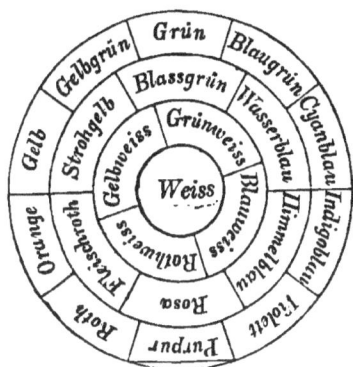

FIG. 12.

the retina by the quickest vibrations must resemble those resulting from the slowest. You may find an analogy in the octave. Although the ground-tone is further removed from its octave than is any other tone of the chromatic scale, yet these two are more nearly related in clang-character than any two others. Indeed, this may be something more than an analogy. The vibration-rate of the octave is twice that of its ground-tone, and the vibration-rate of the light rays at the extreme violet end of the spectrum is approximately twice that of the red rays.

At the same time, the conditions of light and sound-stimulation are in other respects so different that the attempt occasionally made to discover in the series of intermediate colours the exact physical and sensational correlates of the principal intervals of the tonal scale,—fifth, fourth, the two-thirds, etc.,—has in every case resulted in failure.

The great difference between colours and tones, which prevents any such attempt from succeeding, is seen most clearly in the different effects of the mixture of light and sound-waves. When we mix sound-waves, we get a compound sensation, consisting in the last resort of just as many simultaneously distinguishable elements as it contains successively distinguishable tones. But when we mix light-waves, we always obtain a simple sensation. White, which is composed of all the rays of the solar spectrum, is in the last resort just as simple as is any colour which contains but one single wave-modality.

These peculiarities of light-sensations lead us to two facts of general import, which must be taken account of by any theory of chromatic stimulation equally with the subjective similarity of the terminal spectral colours. First, there may be substituted for any simple colour a mixture of the two bordering colours between which it lies. We can obtain orange from red and yellow, yellow from orange and yellowish green, a pure green from yellowish green and greenish blue. Secondly, any colour mixed with its subjective opposite in the right intensive proportions gives us the sensation of white; the two are complementary colours. The first of these facts indicates that the process of retinal stimulation is not continuous, but *graduated*. For the operation of any particular grade of vibration rapidity there can be substituted the mutually supplementary actions of two other grades not too remote from it, one of a less and one of a greater rapidity. These combine to give an intermediate result. The second fact, like the recurrent course of the colour-line, proves that colour-stimulation is a *circular* process, in the sense that every form of it is correlated with an opposite form ; so that when the two excitations concur they cancel one another leaving only the concomitant achromatic stimulation in sensation. We can express this in our diagrammatic construction (Fig. 12) by supposing the colours on the periphery of the circle to be so arranged that the members of every complementary pair are directly opposite to each other, and can be connected by a straight line passing through the centre of the figure. At the centre itself we may put white, and on the area between centre and periphery write in the whitish colour-tones, as they gradually change from complete saturation to absence of colour or brightness.

All the simple qualities of visual sensation are contained in this figure, with the exception, first, of black and the greys intermediate between it and white, and, secondly, of the blackish or greyish colours,—brown, olive-green, etc. If these dark shades of colour and brightness are to be introduced into the diagram, we must first draw a perpendicular to the centre of the circle, white (Fig. 12), and arrange along its length the series of achromatic sensations, and then fill in the dark colour-shades and the intermediates between them and grey or black on concentric circles placed at different heights round the perpendicular line of brightness-sensation (Fig. 13). In this way we get a cone with a circular base, the apex corresponding to black, the centre of the base to white, and the periphery of the base to the saturated colours, while all other qualities are distributed over and through the body of the cone, their exact position being determined by reference to the fixed positions of these fundamental qualities.

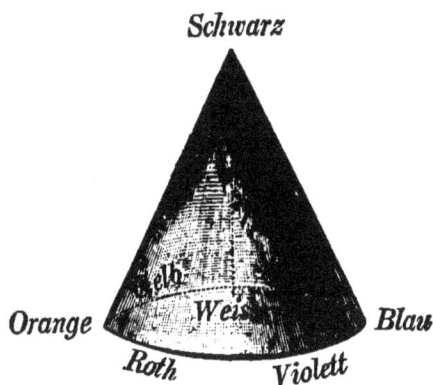

Fig. 13.

LECTURE VII

§ I

IN the light of what we already know regarding the properties of sight and hearing, we can see that the best way to understand the relation between these two senses is to turn to their two essential differences. The first is this : that, whereas a subjectively simple tone-sensation can only be caused by a simple objective movement of the air, in the case of sight any vibration of ether, simple and complex alike, produces a simple light-sensation. And the second consists in the fact that the simple tonal qualities can only be varied in two directions to produce higher and lower tones, whereas light-sensations form two series, the chromatic and the achromatic, each consisting of a number of fundamental qualities and intermediate tints, while the chromatic series, owing to the subjective relationship of the terminal colours of the spectrum, further constitutes a manifold of qualities which returns upon itself. The system of simple tones, therefore, can be represented by a straight line ; that of light-sensations requires a figure of three dimensions for its expression. Besides these general differences, there are further distinctions dependent on various properties which attach to the sense of sight, either exclusively, or to a greater extent than to any of the other senses.

§ II

If we produce a tone (*e.g.*, by striking a string or tuning-fork), and then suddenly arrest the vibrations of the sounding body, the tone-sensation immediately ceases. Even if it actually lasts somewhat longer than the stimulus, the duration of its after-effect is so short as normally to escape our notice. It is quite different with the sensation of light and colour. You know that if a red-hot ember is swung round at a moderate rapidity, you see a complete circle of fire. This phenomenon shows that the light-impression in the eye must persist at least during the time that the ember takes to pass from any point in its course back again in a circle to the same position. We can obtain more exact knowledge of the after-effect of light-stimulation by fixating a luminous object for some time and then suddenly closing the eyes. We then see on the dark field of vision an after-image which resembles the object, but which gradually undergoes very curious changes as regards its light and colour-properties. For the first moment it is exactly like the external object, then its intensity decreases somewhat, and again after a little time its quality alters to just the opposite of the original. If the fixated object were white, the after-image turns black,— darker, *i.e.*, than the dark field of vision on which it is seen,—if the object were black, the after-image becomes white, brighter than the surrounding field of vision. Finally, if the impression were coloured, the after-image assumes the complementary colour, greenish if the object were red, reddish if it were green. In its first stage, therefore, the after-image is called *positive* or *same-coloured*; in its later phases, *negative* or *complementary*. Both phenomena, but especially the more persistent complementary images, are observable with open eyes, if the object fixated is extremely bright. If you glance, *e.g.*, at the setting sun, and then look at the roadway or a grey wall, you may see a clear green after-image of the solar disc at the point of fixation.

The phenomena of after-images prove, in the first place, that the stimulation-process in the retina outlasts the external stimulus by a considerable period, often amounting to several seconds. But they show also that the stage of direct continuance of stimulation, which is manifested in the positive after-image, is followed by an opposite condition of the sense-organ, during which bright objects are seen as dark, and dark ones as bright, and in general every colour as its complementary.

This complementary after-effect may be easily explained on the assumption of a partial exhaustion of the retina. If, *e.g.*, our sensibility to red is exhausted at a particular place, we shall see white light at that place just as if no red rays were affecting it ; in other words, since the subtraction of red from white gives green, a red object will leave behind on the exhausted retina a greenish after-image. The quality of these complementary after-images may be enhanced by contrast with its background. This holds especially of the cases mentioned above, where the after-images of white and black objects were observed upon a dark field of vision.

§ III

These striking differences in the after-effects produced by stimulation of the two sense-organs render it probable that the processes which excite the sensations of tone and light are totally different in character. Indeed, it seems probable that sound vibrations set up corresponding vibrations in the membrane of the cochlear canal, in which the fibres of the auditory nerve terminate ; and that our separate perception of the constituent tones of a clang is due to the tuning of the individual fibres of this membrane to various tones. The stimulation-process would, therefore, here be a mechanical one, and the sensation is consequently interrupted as soon as the vibratory movement of the stimulated fibres ceases. That must, of course, occur very soon after the cessation of the external air vibrations which set up the excitation. It is probable, once more, that the rapid arrest of the vibrations of the membrane is promoted by the action of certain solid cuticular structures, semicircularly arranged, which rest upon it, and whose function must be analogous to that of the damper of a pianoforte. But when a light-stimulus affects the retina, we have obviously a very different state of affairs. There are many observations which go to prove that the sensitivity of the retina to light is like that of the prepared plate in the dark room of a photographer. One of the most striking is this : in the dark, the retina is of a deep red colour ; exposed to light, it gradually bleaches, and finally turns white. This renders it practically certain that light-stimulation is a *photochemical* process. Such processes play an important part in organic nature ; *e.g.*, in the breathing of the green portions of plants and in the production of the colours of flowers. Now a chemical process, even if it is comparatively soon over, always requires a considerably longer time for its completion than does a simple transmission of motion. Regarded from this point of view, the after-image appears simply as a subjective indication of the duration of a photochemical action ; and its two phases point to the fact that there are two processes which run their course during that action. The positive after-image gives us the duration of the chemical decomposition occasioned by the light-stimulus ; the negative or complementary after-image shows us an after-effect of this decomposition. This latter is a phenomenon analogous to those of exhaustion in other living organs,—*e.g.*, in nerve and muscle,—all alike manifest themselves in decreased excitability for stimuli of the same kind as those previously operative.

The senses of sight and hearing may, therefore, be regarded as the principal representatives of two fundamental forms of sensory excitation,—the *mechanical* and the *chemical.* In setting these expressions over against one another, we must be careful not to refer them to processes in the sensory nerves. These consist, probably in every case, of very rapid chemical decompositions. Our purpose is merely to discriminate between the different ways in which external stimuli affect the nervous terminations in the organs of sense. With the mechanical senses may probably be reckoned (besides hearing) that of cutaneous pressure, as is shown by the brief continuance of the after-effect of impression in sensation. To the chemical senses belong (besides that of sight) the cutaneous temperature-sense and the senses of smell and taste. The organ of touch is the earliest, and in the lowest forms of organic life the only, sense-organ. And the fact that it includes both a mechanical and a chemical sense is, perhaps, not without significance for the physiological history of the development of sense-functions.

• § IV

The phenomena of negative and complementary after-images which we have been discussing require little more than purely physiological explanations. There is, however, a very large number of cases in which our sensations of light and colour undergo change where this cannot be derived from the after-effects of stimulation and from the influence of exhaustion, or where at least these phenomena afford only a partial explanation. Of course, wherever after-images manifest themselves at the same time, they may be explained in physiological terms.

If we cause light to pass through a sheet of red glass, and then cast a shadow somewhere upon the red illuminated surface, we ought really to see this shadow grey. For it contains nothing but diffused white light, the intensity of which has been lessened by the shade. As a matter of fact, however, the shadow looks not grey, but green. This same green tint may sometimes be observed in nature, in the shadows cast by trees, when the setting sun gives a reddish light in consequence of the stronger absorption of the refrangible rays by the atmosphere.

An experiment which exactly reproduces the conditions of this subjective colouring of natural shadows can easily be made by means of the colour top and the rapidly rotating discs which we employed to illustrate the mixture of colour-sensations. We take a disc with small coloured sectors on a white background. Somewhere about half-way between the centre and periphery, the coloured sectors are replaced by narrower portions of a black band (Fig. 14). If we let the top rotate at a high speed, the colour of the sectors fuses with the white of the background to a whitish tint ; and where we inserted our black bands, we get a mixture of black and white, *i.e.*, grey. This grey, you see, corresponds completely to a circular shadow cast upon the coloured background. But this objectively grey ring does not appear to us as really grey, but as coloured, and as coloured complementarily to the background. If the sectors are green, the ring looks red ; if they are red, we see it as green, etc.

FIG. 14.

Here is a still simpler experiment. Take a sheet of thin white notepaper and a sheet of coloured paper of the same size ; lay the white paper on the coloured so that it exactly covers it ; and then push a little square of grey or black paper between the two. If the coloured paper which is underneath is green, *e.g.*, its colour shines through the thin notepaper, except at the place where the grey square lies. This should appear grey, but, as a matter of fact, looks red. If our coloured paper had been red, the grey square would have looked green. In short, it always takes on that colour which, mixed with the colour of its surroundings, would produce white. The phenomenon is not so striking, but yet clearly apparent, when you simply lay a piece of grey paper on a coloured background, without covering it with the thin notepaper. Suppose, *e.g.*, that you cut little squares from the same grey paper, and lay them upon red, green, yellow, and blue papers, placed side by side. They all look different ; that on the red is greenish, that on the green reddish, that on the yellow bluish, that on the blue yellowish. And the effect of the background is just as obvious, if you employ colourless lights of various brightnesses, instead of colours. Fasten two similar grey squares upon black and white sheets of paper respectively ; the former appears so bright as to be almost white : the latter looks dark, and under favourable conditions may approach to black.

In describing all these phenomena and enumerating their conditions, we have left out of account one important factor,— the extent of the surfaces employed. If the shadow cast in coloured light is very large, the centre will look grey ; only at the edges will its colour become apparent. We can best show the dependency of subjective colouring upon the space-relations of the surfaces compared by having recourse again to revolving discs. We furnish a disc with sectors in the way shown in Fig. 15 ; the sectors are cut into steps, the absolute magnitude of which is in every case the same. If now the sectors (*B*) are blue, and the background (*G*) yellow, we shall expect, when the disc is rotated, to see a bluish-yellow mixture, which is perfectly uniform within each step of the sectors *B*, but which is changed from step to step in such a way that the yellow

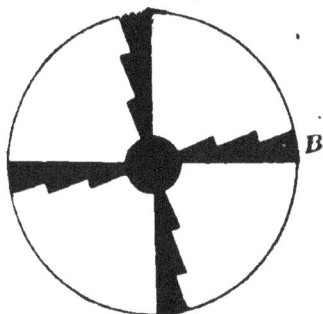

FIG. 15.

1

predominates more and more in the mixture as we approach the periphery; for objectively the colour of each step remains the same throughout its whole extent. In reality, however, the inner and outer borders of each blue-yellow ring are differently coloured, and the intermediate points show a gradual transition from one shade to the other. Each of these constituent colours shows most strongly when contiguous to a ring which contains less of it in its mixture. On our own disc, therefore, the outer edge of every ring is blue, and the inner yellow. We have, that is, a succession of yellow and blue rings outlined upon a background of mixed colour.

The same experiment can be made in a different way. We will take a white disc and paste upon it not coloured, but black, sectors, as has been done in Fig. 15. We should expect to obtain by rotation concentric grey rings, increasing in brightness toward the periphery of the disc, but showing no variation within each ring. We find instead that all the rings look brighter on the inside, where the contiguous ring is darker, and darker on the outside. The difference is so great that the brightness of the different rings oftentimes appears to be the same throughout; so that when rotated rapidly the disc seems to be composed simply of alternating black and white circles.

These experiments illustrate a group of phenomena to which the general name of *contrast phenomena* is given. The name is an obvious one, for their principal condition is clearly the opposition of two colours or of two degrees of brightness. And for the same reason the subjective colour due to this opposition (the green tint of the shadow in red light) is also termed *contrast colour*. We may, therefore, use the expressions 'contrast colour' and 'complementary colour' as synonymous.

The identity of these two concepts has been the chief agent in determining a large number of physiologists to refer the phenomena of contrast to like causes with complementary after-images. Just as the after-image arises from persistence of the excitation in the stimulated region of the retina, so contrast, they say, is a result of the diffusion of excitation over the surrounding portions of the retina. This analogy is, of course, condemned at once by the fact that we distinguish a positive after-image, resembling the original stimulation, from a negative phase of it, which is opposed to the former ; while the effect of contrast is always negative and antagonistic. Moreover, a considerable time is necessary for the development of the afterimage, whereas the contrast change of sensations is instantaneous. But we have still other and weightier reasons to urge against the analogy. A spatial diffusion of stimulation should appear in more intensive form the greater the strength of the original stimulus. This is, however, not by any means the case.

A weak stimulus may under favourable circumstances bring about a stronger contrast than a strong stimulus. This we learned from our experiment with colourless objects laid upon a coloured background. You remember that if we cover them with transparent notepaper, the contrast is much more obvious than if they are left uncovered,—though their colour is weakened by the covering sheet. This suggests that contrast-effects in general may consist not in a direct alteration of sensation, but in changes in our subjective mode of apprehending sensations ; and the suggestion is fully confirmed by other observations. Let us vary our last experiment a little. By the side of the grey square, which has taken on the colour complementary to its background, we place a second square of exactly the same brightness. Now that comparison of the two is possible, the contrast colour disappears, but it reappears so soon as the second grey is removed.

Since the *comparison* of impressions is so plainly important in this connection, it has been maintained, as against the physiological hypothesis mentioned above, that contrast depends upon a *deception of judgment.* A shadow seen in red light looks green to us, it is said, because we are accustomed to regard ordinary diffuse daylight as white, and to judge of colours in comparison with it. So that if in a particular case it is not white, but red, a shadow cast in it must seem to us to be green ; for if reddish light looks white, light which is actually white will be no longer regarded as such, but will necessarily appear to have had a certain quantum of red taken from it, *i.e.,* will seem greenish. If we mix all the colours of the spectrum together with the exception of red, we obtain, of course, a green shade. The same principle of wrong judgment is employed to explain the facts observed in the experiment with transparent paper. If we lay a transparent white sheet upon a dark red one, it looks bright red. Now we push our grey square in between the two sheets. This, of course, renders that portion of the surface colourless ; but we judge that the bright red paper extends over it also, and so see it in the colour which an object must possess if it is to appear colourless when seen through a red medium, *i.e.*, as green ; or, to put it more generally, we see it in the colour complementary to that of the transparent covering.

Now there are weighty objections to be urged against this derivation of contrast phenomena from deceptions of judgment. First of all, it is not right to say that because we are accustomed to see diffuse daylight white, therefore we must do so in the exceptional cases when it is not really white, but coloured. We see clearly enough that the illumination of the setting sun is reddish, and we are not in the least inclined to regard the light that comes to us through a sheet of coloured glass as white ; we sense it precisely in its own proper colour. These instances serve to show that 'the assumption from which the explanation proceeds is not justifiable. And consider, furthermore, how extraordinarily complicated this series of inferences is, in terms of which we are said to sense. In the experiment with transparent paper, *e.g.*, we are not only supposed to take account of the actual colour of the objects seen, but also of the influence of the medium through which the light rays have to pass before they reach our eye. And yet the experiments may be easily performed under conditions which definitely exclude any such influence. We have already seen that the contrast is also apparent when we lay our grey square directly upon the coloured background. Now, if pains are taken to choose saturations and brightnesses exactly corresponding to the modified colour and brightness appearing in the former case through the transparent paper, the contrast is just as clearly observable as before. So that it is not by betraying us into a false judgment that the transparent medium heightens contrast, but rather by producing degrees of saturation and brightness, which are especially favourable for it. Colour contrast always appears most strongly when the coloured object and the colourless one which it tinges by contrast are as nearly as possible of the same brightness. This condition is more nearly realised in all cases where the brightnesses of the objects are originally different, and their relations consequently not so favourable for the production of contrast, by the use of the transparent paper as a covering.

But although the judgment theory is untenable, the expression 'comparison' has a certain justifiable application to the process underlying contrast. We observe that not only is a really grey object modified by its surroundings so as to appear complementarily coloured, but also that this contrast colour disappears as soon as we destroy the influence of the surroundings by introducing another object of the same grey colour. It is, therefore, not amiss to say at least that this removal of contrast is the result of a comparison ; and if the expression is permissible in this case, it must be so also for our original phenomenon. For it is, indeed, conceivable that the result of one comparison could be cancelled by that of another, but hardly that an excitation-process due to definite physiological causes could be annulled by such an act. If we objectively tinge a grey paper with green, no amount of comparison of it with a pure grey of the same brightness can deceive us with regard to the existence of the colour. And we learn from the phenomena of after-images that a colour due to subjective retinal excitation behaves in this connection just like one objectively produced. But apart from these considerations, there is nothing at all in the phenomena of contrast which is not done full justice to by this reference to a relative comparison. A grey paper on a black background looks brighter than it does upon white. This is perfectly intelligible if we assume that our apprehension of a particular light-sensation is not something absolutely unalterable, but is dependent upon other light-sensations which are at the same time present in consciousness, and in relation to which it is, so to speak, measured. The same point of view may be adopted for the explanation of colour contrast. If we gradually decrease the saturation of any colour, it passes over finally into white or grey. We may therefore regard the absence of colour as the lower limit of the various stages of saturation of any given colour. The principle of relativity which we have just been discussing renders it inevitable that the saturation of a colour should be increased, if its surroundings, presented at the same time with it, are complementarily tinged. This being so, the minimum of saturation will, like every other degree, pass over into a more complete saturation ; *i.e.*, a colourless surface will in contrast to a colour appear complementarily coloured.

The notion of *comparison* is, therefore, generally applicable to the mental process underlying contrast. But the judgment theory is wholly wrong in regarding the process as a comparative *judgment*, for the production of which all manner of complicated reflections have to be called into account. When it is said that these reflections take place unconsciously, a suicidal admission is made. At the same time that the mental process is resolved into logical reflections, it is really granted that these reflections do not actually exist, but only represent the translation of a process of an entirely different kind into the form in which we usually cast our reflections. Now such a translation is quite allowable when it is simply used as a means of making intelligible the mode of operation of the elements which have been empirically shown to be present in some particular process. But though allowable to a certain extent, for the sake of clearness, this method of interpretation oversteps its rightful limits when it leads us to ascribe to judgment purely imaginary preliminary stages, such as, *e.g.*, the reflection how a colour ought to appear when it is seen through another colour. That is what it does, you see, in the logical theory of contrast which we have been discussing,—logical : for the theory is really logical, and not psychological. If we wish, therefore, to retain the concept of comparison to designate the processes under discussion, we shall only be able to mean by it an *associative comparison* ; *i.e.*, a connection of two sensations, where the quality of each is determined by its relation to that of the other.

§ V

We have thus referred the phenomena of contrast to the same principle of *sensational relativity* with which we are already familiar as the general psychological expression for the facts of Weber's law. Our apprehension of stimulus-intensity or of tonal pitch is not something absolute, but depends upon the character of other stimuli and tones with which we bring the given sensation into relation. In the same way, the subjective effect of a particular light and colour is determined by the relations in which they stand to other impressions of light and colour which are affecting us at the same time.

We may now subsume all these phenomena—tonal intervals, light contrast, the geometrical increase of stimulus-intensity for equally noticeable sensation-differences—under one general law: the *law of relativity*. The psychological significance of them all is the same. We may formulate this law, which may be regarded as the most general expression of the results of our psychological analysis of sensations, as follows: *wherever there occurs a quantitative apprehension of sensations, whether as regards intensity or degree of quality, the individual sensation is estimated by the relation in which it stands to other sensations of the same sense-modality.*

Sometimes this relative estimation of particular sensations is determined by the impressions which immediately precede or follow it ; this is usually the case in the measurement of equally noticeable stimulus-differences. Sometimes it is determined by sensations simultaneously present, as well as by those coming before and after ; this is the case with tone-pitch. Sometimes finally, only the simultaneous impressions come into account, as in light and colour contrast. Which of these three conditions is fulfilled depends upon the special circumstances of the experiment, and upon the peculiarities of the special sense with which we are dealing. The matter does not at all affect the significance of the law. And this significance is, you see, predominantly *psychological.* For the most obvious interpretation of the law is this : that we never apprehend the intensity of a mental state as if it stood alone ; we never estimate an isolated magnitude ; but measurement implies a direct comparison of one conscious state with another. And so we shall expect to find that the law of relativity is not restricted to the sphere of sensation, but is applicable in every case where the intensity of a mental process is quantitatively apprehended and compared with that of others.

LECTURE VIII

§ 1. Reflex Movements. § 2. Purposiveness of the Reflex.
§ 3. Development of the Reflexes of Touch and Sight.

§ I

WE have seen that sensation is the ultimate source of a very large number of the concrete processes that go to make up our mental life. Everything that we meet with in our world of ideas is derived in the last resort from sensations, and ideas are the raw material of all the higher mental activities. It may be questioned whether the stream of thought could not continue without any reinforcement from the outside world ; but this at least is certain, that it has its source in sensibility, and that sensibility does perpetually interfere to determine its direction. In this way our investigation of sensation leads on directly to the consideration of a second and very important psychological problem,—the problem of the composition of *ideas* from the sensational elements that enter into them in so great number and variety.

In our previous discussions we examined the physical conditions of sensation, external sense-excitations, before we proceeded to sensation itself. Now the further question arises : what is the immediate *consequence* of a sensation ? Observation furnishes us with the unhesitating answer : every sensation which is of sufficient intensity and not inhibited by opposing influences is followed by a *muscular movement.* This muscular movement we term a *reflex* movement, and the name is appropriate. There have plainly taken place a transference of nervous excitation within the central organs of the nervous system from sensory to motor fibres, and a similar transmission of stimulation from these to their particular muscles : the stimulus is, as it were, thrown back, reflected. Reflexes are observed so long as, the nerves of sense retain their connection with the central organs of the nervous system, and these remain united by motor nerves to the muscles. But it is not necessary that the whole central nervous system should be capable of exercising its normal functions. The cord may be severed from the brain, and reflexes still be mediated by the nerves which enter it; indeed, · quite a small section of the cord will suffice for the necessary transference of excitation.

This transmission of stimulus by the sensory to the motor nerves is provided for in the elementary organisation of the nervous system. If we examine brain and cord microscopically, we find, first, a number of nerve-fibres of varying thickness, continuations of sensory and motor nerve-trunks, and secondly, besides these, cells of varying size,—viscous structures, like most organic cells, containing a more solid nucleus and a number of small granules. These nerve-cells are characteristic of and peculiar to the central organs of the nervous system. How important they are functionally you may judge from the fact that they always stand in connection with both incoming (afferent) and outgoing (efferent) nerve-fibres. As a rule each one sends out several nerve-fibres. We may look upon these cells, then, partly as ultimate terminal organs, partly as organs for the mediation of connection between conducting fibres. To make the mechanism of the reflex clear, we have only to suppose a nerve-cell interpolated between two fibres, one of which (*e*) leads from a sense-organ, the other (*b*) to a muscle. We have there the scheme of the reflex. As a matter of fact, however, the nervous connections are very much more complicated and involved, as we should expect, knowing what we do of the complexity of structure of the central organs.

FIG. 16.

The intensity and extent of the movement constituting the reflex response to a sense-stimulus are enormously different in different cases. In general they increase with increase in the intensity of stimulus. The very weakest stimuli do not as a rule excite reflex movements at all; moderate intensities arouse a moderately intensive movement, confined to some particular group of muscles; as the intensity of stimulus is further increased, the reflex answer becomes more general, till it finally involves practically the whole organism. This law of the increase of reflex movement with increase of stimulus remains constant despite the fact that individual and temporal differences may be by no means inconsiderable. These are mainly referable to varying excitability. The more excitable the sensory nerves and nervous centres, the earlier does the reflex make its appearance, and the more quickly does it run through the gamut of its intensive and extensive changes. Again, it may be intensified or reduced by the action of various influences upon the organism. Beheading, removal of the brain, increases the reflexes, until death occurs. Many amphibia may be kept alive for months after beheading, and their reflex excitability is throughout this whole period abnormally great. There are also certain chemical substances the effect of which upon the nervous tissue of the central organs is to occasion a decided intensification of the reflexes. Besides some alkaloids, whose influence is not very great, we may especially mention strychnine in this connection. It produces such an excess of sensibility that the lightest pressure upon the cutaneous nerves, such as normally would fail to arouse any movement at all, brings on reflex twitchings over the whole body. In deep sleep or swoon, on the other hand, the reflex excitability is diminished, as it is by the influence of opium and other cognate poisons.

What the chemical changes within the nerve-cells are upon which the influence of strychnine and similar poisons depends is still unknown. But their general effect may be easily deduced from the universal laws of the action of forces. It cannot be supposed that a material heterogeneous to the constituents of nervous tissue creates fresh nervous force. The only tenable hypothesis is, that it facilitates the actualising of forces already latently present; *i.e.*, that it overcomes certain of the inhibitions making against the transformation of stored into kinetic energy, and so renders the transforming force more effective. Substances of an opposite nature will, on the contrary, increase the number of inhibitions, and thereby increase the amount of external force necessary to release the cell from its state of tension. You will understand this easily enough if you take a simple mechanical illustration.

Suppose that you have a clock whose movement can at any moment be arrested or started by the throwing in or out of gear of some mechanical appliance, say a spring, which stops the works or not according as it is tightened or relaxed. So long as it is tightened, the weight which moves the clock will be exerting a pressure against it. This represents the stresses to which the mechanism is subject. As soon as it is relaxed, these are transformed into energy of motion. To effect the relaxation some small sum of work must be expended. Its magnitude will be proportional to the resilience of the elastic spring, and it can very easily be increased or decreased by varying the tension of the spring.

The movement of the clock in this illustration is the reflex movement, the relaxation of the spring is the operation of a sense-stimulus, and the greater or less tension represents the influence of the substances which are exerting a specific action upon the mechanism of transformation of energy. Just as a greater tension of the spring increases the difficulty of starting the clock, so the alteration produced by opium in the nervous system makes against the release of a reflex movement; and just as a less degree of tension facilitates the starting of the clock, so strychnine facilitates reflex movement. Every clock goes for a certain period, at the expiration of which it runs down and requires to be rewound; in other words, there is in it a certain amount of potential energy which it takes a definite time to use up and transform into energy of motion, and which then requires renewal. Without this renewal the clock cannot go any more. Here again there is a complete analogy with the mechanics of the nervous system. There is a definite amount of potential energy contained in the central organ. This is, partly, only renewed when it has been almost exhausted (as in the clock), the restoration taking place during sleep; but, partly, there is a continual process of renewal, as, indeed, there must be, if the chemical composition of the nervous elements is not to be so radically disturbed that a return to the normal condition is altogether impossible. So that the inevitable result of too heavy a draft upon the potential energy is death. Strychnine and similar poisons bring on death simply by exhausting the energy of the central organs, and particularly of the cord.

The other tissues of the body are left uninterfered with; and even the nerve-fibres, as has been shown by their severance from the cord, remain capable of taking up and transmitting stimuli.

Now the removal of the brain has the same effect upon the reflexes as any one of these poisons, which helps to set the reflex mechanism in action. But it is plain that this effect must be differently produced. The facts are these. The nerve-cells of the cord, which connect sensory with motor fibres, are themselves very complexly interrelated, and, moreover, send out fine nerve-fibres which run to the brain, and there terminate in the central cell plexus. These relations are represented schematically in Fig. 17, where *r r* are cord-cells, functioning as reflex centres, and *cc* central brain-cells. The

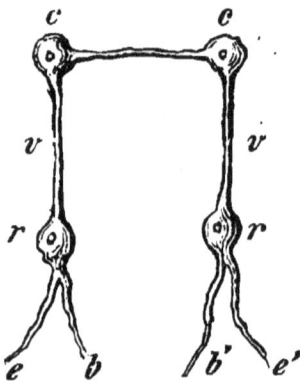

FIG. 17.

stimulus acting on the termination of a sensory nerve (*ee'*) is not simply transmitted by way of the reflex arc to a motor fibre (*bb'*), but is conducted to the higher cells (*cc*), and there diffused —it may be with practically no check or restriction. The figure shows, moreover, that a single sensory fibre is not always correlated with a single motor fibre. The conduction paths between cell and cell are so numerous that every sensory fibre is connected with a number of motor, and even with other sensory, fibres. So that if an excitation arises at *e*, we shall expect force to be released not only in *b*, but also in *b'* and *e'*. To get any idea at all of the organization of the central organs you must multiply these schematic connections indefinitely. A sense-impression will result not only in the movement of a definite group of muscles, but in movements and sensations at quite different parts of the body, extending perhaps to a whole number of muscle groups and to several sense-organs.

There can, indeed, be no manner of doubt as to the existence of *reflected* sensations,—sensations occasioned not by the stimulation of their own sensory nerve, but by that of some other. Normally, it is true, they are very weak; and it is only in conditions of pathologically increased excitability that they attain to a more considerable intensity. They appear, then, to be essentially different from reflex movements in that they are never so intensive as the directly stimulated sensations, but, as a rule, only perceptible at all with strained and careful attention.

Far more important is the other side of the matter,—the extension of the reflex movement due to the transmission of the impulse given by sensation to parts of the body increasingly remote from its origin. Heightening of stimulus-intensity will produce a very considerable diffusion of movement, but the quite universal reaction is only gained in conditions of abnormal excitability, *e.g.*, under the influence of strychnine. Here there is no definite limit set to the extension of the reflex response ; practically all the muscles of the body are violently convulsed during the action of the sense-stimulus.

But, these abnormal states apart, we can assign definite rules to the diffusion of the reflex. It is uniformly dependent upon intensity of external stimulus. When the stimulus is only just intensive enough to call forth a reflex answer, the movement is always restricted to the group of muscles most immediately connected with the sensitive part. If it is the retina that is stimulated, the resultant movement is exclusively eye-movement ; if it is the skin of one of the four limbs, only that limb moves ; if it is some portion of the skin of the head or trunk, there is movement in the nearest muscle group, and generally also in the limb which bears the closest relation to the stimulated spot. Thus a weak stimulus applied to the left cheek leads to a contortion of the left side of the face and a movement of the left arm. The four limbs, which with the eye constitute the most motile parts of the body, are most liable to be thrown into movement by cutaneous stimuli.

If the intensity of stimulus increases, the reflex movement becomes more widely diffused, but is still confined for some time to muscles in the neighbourhood of the organ stimulated. Thus it may extend from one leg or arm to the other. As the intensity increases further, the response grows more and more general ; at the very highest intensity all four limbs are generally in motion together. The movement is at first a flexion, but is changed to extension at the highest intensities. This fact,—

that a stimulus which affects all the motor fibres equally always gives rise to a movement of extension,—seems to show that nerves of the extensor muscles are not so intimately connected with the sensory fibres as those of the flexor muscles. It is not till the stimulus becomes very intensive that the extensors are affected, though when once the reflex response has become maximal, extension tends to get the upper hand of flexion.

§ II

In both of its two principal forms, the reflex possesses the characteristic of *purposiveness.* If the movement is restricted to the muscle group underlying or immediately adjoining the point stimulated, its result is to free the part affected from the influence of the stimulus. If the response is more diffuse, the action of the muscles is primarily auxiliary to the movement of escape ; and it is only in extreme cases, where a whole number of muscle groups are thrown into movements of extension, that all evidence of purposiveness is absent.

The principal reason for this purposiveness of character is that the reflex movement is usually directed towards some *end*, and ceases when it has been attained. The end is contact with the part stimulated. If you stimulate a headless frog by applying a sharp point to the posterior portion of the trunk, one of the legs is violently moved towards the injured spot. Contact with this seems to be the end of the whole movement ; and contact is effected in the simplest possible way,—by that limb and by those muscles which can mediate it with the least expenditure of effort.

Stimulation of the eye produces results analogous to those obtained by stimulation of the skin. If you look at the eyes of a new-born child, you cannot fail to notice the fixity of their gaze. The eye moves, it is true, and especially if light-stimuli fall upon it ; but the movement is altogether irregular, and there seems to be no definite connection between it and the locality of the visual impression. This relation is only gradually built up. If you bring a light into the visual field of a child several days or weeks old, you will find that it turns its eyes towards you and looks fixedly at the light. If you introduce two or more lights, it generally alternates between one and the other. But its gaze is riveted upon the light ; the eye is held to that by a kind of mechanical necessity, and can only leave it when its impression is weakened by fatigue, or when some other stimulus has appeared to oust it. We are in presence, that is, of a similar phenomenon to that of the touching of the stimulated part of the skin by means of a reflex movement. When a light-stimulus makes its appearance in the field of vision, the eye moves towards it, just as the hand moves towards the irritated spot upon the skin.

We must notice, however, that the reflex mechanism of the eye is twofold. On the one hand, there is the connection between light-sensation and the muscles that contract the pupil and close the lids ; on the other, that between light-sensation and the muscles which move the eyeball. The lid reflex may be occasioned by quite weak light-stimuli, if the eyes have previously been in the dark. Thus the first opening of the eye of the new-born child to light is at once followed by a violent and convulsive closing reflex. But the organ very quickly begins to grow accustomed to the light ; and then the connection between light-sensation and muscles of the eyeball makes its appearance. At first, as we have seen, the entrance of a light-stimulus into the field of vision merely produces movement ; we cannot say that the movement is governed or directed in any way. It is still only an uncertain groping for the light. But between the second and fourth weeks after birth some amount of regularity is observable. The child begins to fixate ; and every light-stimulus that appears within the field of vision excites a tendency to fixate it. Fixation consists in the assumption by the eye of the position in which a definite light-impression forms an image upon the most sensitive portion of the retina. This spot lies approximately at the centre of the whole retinal surface, a little to the outside of the point at which the optic nerve enters the eye, and from which it radiates over the retina. It is characterised anatomically by an extremely close packing of the retinal elements which take up the light-stimulus, and by a yellow coloration. On this latter account it is commonly known as the *yellow spot*.

The way in which the reflexes of the eye develop, then, is

this. First of all, light-stimuli excite merely irregular movements. After a time, these take on a definite form, and serve a definite purpose : the eye moves in such a way as to bring the image of the stimulating light upon the yellow spot. If at this stage a light moves to and fro within the field of vision, the eye follows it with an equally continuous movement.

By what steps does this regularity emerge from the initial irregularity ? It is clear that the determinate relation of the yellow spot to the reflex movements cannot be given in the form of an interconnection of the nervous elements conditioning them. If that were the case, the first ray that fell upon the retina would release a reflex movement of the same character as those which we observe later. Now, apart from the evidence against this view that is furnished by the facts of experience, there is nothing to suggest it in the manner in which the force which releases the movement is transmitted within the central organs. The force transmitted from the sensory to the motor fibres is dependent as regards intensity and diffusion upon the intensity of stimulation, and the temporary condition of the organs which subserve its transmission. So that there is no reason why an intensive stimulus in the neighbourhood of the yellow spot should excite only a very slight eye-movement, while a weak stimulus at the periphery of the retina is followed by an extensive one. No ! there must obviously be influences at work during the development of the sense which gradually bring it about that, while the *release* of a reflex movement is effected by the physical mechanism, its *extent* and *direction* depend entirely upon the place of the part stimulated ; so that the greater sensitivity of the organ or the increased intensity of stimulation can only find expression in a greater energy and rapidity of movement. To obtain a complete explanation of how this happens, we must look a little more closely at the structure of the sense-organs.

§ III

The skin over the entire surface of the body is sensitive to stimulation. And the entire retina is similarly sensitive, with the exception of the place of entry of the optic nerve, where there are none of the peculiar end-organs which subserve vision. But the sensations derived from various parts of the skin or retina are by no means entirely similar. So far as the skin is concerned, you can convince yourselves of this very simply. Touch your cheek and the palm of your hand with your finger, being careful to exert the same amount of pressure in each case. The two sensations are quite clearly different. And it is just the same if you compare the palm with the back of the hand, or the neck with the nape, or the chest with the back, or any two portions of the skin which are some little distance apart. More than that, if you observe carefully, you will find appreciable differences in the quality of sensations coming from portions which are fairly close together. As you pass from one point upon the sensitive surface to another, you experience a gradual and continuous change in cutaneous sensation, although the nature of the external pressure has remained perfectly constant. Even the sensations from corresponding places on the two halves of the body, alike as they are, are not altogether the same. If you touch first the back of one hand, and then that of the other, you find that there is a slight qualitative difference between the two sensations.

A similar variation can be shown to exist in the retina. Fixate a piece of red paper held in the hand, and then move it slowly away, without allowing your eye to follow it as it disappears. Its image falls at first, of course, upon the yellow spot ; and then travels across the retina towards the periphery. You will observe that during the lateral movement of the object the sensation of red undergoes a gradual change : the colour-tone becomes darker, appears to take on a tinge of blue, and finally passes over into pure black. Any colour that you choose, as well as white, will show similar alterations. The last stage in sensation is always black.

The obvious explanation of this phenomenon is, that we sense differently with the different parts of the retina ; and that the sensation gradually changes as the impression moves from its centre towards the outlying regions. So far as we can tell, the alteration takes place in just the same manner whatever the direction of movement ; but,—and this is noteworthy,—it takes place with different rapidity in different directions. If the object moves from the centre outwards or upwards, the series of colour-tones is passed through more quickly than if it moves inwards or downwards; so that a body imaged on the outside or upper portion of the retina looks black, while if its image falls on a corresponding portion on the inside or below, it is still seen as coloured.

When you have made these experiments upon the colour-change undergone by small objects seen with outlying portions of the retina, you will ask yourselves with astonishment how it is that you do not always notice the differences; that the blue sky or the red front of a house is not surrounded by a black border. Indeed, we should naturally expect that if a blue or red piece of paper gradually blackens, as it is brought upon the lateral parts of the retina, the blue of the sky and the red of the bricks would also show some change towards the periphery of the field of vision. And we might find considerable difficulty in answering the question, if we had not already become acquainted in our discussion of sensation with a large number of facts which point out the way in which it is to be met. A sensation, we must remember, is not anything determinate and invariable, but the product of a comparison, or, more exactly, of its associative relation to other sensations, which is never conditioned exclusively by the character of the single excitation, but also by that of simultaneous and preceding impressions. It does not stand alone, but is brought into relations. If these co-operative determinants are so strong as to make us, under certain circumstances, see blue red and red blue,—you remember the phenomena of colour-contrast,— why, they may perhaps be able to prevent our noticing the differences in colour-tone which are produced by moving an object over the different portions of the retina.

We have every reason to suppose that the sensations which we get from ordinary contemplation of large and uniformly coloured surfaces are themselves entirely uniform. For if we move the eye, and fixate successively different portions of the colour surface,—*i.e.*, bring them in order upon the yellow spot,— we receive precisely the same colour-impression in every case. There are certainly differences given originally in sensation; but these we have eliminated in thousands of experiences by referring them directly to the spatial distribution of impressions, the association being carried out with the certainty and precision of a machine. So that when we are looking at large and continuous coloured surfaces, we simply do not notice the differences due to the place of the impression: sensation has emancipated itself from them.

This fact serves to illustrate what is a universal rule in the sphere of sense-perception, and one which we shall often have occasion to refer to in the future. We entirely neglect a good many of the different characteristics of a particular sensation simply because they are not directly connected with the objective contents of the corresponding perception. If we are considering a colour as produced by some external object, we take pains to be accurate as regards its quality. But when there is some peculiarity in the colour-tone which has nothing to do with the nature of the external impression, we only perceive it by the aid of special instruments, or by an extraordinary exercise of the attention. We can hardly doubt, therefore, that the sensation-differences in the different regions of the cutaneous surface or of the retina are really far greater than they appear in our experiments. Prepare ourselves as we may, we still tend to notice only those sensation-characteristics which are conditioned by the nature of the external impression; the mere intention to free ourselves from a rule which has been adhered to in the whole course of sense-development, and to which we necessarily and unconsciously conform, does not suffice to abrogate it. We must accordingly not content ourselves with the fact that a peculiarity of sensation, dependent upon the position of the portion of the sense-organ stimulated, can only be demonstrated in the rough or in a general way. But we may certainly suppose that such differences exist and are effective in cases where the deficiency of our observational methods prevents our cognition of them. For the facts mentioned above show that the sensation-difference must have become quite extraordinarily large, if it is to be apprehended as a subjective difference of sensation, and not simply referred to local differences in the objective stimulation.

Now what is the cause of these peculiar differences in sensation? It is plainly entirely local; and it must, therefore, be looked for in the structure of the sense-organs. Differences of tone and colour we referred ultimately to differences in the terminal organs of ear and eye. So these further qualitative differences, peculiar to the sense of sight, must be ascribed to slighter variations in the structure or chemical constitution of the retinal end-organs. The assumption is not by any means without factual support. Observation shows that it is especially the sensitivity to red light that decreases towards the periphery of the retina. This is interesting, because the most frequent type of partial colour-blindness, as we saw before, is red-blindness. So that ordinary red-blindness possibly means nothing more than an extension of the normal sensibility of the lateral parts of the retina to the centre.

In the skin, too, there are many local differences which may serve to explain the qualitative variation of sensation with the place of the portion of the organ stimulated. The bulb-shaped end-organs which take up tactual stimuli are distributed in various quantities, like the retinal elements. There are far more of them, *e.g.*, at the sensitive finger-tips than in the comparatively insensitive back or thigh. And there are further differences in the thickness of the epidermis, and in the nerve-supply of neighbouring tracts of skin, which may bring it about that one and the same impression is differently sensed at different parts of the surface of the body.

And now we have established a fact which may help us to answer the question raised a little while ago. We had asked for the conditions under which a system of reflexes at first entirely irregular could give rise to one of regular and uniform movements. We have found that the skin and the eye, the two sense-organs whose stimulation is the principal incentive to reflex movement, present structural peculiarities which imply definite local differences in sensation. What must our inference be, then, as regards sensation? Evidently, the result will be identical with that which we have when a colour is recognised as the same colour, or a tone as the same tone. Each particular sensation will be recognised in terms of this attribute of dependence upon the place of the impression; and we shall be able to recognise from the attribute the locality of the sensation itself when we have once had experience of its position.

This whole group of facts, then, leads us to a single conclusion : that we have in it the principal condition of the purposive development of the reflex. That development consists, once more, in this,—that a movement which is at first purposeless comes to have a definite object, the object being the sensitive spot which was stimulated by the external impression that released the reflex. That this spot may be discovered by the reflex movement, it is necessary for it to be recognised in each particular case. And just as colour and tone are only apprehended as like or different because their sensations are indistinguishable or distinguishable, so the recognition of the locality of an impression can only be effected by means of definite sensation-characteristics,—characteristics which depend solely upon that locality. We have shown that there are such characteristics. And with that proof we have given the first condition for the regularity and uniformity of the reflex : it must be looked for in the structure of the organs themselves. At the same time we have not yet given an altogether complete and satisfactory account of reflex development. We find that the movements always take the shortest and simplest road to their purpose ; and the structure of the sense-organs cannot, of course, explain that. It can only tell us how it is possible for the reflexes to have a purpose, not how they can attain it. There must be a further explanation of this, to be discovered in the movement itself. We must, therefore, go on to ask whether any such explanation is admissible, and what are the terms of it.

LECTURE IX

§ I

REFLEX movements become transformed from their
original purposelessness and irregularity, so that they
conform to definite ends and follow definite laws. And the
essential factor in this transformation can only be looked for in
the movements themselves. So the question arises as to our
measurement of muscular movement.

When we move the legs in walking, we measure off the length
of each step, without having to follow the movement with the
eye. The practised pianist has acquired such skill in estimating
the distances of the various keys, that his fingers scarcely go
wrong by a hair's-breadth. And we possess an accurate judgment of the force of muscular movement. We discriminate the
magnitude of different weights by lifting them. (It has been
already shown that this discrimination is not made in terms of
the pressure of the weights upon the skin, but by reference to
the act of lifting, You will remember that if lifting is allowed,
a difference of $\frac{1}{17}$ can be distinguished ; while in the case of
simple pressure upon the skin a difference of $\frac{1}{3}$ is only just
noticeable. *Cf.* above, p. 27.) So that we possess a very
accurate measure of the force and extent of movement in the
movement itself. And such a measure can have been acquired
only by aid of the sensation which accompanies muscular movement. For sensations are, as we know, the only means by which
we receive intimation of changes, whether outside of us or within
our own body.

Now, if we attend closely to our movements, we become aware that they are, in fact, always attended by sensations from the muscles. As a rule, it is true, these sensations are so weak that they escape our notice. It is only when we are exerting a certain amount of effort,—*e.g.*, moving a whole limb,—that we observe with any clearness the strain-sensation in our muscles, although much less extensive movements are capable of producing sensations of considerable intensity, if they are frequently repeated, and so occasion fatigue. Fatigue manifests itself by a muscle-sensation, sometimes present while we are at rest, sometimes only appearing (or at least only becoming actually painful) when we move.

The fact that muscle-sensations must be unusually intensive before they can attract our attention depends upon the ultimate character of our sense-perception. We saw above that sensations which cannot be referred to properties of external objects are very easily overlooked. The local colouring of the sensations of sight and touch escapes immediate observation, because we ordinarily direct our attention only upon the place from which the external impression comes. In like manner, we take no account of our muscle-sensations as sensations, but regard only the perception whose instruments they are, the force and extent of the movement made. The sensation calls up at once the complex idea of which it is a constituent ; and we require special experimental methods, or an unusual intensity of sensation, if we are to become conscious of it as such.

The sensations which accompany muscular contraction are probably occasioned by the pressure which the contracted muscle exerts upon the sensory nerve-fibres contained in it. But besides these sensations accompanying actual contraction and the cutaneous sensations of pressure and strain which are always connected with them, there are still other sensations involved in movement, whether executed or merely intended. Our sensations of movement are by no means dependent solely upon the external or internal work performed by the muscles, but are influenced also by the intensity of the impulse to movement proceeding from the central organ in which the motor nerves have their roots. This fact is most clearly indicated by observations on pathological changes in muscular activity. A patient who is partly paralysed in leg or arm, so that he can only move the limb with very great effort, has a distinct sensation of this effort : the limb seems heavier than it used to be, as though weighted with lead ; that is to say, there is a sensation of greater expenditure of force than before, although the work actually done is the same or even less. For the performance of this amount of work there is required an innervation of abnormal intensity. In the same way, the patient will deceive himself, especially in the first stages of the disease, with regard to the extent of his movements. His steps are short and uncertain ; his hand misses the objects which he is reaching for. By degrees, if his condition remains unchanged for a long time, he regains more or less precision of movement ; practice gives him familiarity with his new system of muscle-sensations.

Sometimes this state of partial paralysis is confined to a single muscle group, or even to an individual muscle. In the eye, *e.g.*, the partial paralysis may affect merely the single muscle which turns the eyeball outward, and which anatomists call the external rectus. There then arises a very curious alteration of vision. The patient has a wrong idea of the locality of the objects he sees on the side of the diseased eye : he places everything farther outwards than it really is. If he tries to take anything in his hand, he reaches out beyond it. A day-labourer whose work was stone-breaking, and who was attacked by the disease, began to hammer the hand that held the stone instead of the stone itself. But in these cases, too, it has always been found that, if the condition persists unchanged, the patients gradually become accustomed to their state, and regain the power of accurate movement, the only abnormality being the feeling of greater effort in the diseased part.

These phenomena of partial muscular paralysis render it intelligible that even in cases of complete paralysis there may still persist the idea of active movement of the paralysed muscle. If a patient whose leg is completely paralysed makes a firm resolve to move it, he may have a distinct sensation of muscular strain, and consequently an idea that the leg has really moved. By calling in the aid of sight he can, of course, convince himself that he has been deceived : in the dark the illusion is complete. The same thing happens when it is the eye that is the totally paralysed organ. The idea of an actual movement is connected with the inefficient resolution to perform that movement. The result is an optical illusion ; external objects appear to have moved in the direction of the purposed movement of the diseased eye. This apparent objective movement is evidently a necessary consequence of the subjective illusion. If the eye had really executed the intended movement, the images cast by external objects upon its retina could only have retained their positions unchanged, if the objects themselves had moved in the direction of the eye and in complete accord with it.

It has sometimes been thought that the act of will suffices of itself to explain these subjective movement-illusions. If I will to move an organ which is dependent upon my volition, it is said, there is necessarily connected with my resolve the idea of its actual movement. But it is difficult to see how a resolution can contain in it that peculiar sensation of muscular effort by which the magnitude of movement is measured alike in cases of partial and complete paralysis. It is surely evident that this sensation is a process accompanying the act of will, and capable of varying in degree, for one and the same volition, with variation of its particular conditions. Under ordinary circumstances the sensation is demonstrably caused by the stimulation of the sensory nerves following the contraction of the muscle. How can it arise in cases where the muscle is not able to contract?

We must remember that muscle-sensations always accompany the particular volition. Hence, whenever a volition is repeated, the appropriate muscle-sensations will be connected with it. And since they are familiar to us from numberless previous perceptions, and are inseparably and invariably connected with the will-process, they will be found along with this latter even in cases where the muscle is unable to contract and so to furnish the usual sense-stimulus. Now we give a special name to all those sensations and ideas which, though not occasioned by external, but by internal, stimuli, are yet completely dependent for their determination upon previous external excitation : we call them *reproduced* sensations and ideas. And we may accordingly consider these sensations of muscular effort which are the invariable accompaniments of volition as reproduced muscle-sensations, while we distinguish them from others of like nature by the intimacy of their connection with special processes of consciousness,—volitions,—and by the constancy with which they accompany these. They will, of course, accompany them in cases where the influence of the will upon the muscles is normal and effective; but there they will at once fuse with the actual muscle-sensations occasioned by the stimulus of contraction. In other words, their effects can only be separately followed out when, as in the illustrations given above, partial or total muscular paralysis has disturbed or entirely destroyed the other muscle-sensations which are peripherally excited.

§ II

For any historical investigation into the development of our sense-perceptions, the question of the origin of sensations of muscular effort is of far less importance than the analysis of the phenomena which are brought to light by their disturbance. The gradual adaptation to the diseased condition in cases of partial muscular paralysis, like those which we have noticed, seems in particular to be at least as instructive as the condition itself. It shows what sort of influence muscle-sensations may have exercised upon the development of the senses. Our recognition of the position of an object is normally based upon the sensation of effort attending the movement of the sensing organ to the object. If this power of localisation may be gained afresh after a total transformation of the whole system of muscle-sensations has taken place, there is not the slightest difficulty in the hypothesis that, when sense-perception in general was in process of development, the establishment of a relation between muscle-sensations and the place from which an external stimulus operates was a matter of slow and gradual growth. And this takes us directly back to our original problem. We set out from the proposition that, if reflex movements admit of accurate measurement, the measure can only be looked for in the movements themselves. We found the measure required in sensations varying with the force and extent of movement. We have now proved, by an appeal to experience, that it is by means of these movements that our limbs and organs of touch acquire their accuracy of function. Any alteration in the muscle-sensation does away with this accuracy. And it can only be regained, if at all, by a fresh course of practice.

Our view of the development of reflex movements will, then, be somewhat as follows. They owe their origin in the first instance solely to neural connections existing within the central organ ; that is their only primary condition. The sensation occasioned by a stimulus gives rise to a more or less extended movement, and this in its turn to a muscle-sensation. The movement is, therefore, only a middle term between two sensations : between the original sensation caused by the external stimulus and the muscle-sensation which results from the movement. But there is more in the whole process than this. When we move our limbs, it is either that they themselves may come into contact with the sensitive surface of the sense-organ, or that they may transfer the stimulus from one portion of that organ to another. Suppose that a stimulus operates upon the skin. In the movements which are aroused this or that part of the skin is touched ; in other words, there results a second sensation of contact, beside that already caused by the external stimulus. And this naturally arises in the near neighbourhood of the other, since the excitation-process which underlies sensations of moderate intensity extends only to adjoining nerve-connections, and so sets in movement only the contiguous muscle groups. The total process now consists not of two, but of three, sensations. The last two of these (sensation of movement and secondary sensation of contact) are at first of indefinite extension. But very soon there comes to the front some particular contact-sensation, one which is similar in character to the sensation which stood first in the entire series, and which was directly occasioned by the external stimulus. And this similarity is obviously conditioned by contact with the place upon which the stimulus originally operated. We have seen that there attach to every portion of the skin certain local characters, by means of which it can be distinguished and recognised. The end of the movement will be then the production of a sensation of contact at the place upon which the stimulus operated. This end is easy both of proposition and attainment. For we can recognise not only the peculiar character of the contact-sensation, but also that of the muscle-sensation corresponding to it. If we suppose all this to have happened in a large number of cases, we see that a firm connection will have been established between the two sensations. So soon as a stimulus operates, and a sensation is aroused, the corresponding sensation of movement is awakened, and with it the movement, which is responsible for the final term of the whole series,—a sensation of contact identical in local character with the sensation constituting its initial term.

In the eye we find these phenomena modified by the peculiar structure of the organ. The nervous connections of the retina place it in reflex relation to the muscles which move the eyeball. One portion of the retina is characterised by an especial clearness of sensation. While upon the peripheral parts different colour-impressions are practically all sensed as the same uniform grey, and even as they approach the centre remain for some time indistinct in tone, upon the 'yellow spot' they are clearly and accurately distinguished. Hence the law which governs the development of the reflex movements of the eye :—every impression, upon whatever part of the retina it fall, is brought to the place of clearest vision, the yellow spot. From the whole series of purposeless reflex movements there comes into prominence this particular one, the effect of which is to place the eye directly in such a position, that the stimulus can act upon the yellow spot. Here too, then, a definite movement-sensation, whose purpose is to regulate this definite movement, becomes connected with the local character of the retinal sensation, wherever it may be aroused. And the final term in the total process is always a sensation which is recognisable because characterised by its relation to the yellow spot.

But while we have been describing the development of reflexes, as we observe it in experience, we may, perhaps, have fallen into a grave error. We seem to have been ascribing to the organism, at this early stage of its mental development, definite tendencies and purposive actions. Should not the phenomena under observation be rather regarded as subject only to mechanical laws ? In other words, is it not we who are putting purposiveness into them, while the sensing and moving subject itself knows no more of that than the stone knows of the intention of the boy who picks it up and throws it ?

We cannot, it is true, predicate 'purpose' and 'intention'

of these elementary processes of sense-perception in the signification which the words possess for ourselves. Nor, as a
matter of fact, is anything of the kind presupposed in the above
account of them. The processes which we assumed as necessary
for the regulation of reflex movements are of just the same
nature as those present in the discrimination of sensations according to intensity and quality. They consist in the connection
of sensations which are excited simultaneously or in immediate
succession by the operation of a sense-stimulus. Such connections are termed in general *associations*, and are distinguished as
simultaneous and *successive*. Now the unfailing characteristic of
an association between two sensations, a and b, is this : that
when one of them, a, is given, b is added to it, even though the
external stimulus for b is not present. In other words, the criterion of an association is the spontaneous reproduction of one
of the members of an association complex. Taking this criterion
as our guide, and carefully observing the facts, we find that the
association between definite sensations becomes stable in proportion to the frequency of its repetition. But besides frequency
of repetition, which is involved in all the phenomena of practice
and habituation, there is a second influence of importance at
work in this particular case. If a sensation, a, enters into two
associations, one with a similar or related sensation, b, and one
with a quite different sensation, c, the resulting complexes
possess a different character. In the association $a + b$, the components are apprehended as similar and associated ; in the
association $a + c$, as dissimilar and associated. So that all
associations of sensations may be divided, again, into associations of similar and associations of dissimilar sensations. Two
musical clangs, *e.g.*, which constitute a harmonic interval, form
a similarity-association ; they are related, as we have seen, by
certain common partial tones. But two completely different
sound-impressions, which have no elements in common, form an
association of dissimilar sensations. If now a sensation, a, has
become associated with several other sensations, b, c, d..., of
which b is more like it than c, d... are, the similarity-association
has the advantage, other things equal : it forms more readily
than the alternative associations. This is not difficult to explain. The transition from a to b will be facilitated by the

existence of properties common to both ; *b* is partly contained in *a*, and so is already partly present when *a* is present.

Apply these considerations to the phenomena which we have been examining. It is at once clear that they can all be explained as association-processes. Light-sensations, *e.g.*, form associations with the corresponding sensations of the ocular muscles. And these have become so stable, owing to the functional connection of light-impressions with the resultant reflex movements, that even if actual movement is prevented, the reproduced sensation of movement is still present. Or, again, the relation of light-impressions in the visual field to the spot of clearest vision is a characteristic instance of the association of similar sensations. This association, being preferred to all the other possible ones, is, of course, rendered still more stable by continual repetition. So that if in describing the development of these reflex connections just now we chanced to speak of the 'recognition' of an impression, of its 'being brought upon' the spot of clearest vision, that must not be interpreted to imply that deliberation and reflection are at work, in the general signification of these terms. It was necessary, in order to be intelligible, to translate certain processes of what we may call mental mechanics,—certain association-processes,—into the language of logical thought. Logical thinking is the form of mental activity with which we first become directly acquainted in our internal experience. And so it offers a ready means of making clear the connection of separate elements in a mental process, although the process itself may not belong at all to the sphere of logical reflection. But we must be careful. The logical formulæ which we often find so useful in explaining the connection of mental processes must not be confused with the processes themselves. These associations, into which the processes of sense-perception may be analysed, form the basis upon which all the higher mental activities, including logical thinking, rest. It is always possible to put this farther back, to find it in the elementary processes ; or, as it might, perhaps, be better stated, it is always possible to translate the results of the mental association-mechanics into the language of logical reflection, of which it itself is wholly ignorant.

§ III

The mental associations which we have been discussing are of a very simple kind. And the parallel connections of physical processes are also of a comparatively simple nature. It is not at all difficult to represent the entire complex, which we have been looking at from the psychical point of view, in purely physiological terms. In attempting this, we may, of course, leave entirely out of account the sensations which accompany the stimulation-processes in the organs of sense and of movement. It is true that we are obliged at times to interpolate hypothetical links in the chain of known facts. But these correspond well enough to familiar physical laws for us to be fairly certain that our physiological picture of the functions of sense, determined by the scheme of the reflex process, approximates pretty closely to the truth.

For our present purpose we may regard the general process of stimulation in sensory nerves as some kind of movement. Of the real nature of this movement we know nothing. We will, therefore, not seek to define it; we will only assume that it obeys the universal laws of mechanics. This movement is propagated, as we saw, through nerve-cells to the fibres of the motor nerves; it excites a greater or less number of fibres according to the intensity of stimulus and the degree of sensibility. The weakest stimuli are confined within the particular nerve-channel which is most directly connected with the stimulated sensory nerve; stronger stimuli have more diffused effects. It follows that the reflex process which is set up by the stimulation of a definite sensory nerve is, in the great majority of cases, kept within one definite nerve-channel. That channel will always be employed when reflex activity is awakened; whilst it will only happen occasionally that other channels are occupied. It is an obvious conjecture that this preferential nerve-channel is the one by which a movement is conducted to the stimulated part; that is, that the uniform sequence of events in the reflex is given with the uniform arrangement of its nervous connections. Indeed, this hypothesis may be looked upon as exceedingly probable. Wherever the mind interferes in the series of bodily processes, we find the conditions of its action given in the bodily organisation. Locomotion is determined by the structure of the skeleton and the arrangement of the skeletal muscles, sensation by the character of the nerve-endings in the sense-organs.

Now it is a fact of common observation in external nature that a movement which takes place again and again in the same direction comes by degrees to follow this direction more readily than any other, and will presently be unaffected by influences which at first would have had no difficulty in diverting it. When water is poured upon the ground, it forms a channel for itself. Its initial direction may have been determined by the merest accident; but, once determined, is adhered to, and the more certainly the oftener we pour. When a machine is set in motion, there is always the same resistance of mass to be overcome in its various parts; but friction is lessened by the wearing and smoothing of part against part: so that a machine which has been going for some time usually runs more easily than a new one, or one which has lain for a long time unused. If you let your watch run down, and do not wind it up for a fortnight, you know that it is always liable to stop until it has been going again for a week or so. Now there is good evidence for the view that the same thing holds of neural processes. If we are in the habit of executing some definite muscular movement, we know that it gradually becomes easier, *i.e.*, can be made with less expenditure of force. What we call 'practice' consists simply in changes of this sort. The execution of a practised movement becomes easier because the stimulation-process in nerve and muscle is the more easily set up the more frequently it is repeated. This process is originated by an increased supply of the elements essential to the tissues; so that exercised muscles show an increase in the mass of their contractile substance.

Observation of purposive practice puts it beyond all doubt that this restriction of innervation to a definite channel is a matter of very common occurrence. Most persons are unable to move certain fingers,—the third and fourth,—separately. But a little practice enables one to move either finger independently of the other. At first it requires a very considerable effort to do this; but, as practice is continued, the separate movement becomes so easy that it takes place almost of itself.

The course of practice in cases like these is approximately as follows. The first time that we attempt to move the particular muscle by itself, we are not entirely successful. However great the effort, adjoining muscle-groups are also involved in the movement made. With continued practice, however, this attendant movement becomes weaker and weaker, and finally ceases altogether. The uniform tendency in practice, then, is this : a larger and larger amount of the total excitation follows the channel of the nerve connected with the particular muscle, until, when the process has been often enough repeated, the whole of the excitation is confined to this single nerve-path. And that is precisely what we observe in the development of the constant reflexes. The only difference is, that in the present case this transmission of the major portion of the excitation by the particular nerve-channel is a matter of volition and intention, while in the reflexes it comes about of itself, through the connection of sensory and motor fibres. Moreover, it is obvious that in the present case as well it is really not will, but a frequent repetition of the same physical processes in the nerves, which directly produces the effects of practice. If it were will, we should expect the desired isolation of the movement to be attained at once ; whereas, in fact, however great the effort of will, practice is indispensable. On the other hand, when once isolation has been accomplished, it is not always necessary that the will should intervene for the production of the isolated movement.

§ IV

We have arrived, then, by different roads, at a single result. First of all we considered the development of purposive reflexes as a mental process. Its end proved to be the uniform limitation of reflex movements. Secondly, we attempted to analyse the process in physical terms. And here we are confronted with the same limitation as the result of physiological practice. So that the two investigations together, the psychological and the physiological, furnish a concrete illustration of the principle which we have already found to be universally valid in the sphere of simple sensation-processes,—the principle of *psycho-physical parallelism*. But we must now return to our psychological problem, and ask : what becomes of the reflex movements after they have been uniformly limited in the manner described ? What influence have the associations which have arisen by way of the reflex upon the further development of mental processes ?

We will attempt, first of all, to answer this question with reference to the eye, because the mechanism of movement is there obviously the more simple. Eye-movements are from the first confined to the few muscles which turn the eyeball. And the very special sensibility of the yellow spot brings them into a definite relation to this part of the retina. In the skin, on the contrary, there are numerous tracts of equal importance for sensation ; or, in other words, what is given only once in the eye is here repeated many times over.

The reflex movements of the eyeball are, as we saw, so disposed that any light-stimulus, wherever in the field of vision it may appear, is carried towards the yellow spot, the place of clearest vision. Every light-sensation occasions a movement, which transmits the stimulus by the shortest possible road to the yellow spot, and with which the corresponding sensation of movement is indissolubly connected. Another result of the movement is to change the local colouring of the primary sensation ; this takes on the quality peculiar to the place of clearest vision. The more remote from the centre the original stimulation, the greater the modification of the sensation. Now the intensity of the movement-sensation is also proportional to this remoteness. If I lift a weight two feet, I have a more intense sensation than if I lifted it only half as high. All our sensations of movement are intensively graduated in proportion to the magnitude of the movement made. The qualitative alteration of the light-sensations, therefore, runs parallel throughout to an intensive modification of the sensations of ocular movement. We recognise the relation of a particular light-stimulus to the spot of clearest vision by the local colouring which it takes on ; and we measure this relation quantitatively in terms of the resulting movement-sensation. When a stationary light-stimulus is brought upon the different portions of the retina by a movement of the eye, the character of the aroused sensation varies from point to point. And every such variation is paralleled by a movement-sensation. And so we associate this sensation of movement in the most intimate fashion with the variation, sepa-

rating the subjective sensation-differences from those which are
due to the action of an objective stimulus. This does not mean
that we apprehend them as subjective, that we distinguish them
as something in us from the things outside of us: there can
plainly be no question of any such distinction in these processes
of perception, entirely dependent as they are upon the mental
mechanics of association. Rather are these mental processes of
which we are treating the material from which the subject-
object distinction is gradually built up. They are but the first
step on the road to the conscious discrimination of the self. On
the other hand, there is no doubt that even at this stage a well-
defined distinction exists: the subjective differences form one
group of sensation-qualities, the other characters of sensation
another and a quite different one. And this is the fact which con-
cerns us here. A series of constantly recurring sensation-differences is brought into a relation of dependence with an entirely
similar series of sensations of movement. At the same time, in
saying this we postulate one condition the actual presence of
which might be doubted,—the condition that when once the eye
has brought an impression upon the yellow spot it leaves it again
and turns to another ; thereby, perhaps, bringing the original
one back to the particular portion of the retina which it had
stimulated in the first instance. (That must needs be the case,
if it is to be possible for us to recognise at all that a visual sensation has remained unchanged.) Now there can be no doubt
that the adult human being can move his eye hither and thither,
to this point and that. He takes in any number of impressions,
one after the other, just as he pleases. But can the same thing
be assumed for that earlier stage of development at which the
simple reflex mechanism is still undethroned ? As a matter of
fact, there is one influence already at work there which renders
this variation of fixation possible, and without whose preparatory operation the possibility of voluntary variation would certainly never have been realised. I mean the influence of
fatigue ;—the weakening of light-sensation after long-continued
operation of external stimulus.

Every stimulus which acts upon the peripheral portions of the
retina arouses a reflex movement, which brings its image upon
the place of clearest vision. There the impression is retained

for some little time,—until fatigue sets in and relaxes the mechanism. After this some other peripheral impression of a different kind, for which the retina is still unexhausted, may become the centre of interest, and arouse a second reflex movement corresponding to it. In this way you can see how a large number of external impressions may be successively apprehended at the spot of clearest vision. First of all will always come the most intensive, or that whose place of stimulation stands in the closest reflex connection with the yellow spot ; the others will follow in definite order. Now suppose that there are presented to the eye two luminous points at some distance from one another. Even if the external impressions are perfectly similar, the sensations which they excite will be possessed of a different local colouring. If the eye moves from its original position to another, in which the second luminous point falls upon the place previously stimulated by the first, the second sensation is made qualitatively identical with the first, while the latter has changed. The sensation of movement is there to measure the distance traversed ; that is, the distance of the two luminous points from one another.

Every particular connection of a sensation of movement with the corresponding series of local sensation-colourings is a long-practised association. The number of such associations is very great ; and they enter again into associative connections with each other. The sensations of movement forming a quantitatively graduated series, and the local sensation-differences being qualitatively graduated, there arises at the same time a complete parallelism of the two associatively related sensation-series. And the result of this compound association-process ? We must anticipate a little to state it. Since this process connects together the whole number of sensations excited in and round the eye, it will also systematise those sensory processes which begin with the simple light-sensation ; it will determine the form in which the eye transforms its sensations into perception.

This form is space-perception. So that our observations, even at this stage, lead us to the conclusion that the perception of space, psychologically regarded, is not an innate possession of the mind, but the product of an association of sensations. It will now be our task to test this conclusion in detail by investigating the properties of spatial perception.

LECTURE X

§ I

ALL our previous considerations have been based upon empirical facts. The laws of reflex movement, the muscular sensibility, the local differences in sensations of light and touch, the exhaustion consequent upon long exposure to sense-stimulus,—all these are phenomena which may be verified in experience. But at the conclusion of these considerations we seem to have left the firm ground of experience far behind us. We have ventured upon a psychological construction of *space*, from the associative co-operation of the specified factors. Is that not more than experience can ever achieve? Is not space a connate possession of the mind? Or, if not that, is it not at least an entirely new element in our knowledge, which is *sui generis*, and therefore not a derivative from anything else?

It is certainly true that the perception of space is a new element in our knowledge. But in this sense every psychological fact is new which arises from some particular combination of the elements of our mental life. The laws of this composition are such that the properties of the mental resultants to which they give rise can never be predicted from the properties of the elements which enter into them, although we are afterwards able to see the connection of these elements and their combinations in the complex result. Thus, *e.g.*, after we have completed a complicated process of inference, we recognise that the conclusion follows necessarily from the premises. But still it is, as contrasted with the premises, something new, something which had to be deduced by definite acts of thought. Nothing at all is gained by such general assertions as that the perception of space is a connate possession of the mind, or that spatial quality is an original property of our sensations of sight and touch. Not only are these statements not capable of proof, but those who formulated them have not even taken the trouble of examining the psychological problem before them. For a problem it surely remains—to ascertain whether the laws of the movement of the eye and the organ of touch, and the sensational associations connected therewith, exert any influence upon the perceptions of these senses.

It is a well-known fact that we are able to compare distances fairly accurately by means of the eyes. But it frequently happens that two distances which are not exactly equal are nevertheless regarded as such, just as in the case of simple sensations the perception of a difference only becomes clear when it has attained a certain magnitude, determined in each particular case by the character of the sense involved. Now in the present instance, just as in the sphere of sensation-intensity, we can determine by measurement how great the difference between two magnitudes must be for it to become just perceptible.

We draw two horizontal lines of equal or almost equal length, and ask an observer who knows nothing of their objective relations to say whether they appear to him equal or not. If we begin by taking the two lines equal, and gradually lengthen one of them, we shall reach a point where the longer line is perceived to be just noticeably greater than the other. Here the experiment is interrupted, and the difference between the two lines measured. If this procedure is repeated for various lengths, we obtain a series of different values which tell us how the apprehension of differences of distance varies with the gradual increase of the distances compared.

The experiment is therefore essentially the same as that which we made earlier to determine the dependency of sensation upon stimulus. We have only substituted space-magnitude for stimulus-magnitude. If the two lines with which we begin are one decimetre long, and if we gradually increase one of them, the difference is noticed when the increment amounts to about $\frac{1}{50}$ decimetre, or 2 millimetres. But if the distance with which we set out is only $\frac{1}{2}$ decimetre, the just noticeable difference will be correspondingly smaller; it will now be found to be $\frac{1}{100}$ decimetre, or one millimetre. And this ratio remains constant whatever other standards of measurement we may apply. Within certain upper and lower limits the difference is always approximately $\frac{1}{50}$ of the total distance with which we are dealing. Of the two horizontal lines in Fig. 18 the left is 26 and the right 25 millimetres long. We see at once that the former is the longer;

FIG. 18.

but if it is made just a little shorter, the difference is no longer noticed. You may convince yourselves by experiment that if the lines are drawn twice or three times as long, their differences must also be two or three times as large.

It is at once obvious that we have here the same law which we found to hold for the dependency of the just noticeable sensation-difference upon stimulus-difference. *The just noticeable increment of spatial distance always bears the same ratio to the total distance.* And it is plain that this coincidence may be explained most simply by reference to the fact that we possess in sensation a measure for the perception of spatial relations; and that the sensations which give us this measure most directly are those resulting from the movements of the eyeball, the intensity of which must increase with the length of the path along which the eye travels.

We hold before the face a box, *ss* (Fig. 19), open on one side and having a horizontal slit upon the opposite side, through which both eyes can look towards a white screen, *w*, and see it without perceiving any of the other objects in the room. Now we hang between the screen and the eye a vertical thread, *f*, kept taut by a weight. Each eye will of itself take up such a position that the thread *f* forms an image at the yellow spot, the place of clearest vision. The line drawn in space from this point through the centre of the eye is called the *visual axis*. We may say, therefore, that the visual axes of the two eyes cut one another or intersect in *f*. If we now alter the position of the thread somewhat, by bringing it nearer or removing it farther from the 'eyes, the angle formed by the intersection of the visual axes is changed at the same time; for the eyes always follow the thread and remain directed upon it. If the thread is removed to a greater distance, both eyes turn outwards, and the angle of intersection becomes more acute; if the thread is brought nearer, the eyes turn inwards, and the angle of intersection becomes more obtuse. When we know the alteration in the distance of the thread, we can easily determine how far each eye has turned round its centre. If the thread is moved little by little, the alterations in its distance will not be perceived at all; *i.e.*, the turning of the eyes round their centre is so slight that the accompanying movement-sensation is not noticeable. This sensation of movement only makes its appearance when the alteration in the position of the thread 'has reached a certain magnitude; and then we perceive that the thread has been brought nearer or removed farther off. This limiting point must be determined in a long series of experiments, and with different distances of the thread from the eye. We should find that the eye possesses the finest sensibility for its own movements when the two visual axes are practically parallel, *i.e.*, when the eyes are approximately in their position of rest. Under those conditions we can perceive an alteration of distance if the revolution of each eye round its centre amounts only to about the sixtieth part of an angular degree, or to 1'.

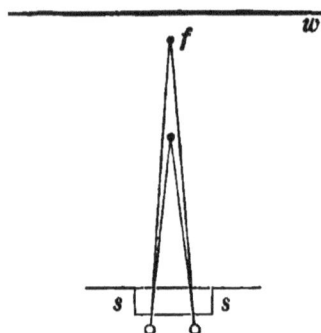

FIG. 19.

But so soon as the eyes have turned inwards to any considerable distance,—which happens, of course, when the thread is brought nearer,—the just perceptible movement is very much larger. And we shall find that the magnitude of this just noticeable movement increases in direct proportion to the distance of the eye from its position of rest.

We are plainly here only dealing with a further confirmation of the universal law of the dependency of just noticeable sensation upon stimulus. The turning of the eye inwards brings about a definite sensation of movement. The magnitude of the movement corresponds to the intensity of the stimulus; the greater the movement already present,—*i.e.*, the greater the stimulus already operative,—the greater must be the increase of movement or the increment of stimulus. And if the apprehension of sensations of movement follows the same law as that of the sensations of the external senses, it is to be expected that the increment of movement corresponding to the equally noticeable increment of sensation will always bear the same ratio to the total movement already present. As a matter of fact, experiment proves that this relation is approximately constant. Even such deviations as occur correspond to the rule which we have found to hold in the case of sensations of the external senses: that is, when the extent of the movement is very great, the fineness of discrimination becomes somewhat less than we should expect it to be according to the law. But the increment of movement which just suffices to produce a noticeable sensation amounts approximately to $\frac{1}{50}$ of the total movement-magnitude. This result is in complete accord with what we have already obtained from the comparison of spatial magnitudes : a longer line can be just distinguished from a shorter when the difference between them amounts to $\frac{1}{50}$ of the length of the latter. But if the perception of a spatial distance is directly proportional to the effort of movement made by the eye in traversing this distance, we must conclude that the effort of movement is the criterion of perception. And since we can only have knowledge of the effort through the *movement-sensation*, the influence of the latter is also demonstrated.

These experiments on the connection of the sensation of movement with the estimation of distance may be supplemented by the following observation. We suspend two black threads, side by side and parallel to one another, at a little distance from a bright background, and fixate them with one eye (Fig. 20). We then move gradually away from them, keeping them constantly fixated as we move. Since distant objects look smaller than near ones, the apparent distance between the threads continually decreases,

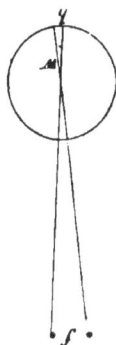

FIG. 20.

until a point is reached where the two appear as one. Now the decrease in the size of an object as we move away from it is due to decrease of the magnitude of its image upon our retina. So that the experiment shows that there is a certain magnitude of the retinal image of two points below which they cannot be perceived as separate. This magnitude of the retinal image (*b*) or of the corresponding visual angle (*w*) may be determined, since the distance between the threads and their remoteness from the eye are known. We find that the two images fuse to one at the moment when the distance between their retinal images has become so small that the eye has to turn only about 1′ in order to bring first one thread and then the other upon the same point of the retina. But that is the same magnitude as we discovered above to be that of the just perceptible movement of the eyeball. It follows, therefore, that the resting eye apprehends the distances of objects in space with the same degree of accuracy as that with which it apprehends its own movements under the most favourable conditions, *i.e.*, when the movement begins with the visual axes parallel. The limit which it can attain to in the cognition of spatial distance is identical with the limit set to its apprehension of the sensations of its own movements.

The dependency of spatial apprehension upon sensations of movement, which we have inferred from these fundamental experiments, is confirmed by many other of the phenomena of vision. The muscles of the eye are on the whole *symmetrically* arranged. Thus one muscle (*a*), the *rectus externus*, turns the eye outwards, and another (*b*), the *rectus internus*, turns it inwards (Fig. 21). The two muscles differ but little in their dimensions, and both lie in a horizontal plane, which passes through the centre of the eyeball. Their position is, therefore, the most advantageous possible for the movements which they are to bring about. This complete similarity of conditions renders it obvious that sensations of movement occasioned by equally extensive revolutions will be of approximately equal intensity, whether these revolutions be made inwards or outwards. We find the same correspondence as regards movement upwards and downwards. The eye is chiefly moved upwards by means of a single muscle (*c*), the *rectus superior*, which runs obliquely forwards in the upper part of the socket of the eye, and is affixed to the upper part of the eyeball, a little outwards from the middle. Its action is assisted by the operation of another muscle, which is hidden by the eyeball in our figure. This muscle, ·the *obliquus inferior*, runs in the lower portion of the socket, from before and within backwards and outwards, connecting with the posterior surface of the eyeball. Equally symmetrical in

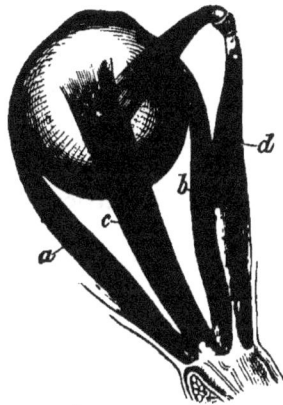

FIG. 21.

their arrangement are the muscles by which the downward movements are mediated. The operation of the muscle lying opposite to *c* on the lower side of the eyeball, the *rectus inferior*, is aided by a muscle, *d*, the *obliquus superior*, which runs forwards and inwards, and pulls on the upper surface of the eyeball. Owing again to the symmetrical distribution of the muscles, the effort of movement is approximately the same whether we turn the eye up or down. On the other hand, there is a very considerable difference between the arrangement of the muscles which turn the eye outwards or inwards, and that of the muscles which turn it up or down. If similarity were required in this relation also, the muscles would have to be so placed that the *rectus superior* (*c*), which moves the eye upwards, and the *rectus inferior*, on the other side of the eyeball, which moves it downwards, should be inserted at the point where they would best subserve the movement which they are to effect. This is, however, as our figure shows, not actually the case. The direction of *c* is somewhat more oblique than that of *a* and *b*. With an equal expenditure of effort, then, the former muscle would move the eye a less distance upwards than either one of the latter pair would turn it in or outwards. For this reason it is assisted by a second muscle. So that the effort necessary to produce a movement up or down is in general greater than that required for an equally extensive movement outwards or inwards, and, accordingly, the movement-sensations are more intensive; and we must expect to find that distance in a vertical direction will appear greater than the same distance in a horizontal direction. This is true as a matter of fact. If we draw a cross with equal arms, it will appear longer in the vertical direction (Fig. 22); while in other figures, such as squares or rectangles, the vertical distances are similarly overestimated.

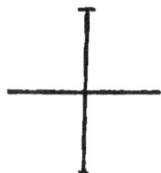

FIG. 22.

§ II

These differences in the estimation of vertical and horizontal distances are the most important, but they are not the only errors made in measurements by the eye. Smaller differences of a similar character may be observed between the upper and lower half of a vertical line, and between the inner and outer portions of a horizontal line. So that, strictly speaking, no one of the four arms of the cross in Fig. 22 appears exactly equal to any other. These lesser differences also correspond in every case to asymmetry in the arrangement of the muscles. When we remarked above that the two muscles *a* and *b*, which move the eye out and in, differ but little in their dimensions, it was, of course, implied that they are not completely similar. As a matter of fact, *b*, the *rectus interior*, is somewhat more strongly developed than *a*, probably because the converging movements of the visual axes predominate in all cases where we are occupied with what is near at hand,—*i.e.*, are fixating near objects : so that *b* is exercised more than *a*. It may be observed, accordingly, that the external half of an exactly bisected horizontal line appears longer than the inner half ; the weaker muscle requires a stronger effort to produce a like movement, and the greater effort is accompanied by a more intensive muscle-sensation. To realise this apparent inequality, it is, of course, necessary to close one eye. For the outside for the right eye is the inside for the left ; binocular vision destroys the inequality. A similar difference, and one which does not disappear in binocular vision, is that between the upper and lower half of the field of vision. If we look closely at the cross in Fig 22, we see that the upper half of the vertical line appears somewhat longer than the lower. And to this difference, again, there corresponds an asymmetry of muscular distribution. The muscles which pull the eye down are more strongly developed than those which move it up,—probably for the same reason which we found to hold in the case of the internal and external *recti.* Since the visual axes are usually directed somewhat downwards,—and this is especially true when we are fixating near objects,—the muscles which move the eye below the horizon get the more exercise, and an upward movement consequently involves a greater strain than an equally extensive movement downwards.

These visual effects of asymmetrical muscular distribution on the eyeball allow us accurately to predict other anomalies in the estimation of distances, which can be experimentally demonstrated. You know that we are more tired if we walk a distance in many short steps than if we take longer and fewer ones. The same holds of the eye. In passing over an uninterrupted path, it moves with less effort than over an equal distance which is frequently subdivided. If we bisect a straight line, then, and divide up one half into numerous smaller sections, the subdivided portion appears considerably longer than the other This experiment may, of course, be varied in all manner of ways. A subdivided angle appears larger than the same angle when open ; a plane figure appears larger when divided up into numerous smaller areas than one which is objectively equal to it, but left undivided, etc. These phenomena, which can be best observed in geometrical figures, have been designated *geometrical optical illusions.* They are all convincing proofs of the co-operation of sensations of ocular movement in the act of spatial vision.

§ III

The spatial perceptions of the sense of touch differ in many respects from those of the sense of sight. The difference may be partly due to the fact that in the normal development of our mental life the eye ranks as a far more perfect instrument than the skin, and that its particular development appears to precede that of the more delicate perceptions of the sense of touch. This does not mean, of course, that the two processes are sharply separated ; they rather cross one other, each influencing and assisting the other. But at least for man and the higher animals vision is the earlier activity, so that the sense of touch is rather guided and educated by sight than *vice versâ.*

If we remember that the pressure-sensations of the skin are always influenced by vision, we shall see that the local relations which attach to them must be primarily visual. But the sensations of touch are of such a character as to be able to throw off this influence to a certain extent. For the skin, as for the eye, the peculiar property of sensation which depends upon the locality of impression varies from place to place. So that this locality may in time be recognised by the local colouring of the sensation itself, without its being necessary to call in the eye to assist in the determination. When once the eye has settled the relation of the locality to its local colouring, we are able to refer a definite sensation to its right place upon the cutaneous sensitive surface.

It follows from this that the spatial discrimination of impressions will no longer depend upon movements, or upon the vividness and comparability of their accompanying sensations, but simply and solely upon the greater or less difference in the local colouring of sensations. If two contiguous portions of the skin differ indefinitely little in this respect, we shall not be able to distinguish the sensations proceeding from them. We shall only apprehend the impressions as spatially different when they affect portions of the skin whose sensational character is really different. And it is plain that this limit is not a fixed and unchangeable one, but that by close attention to our sensations we shall become able to distinguish between impressions lying nearer and nearer together. It is in this way that the great influence of practice observable in experiments of this kind finds its natural explanation.

In the same manner, the differences in power of discrimination, which we find existing at various points on the surface of the skin, will depend on the fineness with which the local sensation-differences are graduated. These divergences are really very large. On the finger-tips we can plainly distinguish as separate two impressions, *e.g.*, two compass points, the distance between which is only one millimetre ; while upon the skin of the back the distance must be 60 millimetres. So that the entire skin may be regarded as a graduated system of sensitive points.

But these points are not arranged uniformly upon it in order of sensitivity, but are at various distances from each other, and variously distributed. Besides the natural character of the cutaneous sensations, their control by the eye may contribute somewhat to this graduation. Not all portions of the skin are equally subject to visual control ; many, like the skin of the back, lie entirely beyond it : others, such as the hand and fingers, are peculiarly subject to it. It must also be remembered that all portions of the skin do not naturally get a like amount of practice. It is again the hands, and especially the tips of the fingers, which are most constantly exercised ; and after them come the lips and tongue. On account of this natural difference in amount of practice, the further development of the spatial discrimination of the skin, which we attain to by voluntary practice, differs considerably for different parts of the organ. On the finger-tips, *e.g.*, it is quite small, on the upper and lower arm strikingly large ; the power of discrimination may be doubled or even quadrupled in the space of a few hours. The advantage of such practice, it is true, quickly disappears : after twenty-four hours it has perceptibly decreased; after a few weeks or months it has entirely vanished. But the result is not limited to the portion of the skin directly exercised. If, *e.g.*, the fineness of apprehension has been doubled on the back of the right hand, the sensibility of the left will have increased by an equal amount, although that hand has not been exercised at all. The same result is obtained from all symmetrical portions of the skin. At the same time, the effects of practice never extend beyond these symmetrical portions. By practising the right lower arm or right cheek, we cannot help practising at the same time the left lower arm and the left cheek ; but there is no practice of the upper arm, or breast, or forehead. This peculiar result must be explained by reference to the psychological processes involved in practice. In practising, we learn to attend to sensation-differences which before escaped our notice. Now the local character of sensations of symmetrical portions of the skin is very similar. If, therefore, we have learned to attend to smaller sensation-differences upon the one side, we shall also have learned to do the same for the corresponding differences on the other. Especially with respect to right and left, there is complete correspondence in fineness of graduation and in the rapidity with which the local colouring alters from point to point. The case is, of course, different when we are dealing with asymmetrical places ; the sensations and their gradations are so different, that experience gained at one place can never be applied to another. Or at most, this previous knowledge can only be valuable because the attention in general has been rendered more keen by practice.

We have seen that the smallest noticeable difference on the skin is probably not determined by reference to sensations of movement at all in the case of the normal seeing individual, but is simply the result of the discrimination of local sensation-differences. In the same way, our judgment of the increase or decrease of the spatial distance between impressions of touch will depend solely upon the knowledge which we possess of the position of each impression in terms of the local colouring of its sensation, or, more correctly, upon the permanent associations into which the two are brought. But this knowledge was acquired with the assistance of the sense of sight. We judge whether a distance on the skin is longer or shorter from the memorial image of the position of the stimulated part which its sensation arouses in us. This memorial image is independent of the movement required to pass over the distance ; it is conditioned solely by the idea which the sense of sight has helped us to form of every portion of the skin as determined by its peculiar sensational character. And it is an obvious inference that the discrimination of spatial distances, whether large or small, always remains unchanged so long as the sensibility of the cutaneous surface itself remains the same. This is the result which we actually obtain by experiment. If a distance of 11 millimetres is just noticeably different from one of 10 millimetres, we can also distinguish 21 from 20, and 31 from 30 ; in short, for cutaneous sensibility in general, it is not the relative, but the *absolute*, just noticeable difference of distance which is constant. Exceptions to this rule are explicable from the fact that in long distances the fineness of our discrimination of neighbouring cutaneous points is considerably altered.

§ IV

The normal development of the sense of touch, then, comes later than that of the sense of sight ; so that the measure which it applies to space is obtained from visual perceptions, and not at all or only secondarily from sensations of movement in the limbs. For this reason the mechanism of the touch reflexes and the laws of its development will not possess the great importance which attaches to them in the development of vision. Their influence must necessarily be diminished to the extent, that is, to which it is destroyed by the predominance of the sense of sight.

But this destruction is only partial. Every influence which makes against that of the sense of sight increases the independence of touch, and helps to develope it to an extent which is never attained under ordinary conditions. *Accidental blindness* shows us striking alterations in this respect : the muscles become much more responsive ; the least tactual stimulus arouses movements which bring the external object into contact with different parts of the skin, and particularly with the most sensitive portions. And much greater still is the part played by tactual movements in those rare instances where the dominant influence of the sense of sight has been absent from the very beginning of mental development,—in cases of *congenital blindness.*

The congenitally blind are forced to construct their entire spatial world from the perceptions of the sense of touch. And they do this with marvellous completeness. That sense which remains throughout the normal life on a low plane of development attains a perfection which in fineness of discrimination may at least be compared with that of indirect vision, the vision of the lateral portions of the retina. In one respect only must the skin necessarily remain inferior to the eye : it requires immediate contact with its impressions.

How now will the congenitally blind acquire ideas of distance in space or of spatially extended objects ? They have at their disposal simply pressure-sensations from the skin, and sensations of movement from the exploring limbs. From these alone they must construct their perceptions of space. The means to this construction is obviously to be found, as it is in visual perception,

M

in the association of the two sensational series by the uniform working of the reflex mechanism. But of course this latter requires a much more complete development in the blind than in the seeing. First of all, each limb is brought into reflex connection with some definite portion of the skin. The local differences of sensation are in consequence associated with definite sensations of movement ; so that there exists for each of these provinces of the skin some central point (although this may perhaps be variable) to which all neighbouring sensations are referred. Then, further, the separate portions of the skin are brought into relation with one another ; and so, by the interconnection of originally diverse sensational systems, the whole mass of cutaneous sensations is united into a single system. This interconnection must necessarily tend to be effected whenever the separate limbs come into contact with one another. For in this way there will be obtained a certain measure, however imperfect, of the distance between the separate organs of touch and their sensation-centres.

The course of this development will undoubtedly require a longer time than the education of the visual sense. The latter was completed in a single act ; but here there is required a great number of successive acts, whose capacity to unite at all in a common effect is simply due to the fact that they are all of a similar nature. That is, the process which gave rise to the space-perception of the sense of sight must be many times repeated for the sense of touch. Now just as we normally fixate with the yellow spot anything that we wish to apprehend clearly, so will the blind be compelled by the great difference in the sensibility of the various parts of the skin to make exclusive perceptional use of those portions which are capable of the finest discrimination. The parts of the organ of touch which possess this capacity in a pre-eminent degree are the hands. The blind are constantly practising their hands in touch, and even more in movement. Touch-sensations alone can obviously never suffice for the exact apprehension of spatial relations. The reason for this is that if the parts of an object do not lie exactly in the same plane, the cutaneous pressure-sense is unable to give any account of them. Hence the slight tactual movements of the hands, and especially of the fingers, which in the blind are wonderfully active, are of very great importance. By their means the spatial properties of objects are more accurately apprehended, partly through successive contact with the parts of the organ of touch which are capable of finest discrimination, partly owing to the continual.connection of sensations of pressure and of movement. But we always find that the blind do not apprehend even fairly simple spatial relations with anything like the rapidity with which the perceptions of sight enable us to obtain an adequate idea of the most complicated figure. Their sensations of touch and movement have to construct the object gradually for them out of its parts.

§ V

Thus the slow and imperfect development of the spatial perceptions of the accidentally and congenitally blind confirms our assumption that the sense of sight normally outruns that of touch. In holding this view, we are in conflict with the opinion which was generally current in the older psychology, and is not yet entirely abandoned,—the opinion that the sense of sight is more probably educated by the sense of touch. What we grasp with our hands, it was said, is more certain to our sense-perception than what affects us at a great distance. It was forgotten that both objects alike make an impression upon the sensory nerves, and that these, in the absence of correlated psychical processes, can have nothing to say regarding the origin of the impression.

But there was an especial circumstance which seemed to support the view that the sense of touch was necessary for the development of that of sight. We see objects in their natural position, and not inverted. But the images which external objects produce upon the retina are reversed. The eye is an optical apparatus composed of a series of curved surfaces, which cast upon the retina a miniature image of all objects lying within the field of vision. The spatial relations of this image, however, are exactly the reverse of those of the object itself: if the latter stands upon its feet, the retinal image stands upon its head, and *vice versâ*. So long, therefore, as it was supposed that the act of vision is concluded with the formation of the retinal image, our vision of objects the right way up necessarily remained a paradox. But what does the mind know of the retinal image? We have only learned of its existence and its inverted position as regards the object from the physicist and physiologist. In order to be able to perceive this image as it really is, we should have to assume another eye behind the retina. And, as a matter of fact, this hypothesis has now and again been considered a probable one. It was never said, of course, that there is a real second eye ; but it was supposed that when the image affected the mind it was again inverted by it, just as it would be by a second eye,—apropos of which an ingenious philosopher has remarked, not unjustly, that, instead of ascribing to the mind this perpetual business of inversion, it would be much simpler to stand it on its own head, so that its inversion might set right again the inverted world imaged on the retina.

From the standpoint of our own investigation of spatial vision this difficulty is capable of a very simple solution. It is merely as a series of locally-coloured sensations that the retinal image affects our mind. Only by the movement-sensations of the eye does the mind learn to connect these into a spatial order. But what do the movement-sensations tell us about the position of objects ? As the eye moves it passes from point to point of an external object. In moving round its centre from above downwards, it passes over an object from top to bottom. It brings all the parts of its retinal image successively upon the spot of clearest vision. Now when the visible portion *a* of the eyeball moves downward in front, the yellow spot *g* at the back will be turned upwards (Fig. 23); as the front point *a* fixates the different parts of the object, the point *g* traverses the retinal image in precisely the same way. So that, if the position of objects in space is inferred from movement, the retinal image *must* be inverted, since only where this is the case is it possible for the movement to correspond with the actual position of the objects. So far from being a paradox, the inverted retinal image is necessary for vision. The retinal image must have been upside down, even if the laws of the refraction of light in the eye had not rendered the inversion physically necessary.

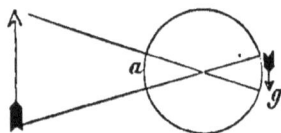

FIG. 23.

Of course the further question may be raised, how we know that we are moving the eye up or down. Are not 'up' and 'down' relative notions, which presuppose the perceiving subject and his position in space? It is really just because up and down are only relative that we are able to introduce order into the world of spatial vision. If we had perception of absolute direction up and down, we should be obliged to think that either by day or by night, as the case might be, we stood upon our heads ; that would follow necessarily from the rotation of the earth. The reason that we do not think so is, that we make ourselves the central point in all space-references. Up and down, like right and left, are terms which only have a meaning when referred to ourselves. In distinguishing an upper and a lower in our spatial perceptions, we make continual reference to our own body : we call that 'down' which for our eye lies in the same direction as our feet ; we call everything 'up' which lies in the same direction as our head.

There still remains one objection, which seems to tell against the influence of movement upon spatial sense-perception, which we have already recognised and indeed proved to exist in numerous cases. Do we really always move our eyes, it might be asked, when we wish to see things spatially? Must we actually turn the eyeball up or down in order to know what is above and what is below? By no means. Without moving our eyes in the least, we can apprehend objects as spatially extended, and assign to each its own spatial position. How shall we attempt to meet this objection? We might, as has been sometimes done, point to the great rapidity of the ocular movements, and our consequent inability to observe them. It might be assumed that though we think our eye is at rest, it is in reality executing very rapid movements. But we cannot escape the difficulty in this way : the rapidity of the muscular movements is by no means so great as we should be obliged to assume that it was on this hypothesis. And, on the other hand, we are able experimentally to reduce the duration of a light-impression so greatly as to entirely exclude the possibility of eye-movement during its operation ; *e.g.*, by illuminating instantaneously with the electric spark. Under these conditions objects are still seen spatially. There can be no doubt, therefore, that movements are not requisite for every single spatial perception.

But there is another point that must not be lost sight of. We must always·distinguish mental *processes* from mental *products*. The latter may depend upon a capacity acquired in the course of previous development. It is not necessary that what was at first a factor in the formation of our space-perceptions, and is still operative to perfect and refine them, should on that account be a persistent and inevitable condition of all vision. The child, taught by its mother to make the first step, learns in time to run alone. Why should there not also be conditions of vision which are operative solely, or at least principally, in the first stages of the development of this sense ?

As a matter of fact, we have already discovered conditions of this kind. The relative position of the sensitive points of the retina is determined by a series of intensively graduated movement-sensations, standing in associative connection with correlative, locally coloured light-sensations. If the impressions once experienced are given a second time, these points can be recognised by their local colouring. So that if two impressions affect two retinal points which were on a previous occasion separated from one another by a movement-sensation of definite intensity, we shall be able, after frequent repetition of the whole process, to distinguish them without the actual occurrence of the movement and its attendant sensation. When once the local sensation-differences have obtained from movement-sensations the measure of the distance which separates them, they retain this measure in independence of its source. A definite place-reference is attached to the local colouring, behind which its true character as qualitative property of sensation entirely disappears. We imagine that we perceive directly the locality of an impression, while in reality we are only perceiving a peculiarity of the sensation, and from it recognising the locality. And when we have extended our power of spatial discrimination by practice, we think that our capacity for the apprehension of spatial differences has been directly increased, whereas in point of fact it is only our ability to discriminate small sensation-differences which has improved. What is true for sight in this regard holds also for touch, only that the latter (even when it has attained an unusually high degree of development) stands in constant need of further assistance from movement-sensations on account of the less definite character of its local sensation-attributes. So that the name 'sense of feeling,' which is sometimes used for touch, is significant. Originally we 'feel for' objects with the eye as we do with the hand. But the hand remains simply an organ of 'feeling,' not only because it must come into actual contact with the objects which it is to perceive while the retinal image is produced by the action of light at a distance, but also because it goes on to 'feel over' after contact has taken place; and a complete perception is only gained by the combination of the two kinds of sensation,—pressure and movement.

§ VI

I have attempted to describe the phenomena of spatial sense-perception in such a way that the theory which is to explain and co-ordinate them results of itself. The theory which I have given you is directly suggested by the facts, and does not attempt to go beyond them. But at the conclusion of our considerations, we must not omit to state that there are both physiologists and psychologists who still believe that they can dispense entirely with any such explanation of the arrangement of our visual and tactual sensations, or, at least, think that one of the factors discussed above suffices to explain all the facts. In the former case, it is assumed,—as was uniformly done in the older physiology,—that the spatial arrangement of our ideas is given directly in the arrangement of the parts of the retinal image; or, as it is put to-day with more show of learning, though without any real improvement in the form of expression, that every sensation of the two senses with which we are here dealing possesses from the first a certain spatial quality. Now we cannot deny that this would be the most convenient hypothesis possible. But it is equally undeniable that it is totally unable to take account of all the factors which we have found to exercise a determining influence upon our apprehension and estimation of spatial magnitudes. Where the attempt has been made to do justice to the factors in question from this point of view, it has been found necessary to set up a number of artificial and complicated secondary hypotheses, some of which even go the length of self-contradiction. These may, perhaps, possess a certain value for the logician, as warning examples of how hypotheses should not be constructed; but they are absolutely useless to the psychologist.

The case is different when an attempt is made to furnish a theoretical explanation of space-construction in terms of one of the influences which determine the act of perception, to the neglect of the others. It has often been thought possible to set up a theory in terms of movement and movement-sensations alone, and either entirely to neglect the local sensation-qualities of the retina and skin, or to regard them as functioning in entire independence of movement, and as being, like the latter, sufficient in themselves for an adequate explanation of the facts. The first of these views inevitably leads to the conclusion that muscle-sensations as such possess a special space-quality; the latter ascribes this quality to the retinal sensations alone, or to both of these sensation series. So that there is indirectly implied a return to the view that the extensive idea in general neither demands nor is capable of any psychological explanation. But it is not enough to grant that the spatial arrangement of impressions is determined on the one hand by movements, and on the other by properties attaching to the resting sense-organ and connected with the place of the stimulus. Experience shows that *these two influences are so intimately connected that neither of them is operative without the other.* The principal proof of this is found in the fact that effects which can only be explained from the laws of ocular movement still persist when the eye is at rest; *cf.* the case of momentary illumination by the electric spark. The above-mentioned illusions with regard to the relative lengths of horizontal and vertical lines and other similar phenomena do not disappear when movement is prevented, although they may sometimes be less striking.

If, following Lotze, we call every constituent of sensation which may be of influence upon the act of spatial ideation a *local sign*, the theories which hold that space-perceptions have been generated by psychological processes, and are neither given *a priori* nor result from a special quality of sensation, may be distinguished as the theory of *simple* and· the theory of *complex* local signs. The first assumes either local signs of movement-sensations, or local signs of the skin, or both, without, however, in the latter case admitting the interaction of the two. The theory of complex local signs, on the other hand, regards the extensive idea as the mental resultant of intensively graduated local signs of the movement-sensation and qualitatively graduated local signs of the sensory surface. Space-perception depends on the uniform association of these two sensation series, although the members of one of them need only be reproduced in order to be effective. This is especially true of the intensive series, whose terms are so intimately connected with those of the other, qualitative series, that every pair of definitely different local signs will be invariably associated with the movement-sensation corresponding to the passage of the organ over the distance between them.

§ I. The Separation of Visual Ideas; Influence of Boundary Lines. § II. Ideas of Depth. § III. Relations between Apparent Magnitude and Distance of Objects.

§ I

WE have now shown at some length how the mind comes to arrange visual impressions in spatial order upon a plane surface. But our knowledge of the formation of the field of vision has given us no idea either of the nature of external objects, or of the visible parts of our own body. The impressions, though spatially disposed, have not as yet been brought into those relations in virtue of which they are arranged as *separate ideas*, each apprehended as a whole of definite spatial form. How does this separation come about? How do we pass from the spatial perception which leaves its objects side by side without distinction or difference to the idea of objects which are spatially separate?

First of all, it is plainly the *boundary lines* of objects which separate them from one another, and further divide up a single object into parts. They afford a definite resting-place to the fixating eye. Whenever we have a series of objects suddenly presented to us, the eye is arrested by the sharpest boundary lines. It thus learns the rough outlines of objects first, and from these passes by degrees to the finer delineations of their parts. This influence of boundary lines on the movement and fixation of the eye may be easily proved by experiment. We hang, *e.g.*, before a white surface a number of vertical black threads of different diameter, and allow an observer to look through a tube towards the screen in such a way that the threads lie in his field of vision. Supposing that the observer knows nothing beforehand of the arrangement and nature of the threads, he will be sure to say when questioned that he saw the thickest thread first, and the others afterwards, for the most part in the order in which their distinctness brings them to consciousness. And a little attention will lead him to discover that in the first moment of looking through the tube the eye turned by a kind of mechanical necessity to the sharpest outline in the visual field, and then, after clearly apprehending this, directed itself successively upon the others in the order of their attraction. This order remains constant if the threads are hung at different distances, except, of course, that the influence of distance upon the apparent thickness of the threads must be taken into account. If two threads of the same diameter are hung at convenient distances for vision, the nearer one is apprehended first. But if their diameters are unequal, that will be first seen which makes the stronger impression upon the eye ; so that the boundary lines appearing in our field of vision determine on the one hand the movements of the eye,—the image of the boundary line being brought upon the place of clearest vision,—and on the other that process within the eye whereby it adapts itself to the distance of the objects viewed. This internal process of adaptation for near objects is also a muscular movement attended by sensation. And the sensation furnishes a measure of the amount of adaptation : the nearer the object we are looking at, the greater the convexity of the crystalline lens under the action of the intra-ocular muscles.

This tendency of the eye to fixate distinct points, or boundary lines, can only be explained by reference to a mechanism similar to that concerned in reflex action. Indeed, it seems to be a justifiable assumption that the relation of movements of the eyeball and of the muscles of accommodation to boundary lines and points is nothing but a further development of reflexes present in the eye from the very first. During the first days of a child's life, every light-impression produces an ocular movement, which brings its image to the place of clearest vision. But as the retina is continually affected by uniformly diffused light, that which is distinct and definitely bounded will very soon be separated out from this indefinite chaos of light-impressions, forming as it does a special stimulus, quite different from its uniform surroundings. To such a stimulus the eye turns ; and when several are present, to each in succession,—the order of fixation being determined in every case by intensity, *i.e.*, by the degree in which each stimulus differs from its surroundings. Even when the sense is completely developed, visual apprehension occurs with the mechanical necessity peculiar to reflex movements. And though we may voluntarily counteract this constraining influence, yet we are always falling again under its sway whenever the unexpectedness of an impression or some other particular reason renders an act of will impossible.

The influence of boundary lines and of points is modified by a third factor, which is dependent upon the same conditions and is operative in the same way as these, namely, the *movements of objects*, by which their position relatively to one another and to the perceiving subject is changed. Since every object which is marked off from its surroundings by boundary lines forms a permanent whole, however its surroundings vary, it becomes the object of a particular idea. If objects at rest are apprehended as similar units, that is only because this characteristic of limitation points to a separation from their surroundings such as is matter of immediate observation in every case of movement. In the series of separate ideas obtained from the original fields of vision in this way, through the medium of movement and boundary line, the first place is taken by the body of the perceiving subject, whose permanence gives it a preference over all other objects,—while, besides that, as the substrate of all sensations and perceptions, it affords the universal centre of relation for the spatial arrangement of the entire external world.

§ II

The sense of sight, then, comes to apprehend objects separately in space by means of changes in their position and corresponding changes in boundary line. Another and a final motive to this spatial separation of objects consists in their reference to points in space at different *distances* from the eye. We can show, even in fully developed vision, how the idea of spatial depth originates, since the idea is of comparatively late growth. This is proved conclusively by the experience of the congenitally blind who have been restored to sight by a surgical operation. In such cases we find a certain power of orientation acquired during blindness by the aid, as we may suppose, of the light-shimmer which is never entirely absent ; but there is no cognition of distance. Remote objects are not infrequently regarded as lying close at hand : the patient shrinks back from contact with them. We may observe the same phenomenon in the first months of childhood : the baby will reach for the moon or for the objects seen in the street through a third-story window.

The development of the ideas of depth is primarily conditioned by ocular *movements.* We let our eye range from nearer objects to more remote, and the path over which it travels gives us a measure of the distance of the objects which we have successively fixated. For a movement-sensation is associated with every movement, and its intensity increases with the extent of the movement. When the relative distances of objects are to be measured, they must not, of course, cover one another. And, moreover, their bases must be visible. If this is not the case, we may quite well estimate objects at different distances from us as situated close beside each other. You may convince yourselves of this by holding before the lower half of the eye a small piece of paper, and so covering the lower part of the objects looked at. If the difference of their distances is small, they are generally regarded as at the same distance ; if it is large, you note, indeed, that one object is nearer, and the other more remote, but you have no approximate idea of the distance between them. That you notice any difference at all in these instances is due to the accommodation of your eye for near and far. Since adaptation also depends upon a muscular movement, we have a rough measure of the focussing of the eye in the accompanying muscle-sensation. At the same time we are obviously less accustomed to attend to this mechanism. Usually we do not employ it in measurement, but make use of the movements of the eyeball, which are much more accurate, and have a far wider range.

When we pass with the eye from the base of one object to that of another we usually begin with the nearer. If we are estimating the total distance of any object from ourselves, we begin from our own feet. So that the foot is the most original and natural unit for measuring distances : the length of the foot is the first spatial distance which occurs to us. Now when we are passing from nearer to more remote objects our eye moves from below upwards. Suppose we are standing at *a* (Fig. 24), and one eye, *o*, is directed successively upon the more and more remote points, *b, c,* etc. Dur-

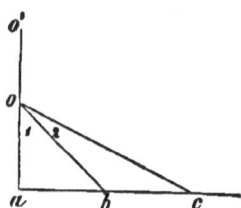

FIG. 24.

ing this process it turns upwards; the visual axis passes gradually from a position where it is directed vertically downwards towards the horizontal, until, finally, when the object is very remote, it completely attains this latter position. This movement is not confined to the eye : the head moves as well, especially for objects lying directly below us, and so assists the ocular movement. For these head-movements, again, we have a measure in sensations of movement ; so that the result is the same, however the movements which carry the fixating eye from point to point are brought about.

Since in these movements both head and eye turn from below upwards, remote objects seem to lie higher than near ones, and the horizon which bounds our view is at the same height as the eye. If the earth were a perfectly plane surface, we should all imagine ourselves to be standing in a depression ; and the surrounding landscape would appear to rise uniformly to the horizon. This phenomenon is, of course, modified in various ways by all sorts of inequalities in the earth's surface, and also in part by the spherical form of the earth. Since, again, distances which are objectively equal require a lesser eye-movement in passing over them the farther they are from us, more remote objects seem to lie closer together than those which are nearer. So that we are frequently unable to cognise a difference of distance in the former case which is quite obvious to us in the latter. If you look at the angles 1, 2, which are subtended by the equal distances *a b, b c* (Fig. 24), you will see that these angles, which give a direct measurement of the magnitude of the ocular movement, become smaller and smaller, and will at last entirely disappear. But if we take a more elevated position, so that our eye is at *o′*, our field of vision is at once extended, and remote distances become visible which were previously hidden from our view. Near distances appear, on the other hand, relatively smaller than they were before. So that if we climb a mountain, or ascend in a balloon, all objects, near and remote alike, appear closer to us.

Objects situated at different distances from the observer present differences not only in the relative position of their bases, but in a whole number of other properties. Originally these are turned to account only when associated with judgments of distance in terms of eye-movement, but later they may become independent indications of distance. To this class belong primarily the *shadows* cast by objects. Their direction and magnitude are dependent on the position of the source of light with regard to the object, and of the object in relation to the point of view of the observer. Then, again, increase of distance means gradual decrease in clearness of boundary lines. And the more remote an object, the paler will be its colouring, which will also vary in quality according to the absorption of light by the atmosphere. All these factors together make up those constituents of perspective in drawing and painting which enable the artist, by proper distribution of contours, light and shade, and colour-tones, to produce upon a plane surface the illusion of actual tridimensional relations.

When the distances are very great, *the apparent magnitude of the objects* comes into play as a further factor. It furnishes the most obvious standard of measurement when we are comparing quite distant objects, and in cases where the above-mentioned factors of perspective are absent is our sole criterion for the estimation of distance in the third dimension. If we compare a tree which is ten feet off with another whose distance is a hundred feet, the former appears larger, even though we know that the two are objectively equal. The magnitude of an object which is directly given in perception we term its *apparent magnitude.* Whenever we have learned from repeated experiences the actual magnitude of an object, its apparent size is employed as a measure of its distance from us. This is not the result of an act of reflection on our part,—that is never involved in the process of perception,—but is due to the direct association of an idea of distance with the impression. The development of this idea, however, is dependent upon experience; and we must therefore explain it itself as the result of an association between the distance of an object whose real size is known and the apparent magnitude of that object. The apparent magnitude of a person approaching from a distance will therefore excite directly the idea of his distance, because in many previous instances we have associated the idea of this particular distance with this particular magnitude by means of other characteristics directly given, and especially by moving the eyes from our own feet to his.

§ III

The apparent magnitude of an object and the magnitude of its retinal image are usually regarded as directly proportional. The obvious reasons in support of this view are that both decrease with distance, and that plainly the magnitude of the retinal image must be the principal condition of our having any idea at all of an object's apparent magnitude. If some one is approaching us from a distance, his image upon our retina and his apparent size (*i.e.*, the size which we ascribe to him in idea) increase at the same time. But since this idea of apparent magnitude is the product of numerous associations, some of which are quite complex in character, we ought not to expect to find any constant relation between the two values,—that of the retinal image (physiological) and that of the idea (psychological). This presumption is confirmed by experience. For we find that while the magnitude of the retinal image remains constant, or, what is the same thing, that of the visual angle formed by drawing lines of vision to the boundaries of the object, the apparent magnitude may be extraordinarily different, owing to its determination by the other factors which enter into the association. Foremost among these stands the idea of distance suggested by other characteristics of the object; and, in the second place, there comes into account the idea of the magnitude of similar objects.

The most striking instance of our seeing the same object at the same distance sometimes as larger, sometimes smaller, is

the one afforded by the sun and moon.   At the distance which
separates the sun from the earth the size of its image cannot
differ at morning, noon, or evening ; its magnitude remains equal
at all times of the day.   But when the sun is at the zenith it
appears smaller than when it is on the horizon at rising or setting.   This is explained in the following way.   We form a definite
idea of the distance of the sun, though, of course, the idea is
very far from the truth.   The sky seems to our eye a solid arch,
which rests upon the earth at the horizon, and closes down upon
the nearest mountains or the towers of the nearest town.   To
frame an idea of the distance of the sun at the zenith, we have
at most only a tower or a mountain as our standard ; to get
an idea of the distance of the horizon, we make use of every
object within our field of vision.   Between ourselves and the line
of the horizon we see large numbers of trees, fields, villages, and
towns : and a distance which contains so many objects must
of course be very large.   So we come to imagine that the
horizon is farther off than the zenith ; the arch of the sky which
rests upon the earth is not semicircular, but is rather shaped
like a very convex watch-glass.   But if our retinal image is
equally large whether that which we are looking at is near by
or far off, the magnitude of the object regarded must be different in the two cases.   The remoter object looks to be actually
larger, just because it seems to be of the same size as the nearer
one.   It is as though a man on a steeple should appear as
large as one by our side ; we could not but imagine that the
former was a giant.   Before we form an idea of the magnitude
of an object we always consider the distance at which we view
it.   Quite frequently we mistake the distance.   But though we
may have convinced ourselves of this error a hundred times, we
cannot free our perception from it, so stable are the associations
through which it has arisen.   Our perception of the size of the
sun rests upon two wrong ideas : in the first place, we suppose
that it is not much farther off from the eye than the nearest
mountain peak or the top of the neighbouring church spire ;
and secondly, we imagine that it is sometimes nearer, sometimes
more remote, according as it stands at the zenith or approaches
the horizon.   We need not be astronomers or physicists to know
that both these notions are false.   But however well we may

know this, and however sure we may be that our distance from the sun does not become alternately greater and less, we still make the same mistake,—the astronomers and physicists among us no less than the ordinary man.

Our perception of objects is, therefore, always dependent upon their distance; not, however, upon their actual distance, but upon the distance as we imagine it. If we could obtain a perception of the actual distance of the sun and the moon, they would appear infinitely large to us. On the other hand, when we try our best to imagine them quite close, they appear smaller than usual. If we look at the moon through a tube, or through the closed hand, seeing nothing but that portion of the sky where the moon is situated, it will seem no larger than a half-crown, whereas it generally looks about as large as a plate. The simple explanation of this fact is that we do not now localise the moon somewhere behind the trees which fill the foreground of our normal field of vision, but close behind the tube or closed hand. In the same way, when we look at the moon through an ordinary telescope, it seems smaller, and not larger, than usual, though the telescope magnifies, and we can see by its aid a number of things upon the surface of the moon which are invisible to the naked eye. That is also because the moon is not localised at a distance, but at the end of the telescope. The same thing happens when we direct our telescope upon distant mountain peaks: we see their outlines more clearly; we observe details which the unaided eye could not distinguish; and yet we note that on the whole the mountains do not appear larger, but smaller. In these cases the magnitude of the retinal image of the moon or of the mountain is increased, yet we see the objects themselves as smaller.

But we have not even yet completed our account of the influences which are here at work. If we look at a man on the top of a tower, he does not appear nearly so small as he should do in accordance with our idea of his distance. When we look at the mirror on the opposite wall, we estimate its distance pretty accurately. But we see it larger than we really ought to, if we compare the size of its retinal image with those of other and nearer objects. Clearly, the fact that we already know the size of the man and the mirror is here of importance. We have seen men close at hand thousands of times, so that we know certainly that there never was a man only a millimetre high, nor a drawing-room mirror only two centimetres square. This experience exerts an influence upon our perception, and serves to modify the idea which we should otherwise have formed regarding the distance of the objects we are looking at. This correction is, as you know, not complete : the man on the roof is much smaller to view than the man by our side, and the mirror on the wall twenty feet away a little smaller than it is when we stand directly before it. There is a kind of conflict between the fact that the object viewed is at a distance, and must, therefore, appear smaller, and the fact that we are acquainted with its true magnitude. Both sides are, as a matter of fact, right in this controversy ; but since it is not possible to grant the claims of both at the same time, we follow the example of that most excellent judge who decided all lawsuits involving money by dividing the sum between the two parties in the suit.

Our perception, then, can only determine the true magnitude of objects where this true magnitude is actually known to us ; and this knowledge must come from direct and often repeated experience. However sure we may be that the moon is immeasurably larger than a plate, we shall not on that account see it a whit larger. We are convinced that a magnifying-glass does not make the objects seen through it any larger ; and yet they continue to be larger for our vision. The sun at midday, we are certain, is not smaller than it is in the morning, yet as we look at it it appears smaller. Vision requires to be convinced in a quite different way. No assertion on the part of other people, no speculation or calculation, is of influence in determining our perception, but only an association of ideas repeated over and over again. Isolated experiences, therefore, make no impression upon our minds. From a window in my room I look directly upon a neighbouring church tower. The face of the church clock appears about as large as that of a moderately large clock which hangs upon my wall. The ball of the steeple looks about as large as the button of a flag-staff. A little while ago the clock face and steeple knob were taken down for repairs, and lay upon the street. To my astonishment, I saw that the former was as large as a church door, and the latter as large as a waggon-wheel. Now the two are in their places again and look to me just as they did before, although I have learned their true size. The workman upon the roof does not seem so much smaller than he actually is, because I have observed the size of my fellow-men hundreds of times. But the ball of a steeple and a church clock are not objects of every-day experience. The button of a flag-staff and the clock on the wall are much more familiar. And so I think of the steeple knob as the flag-staff button, and the church clock as a wall clock. Even that idea seems exaggerated, if I compare these things with objects more immediately around me. For I can just cover the ball of the steeple with the head of a pin, and the tower clock with my watch, if I hold these at a little distance from my eyes. If it were not altogether too improbable that the steeple should carry a watch and have a pin's head as its ball, I should perhaps imagine that that was the actual state of affairs.

We see, then, that our perception of things in space is extraordinarily variable; that it is conditioned by a number of influences which by no means emanate from the objects themselves; that we take into account the apparent magnitude of objects, their distance from us, and finally our experiences of the same or similar objects in other connections. How, then, can we assert that our perception is determined by the objects *outside* us? All these influences are not found in the objects, but in *ourselves*. It is we who involuntarily and unwittingly alter the phenomena, in terms of the ideational elements which are already present in consciousness, ready at a moment's notice to form associations. And this whole variability of our world of perception depends primarily on the *idea of depth*, which gives to the spatial arrangement of the visual field its property of being apprehended as at different distances from the observer. Such a property must necessarily open a wide field for the play of the most diverse subjective and objective influences upon our spatial ideas.

But although the arrangement of objects in terms of spatial distance, under the operation of these influences, must always remain imperfect and incapable of exact measurement, we must not forget that through it alone do we obtain a final form for our world of ideas. With the reference of objects to different distances in space, the world of perception is placed outside of us, and is differentiated into an infinite diversity of content. Although the spatial relations which we ascribe to external objects may at the outset often be incomplete and deceptive, still the decisive step has been taken with the very introduction of those relations. The ceaseless activity of our sense-perception is constantly at work in the endeavour to perfect our ideas. It furnishes us with new ideational groups, and corrects the most serious errors in those already acquired. All the senses co-operate in this work, each revising and supplementing the others. But it is primarily the common action of the two co-ordinate organs of vision to which we owe the greatest part of our ideational development. There are no other organs which so directly supplement and correct each other's perceptions, and which thus give so great an impulse to the fusion of separate perceptions into a single idea, as the two eyes.

§ I

THE two eyes may be compared to two sentinels who, viewing the world from different standpoints, impart their experience to each other, and so complete in idea a common picture, uniting in itself all that each observer has seen separately.

It is not very long since the fact was discovered that binocular vision was different from monocular. The early physiologists thought that the image of an object produced in one eye was not different from that apprehended with both. And it was accordingly supposed that the two eyes were simply equivalent to a single eye,—a conclusion which found an apparent confirmation in the anatomical structure of the optic nerves. At a certain point in their course after leaving the brain, decussation occurs. At this point the nerve-fibres are closely interlaced ; and then again two nerve-trunks appear, one extending to either eye. It was supposed that at the point of decussation and interlacing of the two nerves the fibres were divided. Each fibre, from whichever side of the brain it came, was thought to divide in such a way that one of its parts ran to each eye, and within each eye to correspondingly situated retinal points. It was in 1840 that the English physicist Wheatstone proved that the images cast upon the retina of each eye are very frequently dissimilar, without there being occasioned any disturbance of vision. If we hold an object close before us, and shut first one eye and then the other, we see it a little differently in the two cases. Suppose, *e.g.*, we hold our hand between the eyes, at a little distance from them, so that the surface of the hand is at right angles to the face ; one eye sees only the back, and the other the palm, of the hand. If the point of anatomical decussation were really a place of division, and if the images cast on the two eyes were directly intermixed in the brain, simultaneous binocular vision would only give us a confused picture. For on a portion of one retina there is represented part of the back of the hand, on the corresponding region of the other part of the palm. These two images would therefore be superimposed in the common act of vision, and that would render any clear apprehension impossible. But observation by no means confirms this view of the matter. It is rather the fact that we see the hand more perfectly with two eyes than with only one. It is not merely that we see simultaneously much that in monocular vision could only be apprehended successively ; but we perceive directly that the hand is not a picture painted on a surface, but has extension in the third dimension. The same test may be repeated with the most diverse objects ; we shall always find that the apprehension of the third dimension of objects is intimately connected with simultaneous binocular vision. If we are using simply one eye, we are often unable to decide whether the fixated object is tridimensional, or merely a drawing upon a plane surface. So that in monocular vision illusions are possible in this connection ; drawings in perspective and light and shade in particular may give a very strong impression of the third dimension. If the object is near to us, the illusion vanishes at once so soon as the second eye is opened. Although, therefore, a perception of the third dimension may be gained with a single eye, it is always less complete, instantaneous, and immediate than that given in binocular vision. As a rule we are only able in monocular vision to attain to a perception of the third dimension of objects by degrees, and in terms of the movements made by the eye in passing from a nearer point to one more remote, *i.e.*, by a series of acts following one another in time.

If, then, the immediate idea of depth is always connected with simultaneous binocular vision, it seems obvious to say that we see objects in this way more adequately just because the images cast upon the two eyes are different. We have a direct perception of extension in the third dimension because the two eyes look at things from different points of view. And this fact is, moreover, confirmed by observation. When we remove the object farther and farther from the eye, the perception of depth disappears. But with increase of distance the difference between the retinal images grows less. And at last, when the object is so far off that the distance between the two eyes is practically zero in comparison with it, the two images are precisely alike, and fall upon correspondingly situated portions of the two retinæ. If, *e.g.*, we hold a sheet of paper close before the eyes, so that the right eye sees one side and the left the other, we get a clear idea of its extension in the third dimension. But if it is removed farther and farther off, we come to see less and less of the two sides ; and at last perceive nothing but the front edge, which is just the same for one eye as it is for the other. In other words, perception of depth and difference between retinal images always run parallel to each other.

If this difference between the retinal images of the two eyes is the cause of the perception of depth, it is obvious that this perception may be induced without any actual vision of a tridimensional object, simply by presenting to the eyes retinal images having differences similar to those produced by the perception of such an object. If, that is, we cast upon one retina an image which looks like the back of the hand seen obliquely, and upon the other an image which resembles the palm of the hand seen under similar conditions, an idea of extension in the third dimension will arise in our minds, although the images employed are simply drawings upon a plane surface. The retinal images are precisely the same as they would have been had we looked at an actual hand ; the result as well must therefore remain unchanged.

It is quite easy to test this. And it is best to take for the purpose objects of fairly simple form. Suppose that we are holding before our eyes a truncated cone with circular base, the apex turned towards the face. First, we close the right eye, and draw an exact picture of the cone ; now we close the left, and make a similar drawing. The two pictures are different,

because the right eye sees parts of the object which the left
does not, and *vice versâ*. The left eye sees the cone approximately as *A*, the right as *B* (Fig. 25). Neither of these views
as mere drawing furnishes any incentive to an idea of the third
dimension. The most that we can do is by an effort of imagination to see the small inside circle at will either nearer or more
remote than the larger outside one. But if we let *A* affect the
left eye as though it were an image proceeding from an actual
cone, and *B* act similarly on the right eye, we have just as
definite an idea of tridimensionality as we gain by observing
the cone itself.

It will not, of course, do to view the two drawings with both
eyes directed quite at random. We must look at them in a
way which corresponds to the formation of images by actual

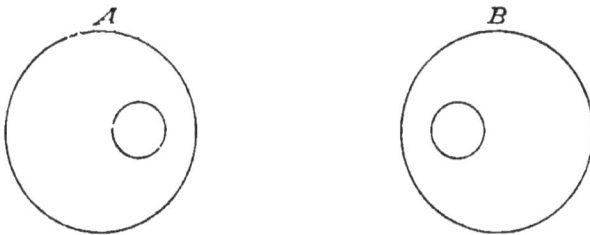

FIG. 25.

objects. The left eye must fixate the smaller circle in *A*, the
right the smaller circle in *B*. Only under these conditions are
the two images in the eyes related as they would be if we
fixated the top of a real truncated cone. But the experiment is
not altogether easy. We are accustomed to direct both eyes
upon one and the same point. Here we have to fixate a
different point with each, the top of *A* with the left, the top of
B with the right. Only long and continued practice can enable
us to command our eye-movements to the extent required for
independent fixation by either eye. Normally the movements
of the two eyes are completely concurrent. The movements
themselves are determined by external impressions; and it is
probable that these were also originally instrumental in bringing about the ·functional concurrence. For, ·as we have seen, it
is a law of the reflex mechanism of each eye that our gaze is

always attracted by distinct points or boundary lines, and moves from one to the other of these according to the intensity of the impression to which they give rise. Since both eyes follow the same law, their movements must necessarily be closely interconnected; the point which leads one eye to fixate it will also arrest the other. Thus there arises an impulse to common fixation on the part of the two eyes, which can only be overcome by practice.

§ II

To obviate this difficulty, which confined observation to a few practised individuals, Wheatstone constructed the *stereoscope*. By means of this instrument any one may readily obtain an idea of the third dimension from representations upon a plane surface. The ordinary form of the stereoscope is that given it by Brewster. It contains two small angled prisms, behind which at a little distance are placed the drawings to be combined. In free vision the eyes must have their axes parallel in order to fixate the drawings *b* simultaneously. But if the prisms *p* are introduced, and their refracting angles turned towards each other, the rays coming from the drawings *b* will be diverted in such a way that these fall upon the place of clearest vision and adjacent parts of the retinæ, although the two eyes do not fixate the drawings *b*, but the point *F*. The necessary result then is that the inner circles of *A* and *B* (Fig. 25) affect coincident points of the two retinæ, and the remaining portions of the figure exhibit precisely the same differences in the retinal image as arise when we directly observe a real object of similar character.

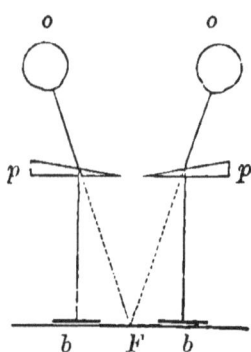

FIG. 26.

The following are the simplest stereoscopic experiments. The perception of depth will arise if two vertical lines at different distances from one another are presented to each eye in the stereoscope; *e.g.*, the lines *a b* to the left eye, the lines *c d* to the right. We obtain in this way a common image of two lines, the first of which, 1, is due to the fusion of *a* and *c*, the second, 2, to the fusion of *b* and *d*. The former lies in the plane of the paper, the second at some little distance behind it. This is in accordance with what would normally be the case. When we fixate binocularly two lines the right one of which is farther off than the left, the horizontal distance between the two lines in the retinal image of the right eye is necessarily greater than in that of the left.

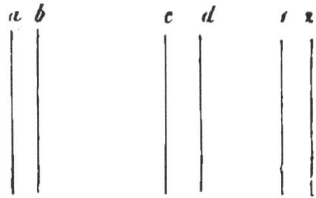

FIG. 27.

The idea of depth arises in the same way when we present to either eye a line drawn somewhat obliquely, and make the inclinations of the two lines a little different. If the lines *l* and *r*, which fall upon the left and right eye respectively, have the inclinations indicated in Fig. 28, we obtain a common image, *s*, which extends into the third dimension, its upper extremity being more remote than its lower. If, on the other hand, the lines are inclined as in Fig. 29, we obtain a common image whose lower extremity is more remote than the upper.

FIG. 28.

These two cases in which oblique lines are differently inclined, or the horizontal distance between vertical lines is different, are realised again and again under the conditions of tridimensional vision with the naked eye, and also in the stereoscope. They constitute the two fundamental experiments of stereoscopic vision. The vertical or oblique lines need not be straight ; the result is the same if they are somewhat curved. All stereoscopic vision depends ultimately upon the combination of these two fundamental experiments. On the other hand, we can never obtain the idea of depth if horizontal lines of different distances are presented. This is very easily explained

FIG. 29.

when we remember that no such condition of tridimensional vision exists in nature. We may turn and twist an object as we will ; its boundary lines are always either vertical or oblique.

The facts of stereoscopic vision prove indisputably that the two eyes perceive independently of each other, and that their perceptions are only secondarily combined in a common idea. Any other view as to the causes of stereoscopic phenomena becomes involved in inevitable contradictions. It is, *e.g.*, absolutely impossible to maintain that the two eyes are really only one ; that every nerve-fibre divides into two branches, which run to exactly corresponding points of the two retinæ. If this were so, the common image which we obtained from our truncated cone would be of the character indicated in Fig. 30, where those parts of the drawing which do not fall upon corresponding retinal points simply cover one another ; but there is no hint of the origin of an idea of a simple tridimensional object.

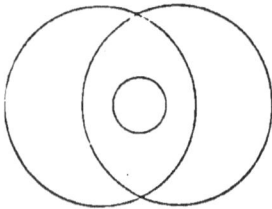

FIG. 30.

If we admit, as the phenomena inevitably compel us to do, that the two eyes are separate organs of vision, which sense independently of each other, we can only look for the fusion of the two visual perceptions in some psychological process. As a matter of fact, the phenomena themselves lead to this conclusion. We see that the idea of depth only arises when the two images exactly correspond to the views which we have of an actual body in space ; and we find that the direct perception of the third dimension always implies binocular vision. Suppose now that the two plane pictures were laid separately before you, and you were told that they were two projections of one and the same object, what inference would you draw as to the nature of this object ? You would say, of course, the object is extended in three dimensions ; and would even possess a tolerably correct idea as to the extent of its third dimension, and perhaps be able to construct an exact model of the whole object. If the perceptions of the two eyes are originally two separate things, it must be by an essentially similar method of procedure that we come to fuse these separate areal images in the common idea of an object extended in the third dimension. We, too, must construct our idea of the model of the object from its areal projections. The only difference is that we do not do this consciously, but unconsciously and involuntarily, by an act of sensational association : so that it is only the result, the idea of the object itself, which appears in consciousness.

The necessity of fusing perceptions in the idea of a single object rests in part in the infinite number of these perceptions. Again and again there are presented to our two eyes corresponding and supplementary areal projections of tridimensional objects. Always and invariably must we perceive such objects from the different points of view of the external world obtained through our two eyes. But this constant motive to connect the two partial ideas is only half the matter. Another and stronger influence is found in the mental endeavour which dominates all perceptive processes,—the effort to obtain a permanent association of simultaneous ideas and ideational elements. We have found this endeavour operative in the processes of areal and tridimensional perception of which we have already spoken. There can be no doubt that the fusion of the two visual images is the result of an act of mental association. But we have still to determine more precisely how this association takes place.

When dealing with the formation of the perceptions of the single eye, we found that sensations of movement furnished a measure of the spatial distance of separate points in the visual field. Similarly in the binocular idea of depth it is sensations of movement which furnish our primary measure of spatial distance. If the common field of vision contains a single bright point, the reflex mechanism which governs the relation of ocular movements to the yellow spot brings about the fixation of this point by both eyes. Its image falls on the yellow spot, the place of clearest vision ; *i.e.*, it is the point of intersection of the visual axes. If other bright points appear in the common field, they are successively apprehended in the order in which their intensity stimulates the tendency of the eye to move. So there results a successive fixation of the distinct points or boundary lines present in the field of vision. But there will necessarily arise at once important differences between the various cases in which the eyes mark out an object in this way point after point. If the points which the eye passes over lie on a plane surface, the image of the points which are no longer fixated,—whose image, that is, falls not on the yellow spot, but on the lateral portions of the retina,—still affects retinal points of approximately coincident position in the two eyes. With this coincidence of position there is also given a certain similarity of the peculiar sensational colouring, dependent upon the place of the impression. If, on the other hand, the points successively fixated lie at different distances from the eye, the image of the point which is no longer fixated does not fall upon points of coincident position and analogous sensational character in the two retinæ. And this divergence will be greater the greater the distance in the third dimension which separates the points. There will necessarily be an essential difference in the association of the two visual images in this case ; and the two series of actual experiences, corresponding to the perceptions of surface and of depth, will be clearly distinguished from one another.

§ III

We must, however, ask how this distinction can give rise to the peculiar idea of a *tridimensional* object as opposed to an areal object. On this point different opinions may still be entertained. It is certain, say many authorities, that only images which fall upon coincident retinal points are seen as simple ; all others are seen as double. We possess, therefore, a sure criterion of the presence or absence of the third spatial dimension in the presence or absence of double images. And the distance which separates these,—*i.e.*, the magnitude of their deviation from a coincident retinal position,—allows us to infer directly the magnitude of extension in the third dimension. So that the perception of depth consists simply in the neglect of double images ; and the idea of depth arises more clearly, the more there is to neglect in order to attain to the perception of a single object.

This view cannot, however, stand the test of experiment. If

Fig. 31 is introduced into the stereoscope, the left eye receives the image *A*, the right eye that of *B*. The lines 1 and 2 fall upon corresponding parts of the retina, lines 1 and 3 upon differently situated portions. The result is that the two heavily drawn lines 1 and 3 are fused to a single idea, the line arising from their fusion giving a clear perception of depth, while the fainter line 2 crosses it in the plane of the paper ; that is, our vision has combined into one the two lines which fall upon differently situated portions of the retinæ, while the two lines which fall upon corresponding parts are perceived separately. It follows necessarily from this that the idea of depth cannot arise from the apprehension and subsequent neglect of double images. Were that the case, it would not be possible for the images of 1 and 2, which are cast upon a series of corresponding points, to appear separately as double images. This experiment also proves therefore that the formation of a visual perception is an ideational act based upon an association of two visual perceptions, and determined not only by the position of the retinal images, but also by other properties which these images possess. The two strongly drawn lines force themselves first of all upon perception, and they alone can be referred to a single object when the images of the two eyes are compared ; while the object itself must extend in the third dimension of space to the position occupied by the images.

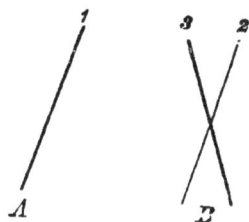

FIG. 31.

The idea of depth is, therefore, not produced by disregarding or by intensively weakening the separate perceptions in the common act of vision, but rather by the clear apprehension of them and their subsequent combination,—this primary association being further associated with other similar ideas. The differences of the two retinal images are by no means to be disregarded as worthless inaccuracies. They give us, on the contrary, an extraordinarily exact measure of the spatial qualities of external objects. And the inference is unavoidable that if we perceive these spatial properties through the difference of retinal images, these must themselves be given in perception, with their characteristic spatial differences.

But some doubt may still remain regarding the manner in which these differences of the retinal images, whose comparison gives us the idea of depth, are themselves apprehended and elaborated into that idea. We must set out from the fact that the movement-sensations of the eyeball which inform us of the spatial relations of the areal field of vision furnished also,—at least, originally,—a measure of distance in the third dimension. The hypothesis then appears probable that the idea of depth has been generated by movement. We have already discussed the signification of eye-movement for the estimation of distance. If we are fixating an object binocularly, any increase or decrease of its distance from us is very distinctly perceived by means of the movements of convergence or divergence, which are made by both eyes in keeping the object constantly fixated. We become conscious of these movements by movement-sensations, and by these latter we measure the approach or removal of the objects. If a spatially extended object lies before us, it simply presents to us simultaneously what is perceived successively when an object moves. At the same time, though the tridimensional object lies before us as a whole, we can only clearly apprehend some portion of it at any single moment. In this case also, then, we pass gradually by ocular movements of convergence and divergence from nearer to more remote points, or from more remote to nearer. In this way we perceive what is nearer or more remote in the object, just as we observe the changes in the position of a single point when it is in motion.

It can hardly be doubted that the idea of depth originally arose in this way by a succession of sensations and perceptions. But it is a different question whether it continues to arise in the same manner, whether each single idea which has been gradually acquired by both eyes continues to be formed by a series of successive acts. We have already discussed a similar question in our investigation of the perception of plane surfaces. There, too, movements play an important part. But we saw that movements are by no means continually operative in every single perception, but that the resting eye is itself able to see things in space, and possesses a fairly accurate measure of spatial extension. And we found that it was the presence of a local colouring in visual sensations which enabled the eye to dispense with the unremitting action of these movements. These local signs are permanent attributes, which, when once their relation to sensations of movement has been discovered, suffice to bring sensations into the extensive form.

And the idea of depth, which in binocular vision comes in to complete our simple apprehension of visual space, may also arise when the eye is wholly unmoved. It often seems to appear at the very moment when the eye is affected by an impression of light, so that there would be far too little time for its formation from a number of successive perceptions separated by movements. This can be shown very beautifully and conclusively if an observer is allowed to look into a stereoscope in the dark, and the stereoscopic pictures are then suddenly illuminated by an electric spark. The duration of the electric spark is so small that there is no possibility of eye-movements during the time of its occurrence. Nevertheless, if the pictures are sufficiently simple, there arises immediately after their illumination by the spark a clear idea of extension in the third dimension.

The idea of depth, then, can arise in an extremely short time, and certainly requires no series of movements for its production. In its case also, that is to say, there must be something among the peculiarities of vision with the resting eye by means of which the sense of sight can be freed from the conditions originally imposed upon it. This something cannot be anything else than that whereby our vision of plane surfaces has also been to a certain extent freed from the co-operation of movements which originally were so necessary. Here again the local character of sensations, which serves to determine their arrangement upon the retina to which they belong, gives the signal, so to speak, at which the mind constructs the spatial extension of each particular retinal image. From the differences which it finds in these images, it measures the extension of objects in the third dimension of space. Just as in the areal field of vision the falling of coincident images upon portions of the retina endowed with a practically common sensational character gives us an indication of the extension of the object in a single plane, the excitation of portions of the retina whose sensations are not of similar character serves as a sign for extension in the third dimension.

Our measurement of spatial distance corresponding to a certain difference between sensations was, as you know, originally made in terms of movement; but after the measure has once been obtained, the indissoluble association of the two sensation-series (sensations of movement and local sensation-qualities) renders it possible for the first series in certain cases to disappear while the measure of spatial distance still remains unaffected. At the same time, observation shows that the connection of the two series cannot permanently be disturbed without causing a disturbance in spatial vision, and this disturbance can then only be gradually eliminated by a new serial association term after term. If, therefore, the eye is freed from the determining influence of movements in particular perceptions of space, its freedom is still nothing absolute; but now and then the renewed control by movement will be found necessary. Only in this way is the firm association of the two sensational series, which are brought into connection with each other by the persistent operation of conditions seated in the sense-organ, preserved in undisturbed integrity.

§ I

THE fusion of the two retinal images to a single idea is only a particular instance of a general law of the formation of ideas. In the visual idea which comes from the two eyes we do not discover any trace of the perception of one eye as distinct from that of the other ; but we blend them at once into a single and indissoluble perception. In this sense it is true that the two eyes constitute only a single organ of vision. That they are really like two independent observers, regarding things from two different points of view, and that we become acquainted with the properties of objects only by combining the result of these observations, are facts which we do not remark ; there is given in consciousness simply the result of this combination. That is to say, it is not until the two perceptions of binocular vision have fused that we have an *idea* ; or, in other words, the fact that the two eyes unite to perform a common function is a necessary result of the nature of the mental processes of association.

This connection of the separate elements of an association to a single idea, which is inevitably conditioned by the nature of the mind, is also furthered by the laws of external perception. Our external perceptions are of such a character, that they can only be referred to an object corresponding to the idea which has been formed of it. So that we can raise the further question : how does the ideational activity behave in the presence of impressions which cannot be referred to one and the same spatially extended object ? These conditions, of course, are never realised in nature; but we can by means of the stereoscope present to the two eyes impressions of the kind supposed. It is just as easy to put into the stereoscope different images, chosen arbitrarily and at random, as it is to bring before the two eyes in this way the planary projections of one and the same object. What does the mind do with these perceptions, which it is unable to combine into a single idea ?

There are scarcely any observations which serve to throw more light upon the nature of our ideational activity than these very experiments, in which something is presented to the organs of sense which is irreconcilable with the laws of their normal functioning, and which bewilders the mind, as it were, with the problem of removing the contradiction involved between conflicting perceptions. One general law may be formulated : however great the difference between the separate perceptions offered to the two eyes, a separate simultaneous apprehension of them is never possible. What happens is either a combination of the separate perceptions, on the analogy of stereoscopic vision proper, or an alternate apprehension of this or that retinal image.

Wherever the two pictures have a certain resemblance to one another,—whenever, *i.e.*, their difference does not too glaringly exceed the differences which occur in nature,—they are fused in a single idea. And even if their differences only correspond remotely to those shown in the pictures of a tridimensional object, the idea of depth will still arise at once. The minor differences are neglected, and the images interpreted in terms of that class of actual objects which they most closely resemble.

But more than this, figures which cannot be combined at all to give an idea of depth fuse to a single idea, if they possess a certain similarity by means of which they can be readily apprehended as pictures of the same object. If, *e.g.*, there are placed in the stereoscope two circles of nearly the same size, there results the idea of a single circle of mean diameter. In the same way, if two horizontal lines whose vertical distance from one another is a little different are presented to each eye, we have as a result the idea of two lines at the mean distance. Now neither horizontal lines nor circles of different sizes can give us the idea of depth. How does it happen that their combination is none the less possible ? We must not forget here that there may be differences between the retinal images of the two eyes even in the absence of the conditions necessary for the perception of depth. If, *e.g.*, we hold the figure of a circle very near our eyes, but a good deal to one side,—so that the figure is nearer one eye than the other,—the retinal image of the nearer eye is greater than that of the more remote, since the magnitude of the retinal image is directly dependent upon the distance of the object perceived. In this case there are in the two eyes retinal images of different magnitude ; and yet, when we fixate the circle, we see it singly. It is just the same with the two horizontal lines, or with any other figures. So that the condition of vision obtained by placing in the stereoscope two figures of somewhat different size is not essentially different from that which sometimes occurs in normal vision. It is, of course, true that in reality we never get images of different magnitudes when we fixate an object lying directly before us, as is the case in the stereoscopic experiment. But that is a secondary circumstance which we may neglect, because when in normal vision we are estimating the magnitude of objects lying very far to one side of us we still do not pay any regard to their different distances from the two eyes.

§ II

Quite different phenomena make their appearance if entirely different objects are presented to the two eyes. If we place in the stereoscope two pictures representing objects taken quite at random, we observe a curious alternation of ideas. We neither perceive two pictures simultaneously and separately, nor do they fuse together ; but first one and then the other makes its appearance. It frequently happens that one picture appears by itself for a time ; then various portions of the other force themselves to the front, and then suddenly the second picture alone can be observed. We notice as a universal rule that there is never any simultaneous superposition of the parts which belong to one picture upon corresponding parts of the other ; and that no complex perception, consisting of parts both of the first and of the second image, can hold its place as a permanent idea.

Such a composite picture is always a transitional stage from one image to the other.   And this transition, or alternation, between two perceptions forcing themselves upon our consciousness at the same time, is very readily occasioned by external influences. In this connection the movement of the eyes is of special importance.   As we move these organs one of them may fixate some sharply drawn boundary line within the first picture, while the second is directed upon some less prominent portion of the second picture.   In this way there arises a tendency for the former to predominate in the resultant idea.   But if we again move the eyes, and this point of fixation changes, the second image may come to prevail in exactly the same way.   First of all there enters into the idea that portion of the picture which forces itself upon our notice with special intensity, and this portion then brings all the rest of the image with it.

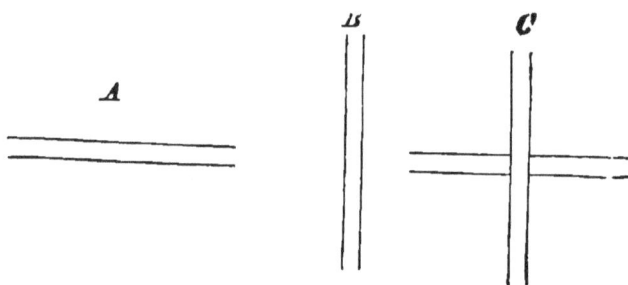

FIG. 32.

For the observation of these phenomena we may employ quite complicated drawings ; but they can be just as well illustrated by means of simple figures, *e.g.*, by letters of different form.   If we present a *U* to one eye and a *W* to the other, or a *J* to one and an *S* to the other, there is never any fusion to a single idea. Frequently we see only one letter ; then it breaks up into parts, and parts of the other letter are added to it ; and, finally, this latter alone is present in our idea.   So that there is no permanence of any one image, but a perpetual alternation, a breaking down and building up of images ; and the eye is greatly fatigued by this alternation of images which it cannot control.

If, on the other hand, the two letters have no conflicting characteristics, they may be united in a fairly permanent idea.   Thus we can unite an *E* and an *F*, or an *L* and an *F*, and in both cases obtain the idea of an E. But still the perception is not quite so steady as is the image of a real E perceived in monocular vision. When portions of the two images perceived are superimposed we observe a curious fluctuation in idea. For some little distance near the boundary the contour is always altogether interrupted ; and this distance varies from greater to smaller.

A similar interruption of boundary lines may be observed in the combination of pictures the lines of which cross one another. If we present to one eye two horizontal lines, A, and to the other two vertical lines, B, each pair being separated by a moderate distance (Fig. 32), we obtain a total image in which the lines in one direction are interrupted by those in the other. It may be either the horizontal lines, as in C, or the vertical lines, which are interrupted. This again depends in most cases on the movement of the eyes. If the point of fixation travels in a vertical direction, the vertical lines are seen as continuous, and similarly with the horizontal lines. It seems as if there is a tendency to see objects in binocular vision as extending in the third dimension, which has arisen from long habit, and which is manifesting itself in these experiments. This tendency is realised so far as the nature of the retinal images permits ; and as a result we simply see one image behind the other. But the resultant idea is not completely explained by this supposition. How is it possible for us wholly to ignore this one set of boundary lines, which is so definitely presented to us ? How does it happen that portions of one retinal image completely disappear ?

§ III

To understand these phenomena, we must familiarise ourselves with a series of facts, which may be observed both in monocular and in binocular vision, and which are of not less importance for the right understanding of the formation of ideas than those which we have already discussed.

It is a well-known fact that we may see reflected in the polished surface of a table the ceiling, furniture, and windows of the room in which it stands. And you all know that not only are the outlines of the reflected objects perfectly distinct, but the colours are given quite truly. Natural as this observation appears, it cannot be directly explained in terms of sensation.

For if the colour of the table is dark brown, one would have thought that the white window, mixing with this dark brown, would produce some shade of light brown. But that is not the case. The colour of the objects reflected in the table is wholly unaltered, while at the same time the colour of the table itself is distinctly seen. We are not able, of course, to apprehend clearly the colour of the table and the colour of the reflected images with absolute simultaneity. But we can accurately observe the two colours in succession without being disturbed by the mixture of light-impressions on the retina.

Suppose we lay a coloured object, *a*, upon a uniform colourless surface, and hold above it a plate of glass, *g*, placing beside the glass plate a second object, *b*, of a different colour from the first upon a similar background (Fig. 33). By looking through the glass plate we see the object *a* directly, and in addition the mirrored image *b'* of *b*. That is, our experiment has exactly reproduced by artificial means the conditions which are present in the reflection of objects in a polished table; we see an object, *a*, of a definite colour, *e.g.*, red, and a reflected image, *b'*, of an object, *b*, which is also of a definite colour, *e.g.*, white. The result in the two cases is exactly the same. The image *b'* is not pale red, but quite unmistakably pure white; and if the attention is directed to the object *a*, it does not either appear as pale red, but we distinctly cognise it as a pure red colour. We are, therefore, able to separate out and to consider in isolation either of the two coloured impressions, in spite of the fact of their intermixture upon the retina.

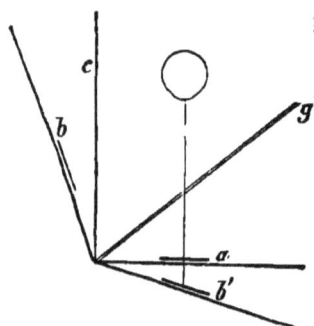

FIG. 33.

But this simple experiment is more instructive than observations of polished tables or other reflecting objects. For in it we can vary the conditions at will, and so gain more exact information regarding the causes of the phenomena. If we turn the support of *b* in such a way that it comes to the position *c*, where it forms with *g* the same angle as that formed by the latter with the support of the object *a*, the mirrored image of *b* falls exactly upon the place where the object *a* is seen. But now that this happens, the two images are referred to one and the same distance in space, and, consequently, fuse together. The result is a colour-mixture; the combined image of *a* and *b* appears pale red, so far as the two are superimposed.

The separation of the colours may be prevented in still other ways. If the coloured objects *a* and *b* are not definitely limited, but are so large that there is no clear perception of their dimensions, we obtain a mixed sensation, just as in the previous case where the mirrored image and the image of direct vision were localised in the same place. But the separation will occur at once if we draw lines to mark out a smaller figure upon each of the coloured surfaces. These boundary lines compel us mentally to assign to each figure its definite distance. And since the distances of the two figures are clearly apprehended as different, there arises the idea of a separation of the two images, with their entire sensation content.

We see here, then, the ideational activity working to effect a disjunction of impressions such as can never take place in the sphere of visual sensation proper. In sensation the impressions are mixed, however different the objects from which they proceed. But since in the idea every impression is referred to its object, there is ascribed to each its own amount of participation in the mixture. Thus the idea corrects, as it were, what is reported by sensation.

Under certain circumstances, once more, an object may appear to reflect when seen with both eyes, while it does not do so for monocular vision. If in Fig. 34 we fixate the object *a* with the left eye *l* alone, we see it in its natural character. If, on the other hand, we look at it with the right eye *r*, we see the mirrored image *b'* behind it. When this image is very bright and covers the whole of *a*, it may happen that the latter is completely ignored, so that the right eye sees only *b'*, and the left eye only *a*. As a consequence, there arises the single idea of a reflecting object, and with that the clear discrimination of

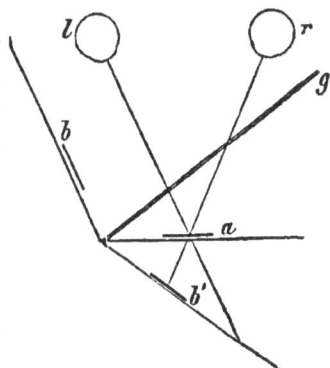

FIG. 34.

the object and the image mirrored behind it. Here we have obviously a case analogous to those already discussed under the head of stereoscopic experiments. Where the position of the object corresponds to that of the reflected image, the former is neglected, just as those portions of one of the stereoscopic images which were covered by lines of the other image were ignored. Since our observations of reflecting objects around us have accustomed us to neglect more or less extensive portions of an image, we carry this habit of neglecting certain elements into cases where the objects seen cannot naturally, and without a forced interpretation, be brought under the head of reflectors. But this is the sole form of combination whereby the two separate perceptions can be fused to a single idea.

The phenomena of reflection, which occur both in free and in stereoscopic vision, are closely allied to those of another class which are important as throwing considerable light upon the nature of the ideational activity,—the phenomena of lustre. Lustre and reflection pass into each other without any very distinct boundary. Since the phenomenon of reflection depends on an ideational activity, we may also infer that lustre will be referable to some mode of ideation. At the same time the popular view is opposed to such a conclusion. According to it, lustre is at least something given directly in sensation, if it is not some quality attaching to the lustrous body as such. But very simple observations will serve to convince us of the falsity of this opinion.

We found that when the furniture of a room is mirrored in the surface of a polished table we are able, despite the mixture of colours which ensues, to analyse our sensation into its constituents, and that in this way we always cognise the reflected objects and the reflecting table in their proper colours. But our cognition of the mirrored objects is only quite clear when the mirroring surface is very uniformly coloured, so that we can abstract from this uniform colour of the table surface at the points where the mirrored images are visible. A good mirror, however coloured, always shows us the reflected objects just as they would be if directly fixated. This is not the case if the mirror is differently coloured in different parts, or if dark and light places alternate on the polished table. Even though each portion of the mirroring surface reflects with equal clearness, the reflected object is not clearly seen. Why? Plainly because it is difficult in such a case to restrict our attention to the apprehension of a single object. On the one hand, the attention is attracted by the boundary lines of the differently coloured portions of the mirroring surface, and on the other by those of the reflected object. And this equally strong attraction by different impressions brings about a conflict of ideas which prevents any permanent or clear apprehension. We cannot see the mirrored images clearly for the mirroring object, nor the mirroring object for the mirrored images. In other cases where a plurality of ideas is simultaneously presented it is still possible to apprehend each particular one distinctly by bringing it in its turn singly before consciousness. Here that is impossible. For the same sense-organ gives us simultaneously impressions which belong to two different ideas. And, moreover, the two ideas are of approximately equal intensity, so that neither the suppression of one by the other nor their alternation is possible.

The correctness of this account of the origin of lustre may be confirmed in various ways by experiment. The phenomenon of mirroring obtained when we produce, by means of a plate of glass, a reflected image behind the place where the object of direct vision lies, can be readily and immediately transformed into a phenomenon of lustre by taking the two objects in the experiments described above (the mirrored object and that directly fixated) in such a way that the ideas which they arouse are of equal intensities. Pure mirroring occurs most readily when the object seen directly is dark, and the mirrored image bright, and when the former is uniform over its whole surface, the latter sharply delineated and obviously situated at a definite distance behind the real or apparent mirroring surface. Wherever the boundary lines of the reflected image are vague, and our judgment of its distance accordingly uncertain, or where the boundary lines of the directly fixated object are prominent and interfere with those of the reflected image, mirroring passes over into lustre.

From this it is clear that the phenomenon of lustre will occur very readily in binocular vision, when one eye sees only the object, the other only the reflected image. In this case we know quite well that we have before us two different things,—an object and an image which is mirrored in the object. In no other way can it come about that the two eyes perceive different colours. But we have no idea of the distance of the image behind the object. We do not even know which of the two perceptions is referable to object, and which to image. And so a very strong lustre may be produced by placing in the stereoscope a strip of coloured paper for one eye and a strip of the same size and form, but of a different colour, for the other. Green and yellow, blue and red, or any colours that are sufficiently different, give an extremely vivid lustre. And in like manner we get lustre by employing very different degrees of brightness of the same colour. Strongest of all is the lustre obtained by the combination of black and white. In this case we do not see a black and a white surface, or a white surface through a black one, but we obtain the same single impression as when looking at lustrous graphite or a lustrous metal, except that the lustre is usually stronger than that which we ordinarily find in natural objects.

Our every-day experience teaches us that wherever we see lustre there is no possibility of a clear apprehension of the objects seen. Too much or too diffuse lustre is therefore unpleasant to the eye, though the luminosity of the object be not nearly strong enough to affect us unpleasantly. A lustrous stimulus is only pleasant for vision when it occurs at rare intervals, and allows the sense-organ to recover itself in the meantime by turning to impressions of the ordinary visual character. Otherwise lustre dazzles us. And this disturbance of vision (which may even affect sensation) is again of a mental or psychophysical character. It makes its appearance wherever there is a conflict of ideas which press upon consciousness with equal intensity. We have observed its analogue in the stereoscopic experiments, where pictures differing so greatly that they could not be combined in a single idea were presented to the two eyes. In both cases we are only dealing with a particular consequence of the principle of *ideational unity*, which we shall have to refer to again in our discussion of consciousness and of the connections of ideas in consciousness. While this principle, in the normal course of our mental life, merely conditions a steady alternation of particular ideas in temporal succession, it leads to such peculiar phenomena as lustre and ideational rivalry, when this normal alternation is prevented, whether by the striving of two ideas for apprehension at the same time, or by the refusal of a simultaneous plurality of ideas to be resolved into its elements.

§ IV

In addition to lustre and ideational rivalry, there exists yet another form of the apprehension of binocular perceptions. If the perceptions of the two eyes do not press upon consciousness with equal intensity, but one of them has a considerable preponderance,—for some reason lying in the nature of the external impressions,—this predominant impression alone becomes an idea, while the other is completely ignored. Here again stereoscopic experiments enable us artificially to reproduce the conditions of the phenomenon. We can do so most simply by employing coloured objects of definite outline. If we place a black background in the stereoscope and lay on this a white square as an object to be perceived by one eye, we obtain, not an idea of white and black mixed, despite the fact that the other eye sees nothing but black, but we imagine that both eyes see a white square on a black ground; and the white is as intensive as that of the object seen with the first eye. That is, the perception of one eye completely suppresses that of the other. The explanation of this is obviously to be found in the fact that the definitely outlined white object, contrasting sharply with its background, possesses a far greater intensity for ideation than does the uniformly black surface. The same phenomenon may therefore be observed when we lay a black square upon a white background, or in general if we present to one eye only a square of any colour we choose, placed upon a differently coloured background.

In like manner one perception may be completely changed by the presence of another if a coloured object of like form and dimensions is presented to either eye, each, however, contrasting at a different degree of intensity with the differently coloured

background upon which it lies. Suppose, *e.g.*, that we lay upon
a white background a dark red object for the right eye, and a
dark green object for the left. The perception of the right eye
entirely suppresses that of the left : we see only the red object,
and nothing at all of the green one. If, on the other hand,
we had taken a black background instead of a white one, we
should see only the green object, and nothing of the red. The
reason plainly is, that dark red contrasts with white more
strongly than does bright green, while this latter in its turn contrasts more strongly with black. That colour which stands out
more clearly from the background is more intense for ideation,
and so we perceive it alone, and entirely ignore the other. On
a grey background we get the idea of a very lustrous object
seen in greenish light. In this case both perceptions come to
consciousness, because they are of approximately equal intensity, *i.e.*, stand out with equal distinctness from the background.
But, as we have seen, this simultaneous presentation of two
different ideas always gives rise to lustre.

It may sometimes be observed that these phenomena of
suppression do not extend to the entire image, but are restricted
to one part of it. This is especially likely to occur when one
retinal image possesses a much greater extension than the other.

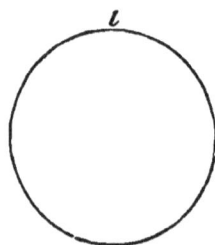

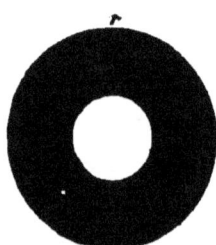

FIG. 35.

If, *e.g.*, we present
to one eye a white
circular surface, *l*,
and to the other a
black circle with a
small white spot at
the centre, we shall
see this last in the
common image as a bright spot surrounded by a very dark
border, which becomes brighter and brighter as we approach
the periphery, and at last almost entirely white. In this case it
is clear that the image *r* entirely suppresses the image *l* at the
centre, but conversely is itself suppressed by the latter towards
the periphery, while between these two portions of the common
image there are continuous transition-stages. The following
experiment is of a similar character. To the eye *l* a uniform
surface, *e.g.*, blue, is presented, and to the eye *r* two coloured

surfaces, which are joined in the median line, *e.g.*, green and red. In the common image we see at the centre, where green and red meet, simply these two colours, while towards the outside they are intermixed with a bluish colour-tone.

But these two last experiments can only partially be subsumed under the phenomena of suppression. This, it is true, plays a part in them, inasmuch as a portion of one of the perceived images preponderates over the other, and thereby causes the latter to disappear entirely. But the preponderance is in their case limited to one part of the image, while in other parts of it it may not infrequently happen, on the contrary, that the image of the other eye is the predominant one. This fact seems to be almost directly opposed to the laws of ideational activity. The fact of the unity of the idea we have seen to be firmly established. It is no contradiction of this that one perception permanently suppresses the other, or that two perceptions alternate in succession ; but that each of the two perceptions should be partially apprehended, and so appear in ideation in the form of a mixed image,—this seems to be scarcely in harmony with our law. We have, however, already become acquainted with a whole class of phenomena in which also two perceptions may combine to form a single idea : these, as you know, are the phenomena of lustre and reflection. In the case of lustre two ideas are presented to us, which we do not succeed in keeping separate. In mirroring we effect this separation, and we can therefore either alternate between the idea of the reflected and that of the reflecting object, or we can unite both of them in a total idea. When we look at the image in a mirror, we usually embrace both image and mirror in a single idea ; the mirror is the frame which surrounds the picture. Now in the experiments which we have just been describing we have obviously the same conditions. Besides the considerable intensification of one portion of a visual perception, the idea of reflection also exerts some influence. And therefore that especially prominent portion of the perception of one eye is alone given in ideation at the place which it occupies in the common image, at the other parts of which ideation has free play, and so tends to apprehend the image of the other eye as a mirror in which the first is seen. Still the conditions in this artificial experiment do not entirely correspond to what takes place in nature. In nature, too, it may happen that we see only the mirror with one eye, and only the mirrored object with the other. We need simply hold the mirror close to our eyes and cause the mirrored image to fall very much to one side. But there are many conditions which we can introduce into the experiment which are never realised in nature. Suppose, *e.g.*, that we place a large blue and a small yellow object in the stereoscope for combination in binocular vision, both lying on a red background (Fig. 36) ; we have a common image, in which the yellow is seen surrounded by blue. So far there is nothing out of the common, for it may also happen in nature that we see a yellow object reflected in a blue mirror. But where this happens we must necessarily see the mirror with the same eye which sees also the mirrored object. For if we are viewing a small object in a very large mirror, it may certainly happen that this image is only visible to one eye, but the mirror itself is never visible to one eye only, least of all to the one which is not looking at the mirrored image. The conditions of the experiment here, then, are not in accordance with nature. How does the eye cope with its perplexity ? Since the right eye sees yellow upon a red background, and the left blue on the same, there is occasioned simply the idea that a yellow object on a red background is mirrored in a blue object ; that is, not only the yellow object, but also the red background which immediately surrounds it, are comprehended in the resultant image. Farther towards the lateral parts of the image, however, there arises in ideation the perception of a blue surface distinctly outlined against the red ; and so the blue sensation gradually comes to predominate. Thus we obtain as a final image that of a large blue square upon a red background, in the middle of which is a small yellow square surrounded by a fringe that shows a deep red on the inside, but becomes more and more tinged with blue towards the outside.

All these phenomena, which can be varied in many other

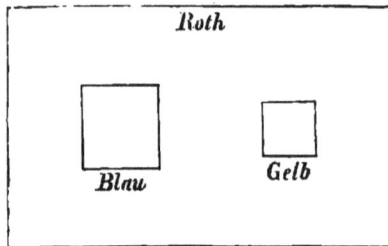

Fig. 36.

(labels in figure: *Roth*, *Blau*, *Gelb*)

ways, show that a single idea is always formed from the perceptions of binocular vision, and that this always takes place on the analogy of the conditions of vision found in nature. The process by which the two visual perceptions unite to form one idea depends, therefore, upon the formation of numerous associations, some acting in harmony with, and some in opposition to, one another. In this latter case, phenomena of suppression, or ideational rivalry, make their appearance. The separate perceptions of the two eyes are themselves composed of sensations which combine with each other in a quite different way from the light-impressions which fall upon one retinal point in the same eye. We have rather to regard the binocular visual idea as a *mental resultant of the originally separate perceptions of the two eyes.*

§ I. THE FEELINGS. § II. SENSE-FEELINGS. § III. COMMON FEELING AND THE OTHER TOTAL FEELINGS. § IV. RELATION OF FEELING TO IDEA.

§ I

THE mental phenomena with which we have been concerned hitherto have represented stages in one and the same great process. We have seen that ideas are derived from sensations in the regular course of development, and that both alike have a single end,—*knowledge of the external world*. But we have purposely neglected one very important side of our mental life. We never actually find a mind which apprehends things without joy or sorrow, and contemplates them with absolute indifference. In cognising objects we feel ourselves attracted to or repelled from them, or incited to the performance of some kind of action, according to their nature. We can, therefore, comprehend all those phenomena which are not included in the ideational process under the two words 'feeling' and 'will.' Feeling and conation always accompany our sensations and ideas; they determine our actions, and it is mainly from them that our whole mental life receives its bias and stamp of individuality.

Feeling and will are closely interconnected. And both are again connected with ideas. The separation between these processes is one that exists only in psychological abstraction, and has no basis in reality. Feeling passes over into impulse, impulse into voluntary action, and voluntary action has reference to objects which are given to us as ideas.

In ordinary language we employ the word 'feeling' in various senses. We call hunger and thirst 'feelings'; we speak of the 'feeling' of pain, and of 'feeling' external objects with our hands; we call love and hate, joy and sorrow, hope and anxiety, 'feelings'; we talk of our 'feeling' for the beautiful and the ugly, and even of ' feeling' that something is true, or honourable, or virtuous. What is our justification for bringing under one and the same concept mental processes which are so diverse in their nature, and belong to such different stages of development ? It may be mere chance ; language may somehow have come to apply the same name to a number of totally different phenomena. Or it may be purpose ; perhaps these processes possess something in common, notwithstanding all their divergences.

As a matter of fact, there is one point in which all ' feelings ' agree, however different they may be in other respects : they all imply a condition of the feeling subject, an affection or activity of the self. Feelings are always subjective, while the idea always has an objective reference. Even when the matter of ideation is some one of our own conscious processes, this is regarded objectively. So that the term 'sense of feeling' has been reserved for the sense whose impressions are most obviously connected with subjective states of pleasure and un-pleasantness. What language implies in calling joy and care, love and hatred, 'feelings,' is simply that they are apprehended as exclusively subjective states, and not as properties of objects outside us.

The attempt has sometimes been made to limit the meaning of the term 'feeling,' on the ground that the processes compre-hended under it are entirely too diverse in character. And in particular it has been thought necessary, from the psychological standpoint, to strike out from the category of feeling all those subjective excitations which are directly connected with sen-sations. Hunger, thirst, bodily pain, impressions of touch in general, are, it is said, sensations ; they are accompanied by physical processes in the nervous system : but feeling is a purely mental condition. The term should, therefore, be confined to mental states which are independent of bodily affections and arise solely from some kind of reciprocal action among ideas. But as soon as we give up the reference of feeling to a subjec-tive condition of pleasure-pain, or of some similar pair of sub-jective opposites, we have no reason for uniting affective states in general in a common class. If, on the other hand, we retain this principle of classification, we cannot exclude from ' feeling' the sense-feelings attending the simple sensations. It seems to have been regarded as a difficulty that one and the same simple process should be called both sensation and feeling. But it has been forgotten that joy and sorrow, hope and anxiety, and all the other 'feelings' are really states of mind which are affective only so far as they have reference to the feeling subject ; while in other respects they depend upon ideas which objectively regarded are entirely empty of feeling. The only difference is, that in the case of these more complex feelings we attach a greater value to the feeling as such, and therefore give a separate name to each of its particular forms. A very complex ideational association may give rise to a feeling which is simple and uniform throughout. We are, therefore, inclined to substitute the subjective result, the feeling, for those various and complex processes which serve as its objective background. There is no need for this in the case of the simple sense-feeling. Its objective substrate is a sensation which is equally simple with it, and more readily discriminated by virtue of its relation to an external object.

If the feeling is characterised merely by its relation to the feeling subject, it is clear that the distinction between it and the sensation or idea cannot be in any sense original. The simple sense-feeling in particular is contained *in* the sensation ; and it is just as incorrect to say, ' Feelings alone are primitive,' as to say, ' We have at first simply sensations.' The ultimate fact is, that we sense and feel. The logical separation of feeling from sensation can only come about after we have distinguished subject from object. Then, and then only, is the elementary process of sensation analysed into a subjective factor, the feeling, and into an objective factor, the sensation. The sense-feeling may in this way be considered as an integral element of the sensation itself ; and for that reason it is also termed the *affective tone of sensation.*

Since the aim of these lectures is to give a general description of the elementary constituents of our mental life before dealing with their connection in consciousness, and the complex processes to which this connection gives rise, it will be best for us to confine our attention here to the *sense-feelings*. The simplicity of the conditions upon which they depend makes it easy to examine them without any detailed reference to their connection in consciousness. For the same reason they will give us the most efficient aid in our endeavour to understand the general nature of feeling and its relation to will. But we shall be obliged occasionally to glance at the higher feelings,— especially the intellectual and æsthetic. There are certain problems to which these furnish more' definite answers, just because of the more complex associations which they involve.

§ II

There are some sense-organs which require quite intensive stimulation for the production of a sense-feeling possessing any considerable degree of strength. This is especially true of the eye and the ear. A moderate visual or auditory stimulus seems to give rise to scarcely anything except the objective sensation ; a moderate light-impression, *e.g.*, is referred simply and solely to the external illuminating object. But it must be admitted that careful introspection enables us to recognise a certain affective tone even in the sensations produced by weak visual and auditory stimuli. We notice this especially when the impressions are given without direct reference to external objects with clearly defined boundaries. Thus the different spectral colours,—red, green, blue, etc.,—as well as white and black, have characteristic, if weak, feelings attaching to them ; while every single musical clang is also attended by an affective tone, determined in each case by tonal pitch and clang-character. But the intensity of these feelings is very slight. Their importance is, however, increased by the fact that they enter as intensifying elements into the *æsthetic* feelings, which are more elaborate affective states, connected with entire ideational complexes. On the other hand, a dazzling light or a deafening sound will directly occasion a feeling of pain, behind which the objective significance of the sensation may in its turn entirely disappear. These intensive stimuli disturb the normal functioning of the organ ; and it is, therefore, the subjective factor which comes most predominantly into consciousness.

Cutaneous sensations are also as a rule referred wholly to external impressions, so long as they are not painful. But there are certain stimuli which, although truly tactual, produce sensations with very intensive affective tones. Stimuli of quite weak intensity, which only lightly touch the surface of the skin, excite tickling or itching. Both of these may also arise independently of external impressions. They are always characterised by a tendency to diffusion of effect. The mechanism of the matter probably is that weak tactual stimuli call into reflex activity the unstriated muscles which lie directly beneath the skin, and to which cutaneous movement is due. When these muscles are contracted, we have also the peculiar feeling of shivering accompanying the muscle-sensation proper ; and this readily combines with the feeling of tickling. The reflex excitation of the cutaneous muscles frequently extends to other muscle groups ; and when the excitability is great, may lead to general reflex convulsions, which are greatly exhausting to the organism.

A fairly low degree of temperature has an analogous result. If we let a cold stimulus of slight intensity act upon the skin, we have first a sensation of cold,—*i.e.*, the cold is perceived as an alteration in the condition of excitation of the cutaneous organs,—then the smaller cutaneous muscles are reflexly excited, and so the feeling of shivering is occasioned. The same effect may be produced by the action of internal causes in producing a sudden loss of heat, and consequently a cutaneous sensation of cold. This happens in fever chills, where the effect is much intensified by the abnormal reflex excitability of the cutaneous muscles. Very high or very low temperatures, finally, have the same result as very strong pressure-stimuli : they do not produce sensations of heat or cold, but only severe pain. And it is of the very nature of pain that its character is always constant : a prick, a grinding pressure, intense heat, and freezing cold all excite pain of the same strongly affective quality.

It is different with impressions of smell and taste. Even at weak intensities, these are accompanied by distinct feelings of pleasure and unpleasantness. And the feeling is so closely fused with the sensation that it seems impossible to think of the two as even temporarily dissociated. It is only the fact that there are impressions which are relatively free from feeling which leads us to believe that the quality and intensity of the feeling depend upon other conditions than those of the sensation. It is, perhaps, scarcely necessary to point out how important the strong affective tone of the sensations of smell and taste is for our physical life. Here, more than in any of the other senses, the feelings of pleasure and pain serve as subjective indications of the impressions which we should seek and of those which we should avoid. These indications may, of course, occasionally lead us astray; but the adaptation of natural unperverted feeling to the beneficial or injurious character of stimuli is on the whole marvellously complete.

But peripherally excited sensations do not constitute the only material for the formation of sense-feelings. There are a large number of sensations which are not caused by external impressions, and which do not imply the apprehension of external objects, but which may in all other respects be co-ordinated with the sensations from the sense-organs proper. To this group belong, in the first place, the muscle-sensations, which we have already described in our inquiry into the processes concerned in perception. Moderate muscular exercise is connected with a more or less distinct feeling of pleasure; while exhaustion, excessive effort, or a pathological condition of the muscles not only alters the quality of muscular sensation, but also brings with it a very intensive feeling of unpleasantness. Secondly, there belong to this class sensations from the various tissues and organs of our body. These are generally of slight intensity, and therefore easily overlooked. But under special conditions, especially where the state of the particular organ or tissue is pathological, they may become so intensive that the feelings connected with them dominate consciousness almost exclusively, and produce general discomfort throughout the organism. These sensations, therefore, are only familiar to us at their highest degree of intensity, *i.e.*, when they have become painful. But the affective character of pain is, as we have already remarked, essentially the same in every case. For that reason, the specific differences of the organic feelings are not generally apprehended by us. Nevertheless, observation shows that such specific differences really exist. Language uses different terms to denote the pains coming from different organs. We speak of 'stabbing' and 'gnawing' pains in our bones, of 'pricking' pains in the porous serous membranes, and of 'burning' pains in the mucous membrane. Here, as in the external sense-organs, pain is simply the sensation raised to its highest degree of intensity; and that peculiar property of the pain which is dependent upon the structure of the organ is prefigured in the pure sensation. This is especially noticeable in intermittent pains. There are times when sensation is present which cannot be called painful; and during these intermissions the peculiar sensation-colouring which gives its special character to the ensuing pain is usually not at all affected.

We may, therefore, regard these sensations from the bodily tissues and organs as originally equal in value to those from the organs of special sense. Gradually, however, these latter acquire a paramount position, through their importance for the development of ideas; while the great majority of the organic sensations pass unnoticed until their unusual intensity in a particular case announces some important change in the condition of the organism, of which consciousness is obliged to take account. That is to say, in proportion as sensations, originally given as undifferentiated, are divided up into sensations of special sense (relatively free from affective tone) and organic sensations (strong in affective tone), there is a tendency for this latter group gradually to disappear from consciousness. Hence the entire life and action of the child is determined by the sense-feelings; while the more developed and better furnished the mind becomes, the more independent is it of their domination, and the greater is its measure of success in permanently repressing the weaker sense-feelings and at least temporarily subduing the stronger ones. The only exception is the hypochondriac, who delights in observing his own bodily symptoms and states. By dwelling anxiously upon all those weak sensations which pass unnoticed by the normal consciousness, he gains a great deal of training in the apprehension of his sense-feelings. The physician will often laugh at his 'pains' and 'aches' as illusions; but generally they are real enough. The abnormality of the hypochondriac does not consist in his perception of feelings which have no existence, but in his distinct apprehension of, and anxious reflection upon, feelings which a healthy man usually takes no notice of.

To the specifically organic feelings belong also those of hunger, thirst, and shortness of breath. They, too, are connected with sensations which normally recur with moderate intensity at definite intervals, but which will increase more and more in affective intensity, if their demands are not satisfied. Hunger, thirst, and respiratory excitations are sensations centrally aroused, but peripherally localised,—thirst in the mucous membrane of the palate and throat ; hunger in the stomach ; the respiratory sensations in the respiratory organs, and especially in the muscles of the chest, which subserve the process of breathing.

§ III

Our general condition of bodily comfort or discomfort is dependent upon the mass of organic sensations which are always present, but in varying degrees of affective intensity. The totality of feelings acting upon consciousness at a given moment is termed the *common feeling*. This has been defined in terms of its origin as the sum total of simultaneously present organic feelings of whatever quality. But the definition overlooks the fact that our state of feeling is always qualitatively single. We can never be moved simultaneously by a number of diverse and independent feelings ; they combine to form a resultant possessing the character of a feeling of definite quality and intensity. At a given moment we feel 'well,' or 'ill,' or indifferently. If we should ever say, in speaking of our general bodily condition, that we are feeling at once 'well' and 'ill,' it may always be proved by introspection that we have been uniting *successive* feelings in the single judgment. But this is the result of reflection upon our feelings ; the fusion is not given in the feelings themselves.

This qualitative unity of feeling seems to correspond to the ideational unity of our consciousness. The sensations excited at a given moment by external and internal stimuli are not perceived as a mere medley of impressions, but are associated to form ideas, which are then brought into spatial and temporal relations with each other. In the same way all the particular feelings are united in one total feeling, into which each enters as a constituent factor. But the analogy cannot be carried any farther ; there is an important difference between the two processes. We can prove that an idea is a compound process by analysing it into sensations. A clang, a compound clang, and a visual object are single, but not simple, mental facts. We can analyse each of them into a number of simple sensation-elements. But no feeling is capable of analysis in introspection, whether it is connected, like the sense-feeling, with a single sensation, or, like the elementary æsthetic, intellectual, and moral feelings, with a complex ideational group.

This simplicity of feeling, together with its subjective character, which makes it impossible for us to refer it in every case to external objects, as we do with sensations and ideas, is doubtless the cause of that 'obscurity' which has been so often emphasised, and which consists simply in the indefinable nature of affective quality. It is this 'obscurity' which has led to the attempt to substitute for the impossible definition an enumeration of the objective conditions under which feeling arises, and a description of the relations obtaining between the ideas present at the moment of its appearance. These supply the only means at our disposal for the production in others of feelings similar to those which we experience ourselves under particular circumstances ; and they are perfectly justifiable so long as we do not mistake them for an account of feeling itself. But psychology has fallen into this very error again and again. That is, it has 'explained' feeling by reflecting upon the ideas among which it arises, and by which it is attended. Sense-feeling has been defined as a furtherance or inhibition of our bodily well-being, or even as a direct cognition of the usefulness or danger of sense-stimuli. Æsthetic feeling has been said to consist in the idea of definite mathematical proportions, moral feeling in reflection upon the useful or hurtful consequences of our actions, and so on. Leaving out of account the objections to these theories on other grounds, we see that they are all overthrown by the consideration that feeling is not itself an intellectual process at all, although it is always connected with intellectual processes.

Every feeling is a qualitatively simple and undecomposable mental state. This fact does not, of course, exclude the possibility of there being in consciousness several simultaneous feelings. Only, these simultaneous feelings always combine in a *total feeling* which possesses a unitary character, and cannot, therefore, be regarded simply as the sum of the original particular feelings. The *oscillatory* and the *discordant feelings* are, perhaps, the most instructive examples of these complex affective states. In the first group, opposing feelings alternate with each other in rapid succession. But there is also a continuous modification of one affective phase by the other, so that a new feeling with a characteristic quality of its own arises alongside of the primary changing feelings. Its quality is, of course, dependent upon those of the original feelings; but it cannot be analysed into them. Its intensity is constantly altering, so that at one moment the primary feelings, at another this new permanent feeling which is characteristic of affective oscillation, predominates in consciousness. Affective discordancy is directly derived from affective oscillation. It occurs when the oscillations of feeling follow each other very quickly, and the successive feelings themselves are strongly opposed. We have an example of this from the sense-feelings in tickling, and from the intellectual feelings in doubt; while the dissonance of two clangs may be taken to exemplify it in the field of the elementary æsthetic feelings.

The statement that doubt is a compound of the feelings of acquiescence and repugnance is certainly a true description of the alternating affective states which go to constitute the entire mental process. But there seems to be present in addition a resultant total feeling directly corresponding to the dissension in the emotional condition. There may be moments of doubt when neither the feeling of acquiescence nor the feeling of repugnance is in consciousness at all; and these moments possess a unique affective character which does not appear to be analysable into either of the other two feelings which displace it from time to time; but it may continue to exist alongside of them. At such moments, therefore, there exist *three* feelings,—those of acquiescence and repugnance and the total feeling resulting from the two, but qualitatively different from them.

Doubt always involves a strong opposition between the constituent feelings. In the feelings of tickling and of dissonance, which are formally related to it as 'discordant' feelings, although their ideational connections are entirely different, the affective state is more homogeneous. In tickling, which is due to a continued weak cutaneous stimulation, we can clearly distinguish two original feelings, either of which may predominate according to circumstances,—a pleasurable feeling, which probably accompanies the weak touch-sensations ; and a painful feeling, which appears to be connected with reflexly excited muscle-sensations, to which, *e.g.*, those of the diaphragm belong. If the tickling is slight, the direct effect of the stimulus, and therefore the pleasurable feeling, is most prominent ; if it is more intensive, the reflex effect, and therefore the unpleasant feeling, gains the upper hand. The specific total feeling of tickling is the resultant of these two. It, again, may be perceived with especial clearness when the two opposing factors are of approximately equal intensity. But in general, and perhaps as a result of the strength of the two factors, the total feeling in tickling is of relatively slight intensity. The opposite holds of the dissonance of two clangs, when we can always distinguish the feelings which are connected with the separate clangs from the total feeling of dissonance itself. As the dissonance increases, the total feeling prevails more and more over those excited by the separate clangs.

We see, then, that the total feelings arise from the union of particular feelings, but that they constitute new and simple feelings of definite quality, quite distinct from their constituents. Plainly the *common feeling* must be regarded as belonging to this group. We do not intend to denote by the phrase the medley of miscellaneous separate feelings present in consciousness at a given moment, but rather a new feeling to which they give rise, and whose quality they all help to determine. The entire sum of separate feelings combines to form a complex unity, the trend of which finds its expression in the resultant total feeling. Similar total feelings with accompanying particular feelings constitute the 'higher' intellectual, æsthetic, and moral feelings. In all these cases, every particular feeling and every total feeling have their own characteristic quality, in

virtue of which they stand in relations of agreement and dis-
agreement to other feelings, though they are never analysable
into them. So that nothing can be more erroneous than the
opinion sometimes held that the entire world of feeling is composed of a certain sum of elementary feelings,—perhaps sense-
feelings,—of approximately constant quality. The essential
characteristic of feeling, especially of the higher feelings, is
rather an inexhaustible wealth of qualities ; new qualities arise
from the mutual influences of simultaneous feelings, and from
the induction of present by antecedent feelings. And to this
we must add that the worth of the feelings constantly increases
as the relations in which they stand become more complex, for
it is these relations which determine the influence of any particular feeling upon our entire mental life.

§ IV

Finally, the existence of total feelings, and especially of those
which reflect an oscillatory or discordant affective state, leads
us to an important fact, without mention of which everything
that we have said about feeling, and especially about the origin
of resultant feelings, would be incomplete. We have considered
feeling primarily as a process which accompanies ideation.
Since we have been occupied hitherto with the analysis of ideas,
they naturally suggest themselves as points of departure for our
investigation of feeling. But ideas are not the only mental
processes, even if we abstract from feelings and the other subjective processes connected with them. All the changes which
occur in a given ideational content are as such also mental
processes characterised by a particular rapidity and manner of
occurrence, and, like the ideas themselves, connected with feelings. So that even from the standpoint of an objective observation, which only takes account of the ideational, and not of
the affective, side of mind, we are obliged to distinguish these
processes of change in ideational content from the ideas which
are altered. Introspection of such changes is also sometimes
called 'ideation.' We are told, *e.g.*, to 'form an idea' of some
change ; that a content is appearing or vanishing, or that contents are passing through consciousness with greater or less

rapidity. It is scarcely necessary to remark that we are not really dealing here with two different things,—ideas and changes in their condition and arrangement. The analogy with physical bodies and their changes of position which is usually thought of is entirely misleading. The ideas themselves are, as you know, not unchanging objects, but *processes, occurrences*, whose existence is necessarily bound up with that of the changes supposed to take place in them. If, *e.g.*, an idea disappears, that means that the mental process which we call an idea ceases to exist. So that when we speak of 'ideas with which feelings are connected' our language is at least subject to misinterpretation. We should rather say: all ideational processes, whatever their nature, whether they consist of the idea of an external object, or of some internal change in this idea, are at the same time affective processes. Affective discordancy furnishes an obvious proof of this: in doubt and in dissonance the resulting feeling is determined to a far greater extent by the characteristic alternation of ideas than by the nature of the ideas themselves. The total feelings in particular are always essentially dependent on some peculiarity of the alternation and succession of ideas. We shall return again to this point in our consideration of the *emotions*, of which total feelings are important constituents.

LECTURE XV

§ I

IN the previous discussion we took as our starting-point the fact that the affective side of consciousness at any moment seems to exhibit a unitary interconnection similar to that presented by its ideational contents. But further investigation convinced us that the affective unity of consciousness differs in important respects from its ideational unity. The latter appears to be *external*, in the sense that the particular ideas are united into a whole, more especially by the spatial relations in which they stand to each other, without the constituents of this whole being necessarily brought into any internal relation. In feeling it is quite different. It is true that several qualitatively different feelings may exist side by side, but they always give rise also to a total feeling which endows the entire group of separate feelings with an internal coherence.

We shall best understand this *internal* unity of feeling if we look somewhat more closely at the connection of *feeling* with *will*,—a subject to which we referred in a general way at the beginning of the last lecture. This connection may be regarded from two points of view. In the first place, feeling is only thinkable as a mental state of a being endowed with will; pleasurable and unpleasurable feelings tend to direct the course of the will. Whether or not they pass over into actual volition is determined by internal and external conditions. But without the capacity to will the alternatives could not possibly exist. Secondly, will is an internal process, distinguished from other mental activities through the fact that in it we are conscious of definite *motives*. But motives are always accompanied by feelings, and the feelings further appear to us as those elements of the motive which contain the real reason for the activity. Without the excitation which feeling furnishes we should never will anything. A mind which contemplated things with entire indifference as 'pure intelligence' could never possibly be roused by them to volition or action. Feeling, therefore, presupposes will, and will feeling. In a concrete voluntary action the two are not different processes, but part-phenomena of one and the same process, which begins with an affective excitation, and passes over into an act of will. It often happens, however, that the final term in this series is wanting : the intensity of a feeling may become lessened, or it may be displaced by another feeling without leading to an act of will. We may therefore divide feelings at once into two groups,—those which form constituents of a voluntary act and those from which no definite volition results. The latter class, again, contains different degrees. If the subjective condition is, and remains, simply a pleasant or unpleasant mood, we speak of a feeling proper. When there is added to this a definite tendency towards a willed result, we term the internal process an *effort* or an *impulse*. If in this effort we are further conscious of some inhibition, which prevents it from passing over directly into volition, we call it a *desire*.

§ II

It is in the doctrine of feeling and will more than anywhere else that psychology still wears the fetters of the old faculty theory. And so it has usually taken a radically false view of these intimately connected part-processes, regarding each constituent as an independently existing whole, which might incidentally, but need not necessarily, exert an influence upon the constituents of the other. Thus first of all feeling was considered apart from its connection with will, and then desire was treated as a separate process, sometimes found in connection with feeling. Further, impulse was opposed to desire proper as an obscure desire, in which the subject is not conscious of the desired object ; or, perhaps, as a lower desire, referring exclusively to the needs of sense. (That is why many psychologists hold that impulses only exist among animals.) And finally these processes are still further supplemented by the postulation of will as an entirely new and independent faculty, whose function it is to choose between the various objects of desire, or in certain circumstances to act in accordance with purely intellectual motives and in opposition to impulses and desires. According to this theory, that is, will consists in the capacity for *free choice*. Choice in this sense presupposes the possibility of decision between various objects of desire, and even of decision against the desired object on the ground of purely rational considerations. It was therefore supposed that desire is a condition which precedes volition, and that at least in many cases this latter is only the realisation of desire in action.

We must pronounce this theory a purely imaginary construction from beginning to end. It has taken its facts from every possible source except an unprejudiced introspection. Feeling is not independent of volition, as alleged ; impulse is not a process which can be distinguished from will, still less opposed to it ; and desire is not the uniform antecedent of will, but rather a process which only appears in consciousness when some inhibition of voluntary activity prevents the realisation of volition proper. Finally, to define the will as the capacity of choice is to render any explanation of it impossible from the outset. Such a capacity presupposes volition as its antecedent condition. If we could not will without choice,—*i.e.*, as directly determined by internal motives,—a volition involving choice would necessarily remain impossible.

This confusion of volition and choice brings another error in its train. Will is supposed to arise from all sorts of involuntary activities. Generally this view is applied exclusively to external voluntary acts, which many psychologists regard as the only ones. Both the human and animal body, it is said, were originally, before the appearance of will, the seat of reflex movements of the most diverse character. These were for the most part purposive, owing to the teleological connection of sensory with motor fibres in the central organs. Thus a stimulus which caused pain would give rise to a reflex movement of defence, resulting in the removal of the stimulus. It is further supposed that the mind perceives the purposiveness of these reflex re-actions, and so the thought arises in it that it might possibly undertake similar movements itself, and attain the same purposive result. The next time that the stimulus approaches, therefore, the mind will be on the alert to execute the defensive movement, and so remove the stimulus before it has any painful consequences. The most remarkable results of this kind of re-flection are those obtained as regards locomotion. It may per-haps happen that the body gives a sudden spring, in response to a strong reflex stimulus. 'Eureka!' says the mind to itself. 'Why should I not cause my body to spring when this un-desirable stimulus is not there?' But when the will has once discovered that its voluntary muscles enable it to do almost any-thing it wishes, it, and not the reflex, is master. The reflex has played its part, and is henceforth restricted to the sphere of the absolutely necessary.

You will not find, of course, that this description is in literal agreement with that given in any of the works treating of the development of voluntary acts from reflex movements, but in substance there is no difference. You will even find such ex-pressions as ' The mind takes note of this and that,' or ' It now executes movements voluntarily which it formerly observed taking place in the body involuntarily.' And indeed there is no reason why the mind should not act in this way, if it were 'pure intelligence,' as these writers palpably assume, or even if it only had at its disposal a small number of feelings to occupy its leisure moments.

But the matter assumes a different aspect when we look at it without preconceptions, and refrain from reading into the facts of observation notions and reflections which exist only in our own minds. In the first place, there is not the slightest con-firmation to be found for the assertion that the lower animals, and children in the early days of life, are merely reflex machines, which make certain movements with mechanical certainty as soon as we press the spring. Even such of the protozoa as un-doubtedly belong to the animal kingdom give plain evidence of voluntary movement. The chick just out of the shell executes movements which are in great part at least of the nature of voluntary actions. No one will of course deny that reflex movements may also be observed from the first especially among the more complexly organised animals. We have ourselves referred to the reflex movements of the eye and the organs of touch, and the part which they probably play in the formation of our space-perceptions (pp. 126 ff.). It must not, however, be forgotten that these purposive reflexes have become possible through an organisation acquired in the course of countless generations. What are the conditions which have been operative during this development to increasingly modify the organisation of the nervous system, so that the movements which constitute its mechanical response to external stimuli may be as well adapted as possible to subserve the immediate ends of the life of the organism ? There is only one intelligible answer to this question. It consists in a reference to those processes which even during the individual life mediate the formation of purposive `reflex and automatic movements, to the processes underlying *practice*. Practice always implies that an action which at first was performed voluntarily has gradually become reflex and automatic. Thus when the child learns to walk, the taking of each single step is accompanied by a considerable effort of will ; but after a time and by slow degrees it becomes able to initiate a whole series of movements without attending to their execution in detail. In the same way, we learn to play the pianoforte or to execute other complicated movements of the hands by frequent repetition of particular and connected acts, and their consequent transformation into a chain of effects which follow each other with mechanical certainty when once the appropriate impulse has been given. Now the modifications which the nervous system undergoes during the life of the individual in consequence of the mechanising of these practised movements must naturally, like all other modifications of the same kind, be summated and intensified in the course of generations. The purposive character of the reflexes becomes then readily intelligible, if we regard them as resulting from the voluntary action of previous generations ; while, on the contrary, the view which sees in them the starting-point of the will's development fails to explain their existence and purposiveness, and is further in disagreement with the results of objective and subjective observation,—with objective : for the observation of animals, and especially of the lower forms of animal life, never goes to prove the primitive character of reflexes which the theory assumes ; and with subjective : for it remains completely unintelligible how a decision of will can arise from purely intellectual processes. Introspection invariably points to feeling as the antecedent of will ; but feeling, as we saw above, is not separable from it, since it always implies a certain tendency to will in one way or the other.

Moreover, this theory usually occupies itself exclusively with *external* voluntary actions, and entirely overlooks the fact that there is an internal volition manifesting itself only in the form of conscious processes. We direct our attention voluntarily upon any object which appears in our field of vision ; we are clearly conscious of an effort of will in trying to recall a word or a fact that we have forgotten. We voluntarily turn our thoughts in different directions in selecting out of a number of ideas which are passing through consciousness those which have most direct reference to the general trend of our thinking at the time. It is quite impossible to derive these internal voluntary processes from external voluntary actions. It is surely obvious that the contrary is true,—that every external voluntary act presupposes an internal volition. Before we voluntarily execute any particular movement, we must have formed the decision to make it. And this decision is an internal voluntary process. Internal voluntary actions, then, are possible without external ; but external acts always require antecedent internal ones.

§ III

External voluntary actions, therefore, presuppose as their condition internal volitional processes. And in like manner reflection and choice between various possible actions (which are usually and wrongly held to be the essence of will) imply the pre-existence of *simple* voluntary actions. In these latter some definite object, whether it be something external or an internal idea, is willed, without any reflection or choice at all. Choice is nothing but a *complex* voluntary process. At first several motives to will are present simultaneously. Later some one of these, which accords with the decision we have formed, gains the predominance over the others. If this predominance is exclusive enough to allow one definitely directed volition to prevail over the others, but not strong enough to give rise to an external voluntary action, we have simply a *desire*. If the inhibitions due to conflicting voluntary impulses are gradually overcome, desire subsequently passes over into a voluntary action. This explains the fact that desire may exist in two forms,—first as the state of mind preliminary to a voluntary action, and secondly as a permanent conscious process which does not give rise to any such action. If in the latter case there is connected with the desire the idea that it cannot be realised either for the present or at all, we have what is called a *wish*. Desire, then, is mainly a matter of affection and conation, while in wish there is present besides these an intellectual process of considerable intensity. But the popular view that desire is the uniform and necessary antecedent of volition is the result of the erroneous doctrine of will which we have just been discussing, and is entirely without foundation. It follows from the conditions which we have enumerated that desire *may* be present in the mind before a voluntary act occurs, but it is not *indispensable* ; indeed, it is probably absent oftener than it is present. Even in complex voluntary processes the action may take place before the state of desire has had time to develop. And in simple processes the possibility of desire is altogether excluded, since the internal voluntary action gives rise directly to the external without finding in consciousness any resistance to be overcome. Feelings, of course, occupy an entirely different position. They are always present alike in simple and complex conative processes, the only difference being that they are more complicated in the latter case than in the former. Before volition is realised in action, the tendency of will is known, and this tendency is simply a matter of feeling. Feeling, therefore, is not a process different from volition, but simply a constituent of the complete voluntary process. It is only because we have so often had experience of feelings from which no voluntary acts arise that we are able to separate the two processes. The reverse is altogether impossible : voluntary activity always presupposes an antecedent voluntary tendency, *i.e.*, a feeling.

What is it that must be added to feeling in order that a

volition may result ? This question has really been answered in what has just been said. The tendency to will contained in the feeling passes into a voluntary *activity* of the same direction. What exactly are we to understand by this activity, which, together with feeling, constitutes the chief characteristic of volition ? The concept of activity contains two factors,—in the first place, activity implies a process or change in the given condition of an object, and, secondly, the reference of this change to some subject as its immediate cause. Thus in the physical sciences we speak of the chemical action of the electric current, of the mechanical action of wind and water, etc. The chemical decomposition of a liquid into its constituents, the movement of a mill-wheel, etc., are in these instances the observed changes ; the electric current, the moving water, and air are the subjects to which these changes are referred. So that we may ask in this matter of voluntary activity, what is the change that occurs, and what is the subject which we postulate to explain it ? In the first place, the change is always an alteration in our states of consciousness : an idea may arise which was not previously present, or an existing one may disappear ; or, again, the change may consist in an obscure idea becoming clearer, or a clearer one more obscure, etc. These ideational processes are further always connected in consciousness with various feelings and emotions. In external voluntary actions the changes which refer to move-ments of the body play the most important part. If we abstract from the active subject, it is muscle-sensations, and perceptions of movements and their results, which form the chief conscious constituents of an external voluntary action ; and all or some of them have now and again been regarded as the exclusive characteristics of volition. But it is surely evident that they do not exhaust the psychological analysis of will ; every one of the changes in ideational content to which volition may give rise can under certain circumstances occur inde-pendently of it. Ideas which are brought to consciousness by voluntary recollection may also crop up through involuntary association ; and muscle-sensations may be produced by re-flexes, or, as you know, by means of external and artificial stimulation of the muscles. What must be added to all this, therefore, is the reference to an active subject, which introspection teaches us to regard as the direct cause of the ideational changes. But what is this 'active subject'? The most obvious answer appears to be : the willing subject is our own self. But that answer does not in any way assist our psychological analysis. For what, again, is this 'self' which we are led to look upon as the author of our voluntary actions? When we examine it closely, we see that it is only another expression for the old phrase 'willing subject.' We perceive changes in our conscious content, and refer them to a single subject ; then we go on to name these changes 'voluntary actions,' and the subject brought in to explain them our 'self.' The only means of determining more exactly the nature of the 'self' is to analyse out what we regard as the cause of our voluntary action in each particular case.

Now the willing self is usually regarded as the immediate cause of voluntary actions, but by no means as their final and only condition. We suppose that the will is determined by *motives*. We assume, of course, that a motive cannot be effective without a willing self ; but, on the other hand, we regard it as equally obvious from the facts of our immediate internal experience that a willing self cannot act without motives. The connection between motive and will is, therefore, just as necessary as that between will and active subject. A reflex, or a passive movement which some external force compels us to make, is not conditioned by motives, although they have causes of their own just as certainly as voluntary actions. Motives are therefore *causes of volition* ; and since volition always arises from internal processes, it is at once clear that they must be *internal*, psychical causes.

Now what is a motive ? It is customary to make a distinction between simple and complex motives, and to comprehend under the latter rubric complex groupings of motives, where the constituents may to some extent operate in different directions. But in giving an account of the particular causes which determine volition, we shall only recognise as *determinate* motives those which give it a definite direction, and which act like simple forces, incapable of further analysis. In this sense every motive is a *particular idea with an affective tone attaching to it.* And since feeling is itself simply a definite voluntary tendency, this combination of idea and feeling in motives only means that an idea becomes a motive as soon as it solicits the will. Hence it is tautological to say that only ideas with a strong affective tone can operate as motives, since it is just the affective tone of an idea which gives it the power of acting as a motive.

Nevertheless, introspection can show the *conditions* in virtue of which some ideas become motives and others do not. These conditions are of two kinds,—they consist partly in the immediate attributes of sense-impressions, partly in the nature of our previous conscious experiences. All those attributes of sensation which endow it with a vivid affective tone serve also to make the impression effective as a motive to will. In this case it generally happens that the impression, with its strong affective tone, is the only motive present in consciousness : the voluntary action is a *simple*, or, as it is ordinarily expressed, an *impulsive*, action. There can be no doubt that the majority of the actions of animals are of this character. But impulses make up a large part of human action also, and especially in the earlier stages of its development. All sense-impulses are simply tendencies to will connected with definite sensations ; *i.e.*, they are feelings which have a strong tendency to pass over into actual volition.

But in course of time the mind acquires various dispositions toward the renewal of previous ideas which are themselves connected with definite voluntary tendencies. An external stimulus will not any longer simply call out the impulse corresponding to it ; but this impulse will increasingly tend to influence and be influenced by the dispositions already existing in the mind. These, again, may be transformed into conscious motives to will either by the external impression or by secondary influences. So that the chief motive of actual volition is henceforth not some particular sense-impression which happens to be there, but the entire trend of consciousness as determined by its previous experiences. This trend or disposition does not, of course, come directly to consciousness as such. We can only give an account of those dispositions which enter into the conflict of motives in virtue of their perception as ideas to which a strong affective tone attaches. And even of these many remain so indistinct, that though they may be factors in the resultant total feeling, and therefore in the act of will which finally results, there is still no clear perception on our own part of their independent existence. . On the other hand, we can know nothing whatever of the influence which may be exerted by the dispositions that never become realised in idea at all upon the changes in our ideational content, and so upon the final act of will. The links which join the actual current processes with the past history of consciousness simply serve to bring out with unmistakable clearness the general fact that the determining ground of action has not been any single impression, nor any particular motive, whether called up by association or arising ' of itself,' but the entire trend or tendency of the mind, which has its roots in the original nature of consciousness and the accumulated experience of the mental life. A more or less intensive feeling is connected with this general idea, and becomes an essential element in the common feeling of the moment. The action which results from this plurality of conflicting motives we call a *complex* voluntary action or a *volitional* action. It possesses two distinguishing marks in consciousness,—first, the feeling of a *decision*, preceding the action and based upon the connection of the present impression with past experiences ; and secondly, the idea of the voluntary act as determined by a *choice* between different and conflicting motives. Either one of these characteristics may be more or less distinct. The clearness of the perception of either usually stands in inverse ratio to that of the other. The feeling of decision is predominant where the voluntary act occurs at once and with complete certainty ; the feeling of choice prevails where there is a long preliminary conflict of motives.

These facts make it obvious that simple voluntary actions are the necessary presuppositions of the more complex. Even in the former the impression does not cause the action of itself ; its effect depends upon the state of consciousness at the time. But as this effect is relatively simple, the directly given stimulus is the principal motive in the decision, and other motives have no appreciable significance as compared with it.

§ IV

If we bring together once more all the essential elements of a voluntary action, we see that it consists, in the first place, of a *feeling*, in which the tendency of the will is manifested; secondly, in a *change in ideational content*, which may be accompanied by an external effect mediated by the organs of movement; and thirdly, in the general idea of the *dependence of this change upon the whole trend of consciousness.* This last, like all secondary ideas, finds its principal expression in a *feeling*, which partly precedes the decision of will (in the form of the above-mentioned feeling indicating the volitional tendency), and partly accompanies it. To these three constituents must be further added the feelings which arise subsequently as a result of the internal and external effects of the action, but which exert no influence upon its performance.

One very important attribute of volition, which affects all the elements of voluntary action which we have here cited, is its *unity*. Despite the conflict of motives and the oscillations of feeling conditioned by it, the voluntary act itself at any given moment must be single and unitary. This fact is the basis of the unity of the self. By a *hysteron proteron* which often recurs in psychology we tend to regard the latter as the cause of the unity of volition. But, as a matter of fact, what we call our 'self' is simply this unity of volition *plus* the univocal control of our mental life which it renders possible. Furthermore, this unity of volition enables us to explain directly another fact to which we have already referred,—the fact that the feelings of each moment unite in a single total feeling, whatever oppositions may exist between them. This total feeling is the resultant volitional tendency. And it is just as impossible for it to be resolved into a number of independently coexisting feelings as it would be for us to will several different things simultaneously.

In virtue of the attributes reviewed in these lectures, feeling and will react upon the ideational side of our mental life; and thus help to determine the entire content of what we call, by an arbitrary distinction, but one which is of service in the analysis of the facts, *consciousness*. Now that we have described the various constituents which go to make up the mental life, we will turn for a time to the phenomena which result from the combination of all of them.

§ I

WHAT is 'consciousness'? Much attention has been devoted to this question in modern times both by philosophers and psychologists. There could be no doubt that the word denoted some phase or aspect of our mental life, and was not identical with any of the other concepts, like 'idea,' 'feeling,' 'will,' etc., which we apply to particular mental processes and states. So that the view naturally suggested itself that consciousness is a special mental condition, requiring to be defined by certain characteristic marks. And the feeling that it was necessary to oppose to consciousness an *unconscious* mental existence promoted this opinion. Ideas, affective processes, may vanish and then again appear. It is therefore inferred that after leaving consciousness they have continued to exist in an unconscious state, and at times return to their former condition.

From this point of view, nothing is more natural than to think of consciousness as a kind of stage upon which our ideas are the actors, appearing, withdrawing behind the scenes, and coming on again when their cue is given. And the notion has become so popular that many philosophers and psychologists consider it much more interesting to learn what takes place behind the scenes, in unconsciousness, than what occurs in consciousness. Every-day experience, it is supposed, has made the latter familiar to us ; but we know nothing of the unconscious, and to learn something about it would be a really interesting addition to our knowledge.

Nevertheless this comparison of consciousness to a stage is entirely misleading. The stage remains when the actors have left it ; it has an existence of its own, which is not dependent upon them. But consciousness does not continue to exist when the processes of which we are conscious have passed away ; it changes constantly with their changes, and is not anything which can be distinguished from them. When the actor has left the stage, we know that he is somewhere else. But when an idea has disappeared from consciousness we know nothing at all about it. Strictly speaking, it is not correct to say that it subsequently returns. For the *same* idea never returns. A subsequent idea may be more or less similar to an earlier one ; but it is probably never exactly the same. Sometimes it has constituents which the earlier idea had not ; sometimes certain of those which belonged to the latter are lacking in it. There is scarcely any view which has been a greater source of error in psychology than that which regards ideas as imperishable objects which may appear and disappear, press and jostle each other, objects to which, it is true, additions are at times made through the action of the senses, but which, when once they have come into being, are only distinguished by the variation in their distribution in consciousness and unconsciousness, or at most, by the different degrees of clearness which they possess in consciousness. As a matter of fact, ideas, like all other mental experiences, are not *objects*, but *processes, occurrences.* The idea which we refer back to a previous one, when we apprehend it as similar to that, is no more the earlier idea itself than the word which we write or the picture which we draw is identical with the same word which we wrote previously or the similar drawing which we made some time ago. Indeed, you will see, if you consider the complex conditions under which our inner experience arises, that nothing like the same degree of similarity between the earlier and the later product can be expected here as may be found under certain circumstances in the field of external actions like writing and drawing. The circumstance that new processes exhibit relations and similarity to others previously existing, can no more prove the continued existence of the idea as such, than it can be inferred from the similarity of the movement of the pen in writing a definite word now to that involved on a former occasion, that this movement has continued to exist in an invisible form from the time it was first made, and has simply become visible again when we have written the word anew. If ideas are not imperishable facts, but transitory processes which recur in more or less altered form, the whole of this hypothetical structure falls to the ground. And at the same time the unconscious loses the significance ascribed to it as an especial kind of mental existence, which, though not itself consciousness, might at any rate enable us to determine the characteristics or conditions which must attach to the objects of mind in order that they may become conscious.

In the same way, all attempts to define consciousness as a particular mental fact co-ordinate with our other internal experiences have proved fruitless. It is obvious that those who would regard it as the capacity of internal observation, as a kind of 'internal sense,' commit in this analogy an error similar to that involved in its comparison to a stage. The perceiving organ and the perceived object are two different things; consciousness and conscious process are not. The activity of observation, of attention, is of course found among what we call conscious processes. But it is just one conscious fact, co-ordinate with the rest, a fact which presupposes the existence of consciousness, not one which renders consciousness possible. The same criticism applies to yet another explanation which is sometimes given. We distinguish in consciousness, it is said, a whole number of ideas. Therefore consciousness must possess the capacity of *discrimination* ; the word must be equivalent to discriminating activity. But here again the question arises whether the discrimination of processes directly perceived is the antecedent condition of these processes, or whether it is not rather a result to which they are essential. In the first place, the objects must be there to be distinguished. The child runs together a number of separate objects into a single idea, where the developed consciousness keeps them separate. Discrimination, then, like observation, consists in processes which presuppose consciousness, and which consequently cannot constitute its essence. And consciousness itself is not a particular mental process, co-ordinate with others; it consists entirely in the fact that we have internal experiences, that we perceive in ourselves ideas, feelings, and voluntary impulses. We are conscious of all these processes in having them ; we are not conscious of them when we do not have them. Such expressions as 'the limen of consciousness,' 'appearance in and disappearance from consciousness,' etc., are pictorial ways of speaking, useful for the brief characterisation of certain facts of internal experience, but never to be regarded as a description of these facts. What really takes place in the raising of an idea above the limen of consciousness is, that something occurs which had not occurred previously. And what really happens when an idea disappears from it is that some process ceases which has hitherto been in progress. In like manner we must think of the range of consciousness as denoting simply the sum of mental processes existing at a given moment.

Although, therefore, consciousness is not an especial kind of reality co-ordinate with the particular facts of consciousness, modern psychology still finds the concept indispensable. We must have a collective expression for the whole number of mental experiences, given either simultaneously or successively. As simply denoting the existence of internal experiences, while leaving their nature altogether undetermined, the concept is especially serviceable for the treatment of the interconnection of the mental facts, of all those processes with which we have already become acquainted in isolation. It has no meaning apart from its reference to this interconnection of simultaneous and successive mental processes ; and the problem of consciousness consists in determining how the particular phenomena are interrelated, and how their relations and connections again combine to form the totality of mental life. For the sake of simplicity in treatment, it will be convenient to confine ourselves at first to the ideational side of consciousness, and then, when we have discussed the problem just formulated from this point of view, to supplement our results by reference to the affective and conative elements. This is of course the plan which we have followed in our analysis of particular mental processes. It will, however, soon become evident that in dealing with the interconnections of mental processes we cannot carry our abstraction through to the end, since the affective side of our mental life constantly exercises a determining influence upon the combinations and relations of ideas. In certain instances, therefore, we shall not be able to avoid at least a passing reference to the affective and conative factors.

§ II

The first question which may be raised within the limits of the conditions laid down runs of course as follows : *how many ideas may be present in consciousness at a given moment?* The content of this question is not quite so precise as its wording seems to imply. The estimation of the number of constituents which a whole contains is, naturally, dependent upon what we regard as the constituent unit. Now, even if we neglect the continual change among ideas, their combinations in consciousness are enormously complex. So that it may easily be a matter of doubt whether some given portion of conscious content is to be considered as an independent idea, or as a part of a more comprehensive idea. We may here dispense for a time with any final theoretical solution of this difficult preliminary problem. For our present purpose it will suffice if we can furnish a practical criterion. We shall accordingly regard an idea as separate and independent when it is not connected by customary association with other ideas simultaneously present. If, *e.g.*, there is placed before the eye a number of letters in serial order (say, *x v r t*), we shall consider that each one forms an independent idea by itself, in spite of its spatial association with the rest. For, regarded as a whole, letters form no new complex idea, capable of entering into definite connections with other ideas. If, on the other hand, we perceive four such letters as *w o r k*, we shall not hold them to be independent ideas,—at least, for one who reads them as a word,—but shall look on them as combined into a single complex idea. From these considerations there follow two results, which should never be lost sight of in experiments made to determine the ideational range of consciousness. First, we must always decide from the objective and subjective conditions operative in each particular instance what ideas may pass as independent units and what not. It is, of course, obvious that the same objective impression may in one case be apprehended as one idea, and in another as more than one, according to the subjective conditions involved. Secondly, the conclusions derived from ideas of one kind will not necessarily hold for ideas of any other kind. In particular we may expect to find that the range of consciousness will be smaller for complex ideas than for relatively simple ones.

When the question of the range of consciousness was first raised, these conditions were entirely overlooked, and the general method of investigation pursued was not one which could lead to any certain results. Conclusions were either deduced from certain metaphysical postulates,—*e.g.*, that the mind, as a simple being, could only contain a *single* idea at a given moment,—or the investigations were based solely on introspection. Any one may convince himself of the fruitlessness of this latter procedure by seriously asking himself the question : how many ideas do I now find in consciousness ? And at the same time the experiment shows him the reason why his efforts at an answer are without result. The question is scarcely raised before the moment to which it refers has passed, nor can the following moment be fixed any more successfully. It thus becomes quite impossible to distinguish what is simultaneously given at a particular moment from that which comes later. This defect of direct introspection, however, itself shows us how we should endeavour to supplement it by experiment. It is only necessary so to arrange the conditions of experimentation that the confusion of simultaneous impressions with successive is less easily possible. This we can do by momentarily presenting a number of sense-impressions, which are capable of becoming independent ideas, at a given signal, and then trying to determine how many of them have been actually ideated. Now it would be wrong to suppose that the running together of the momentary impression with subsequent ideas is here altogether precluded. Suppose, *e.g.*, that by a momentary illumination we present to the eye a number of visual objects. The perceptions of the first moment will naturally be supplemented by others of which we do not become conscious until later. You may easily convince yourselves of this by holding a book in the dark at a convenient distance from the eye and illuminating the room for an instant by an electric spark. Even if in the first moment you only cognised a single word, it may very well happen that you will subsequently, by the aid of memory, be able clearly to apprehend others. Indeed, what is read subsequently in this way is often more than what was recognised in the first instance. But these experiments, again, lead us to a further fact, which shows that it is possible to draw valid inferences from them with regard to the condition of consciousness at a given moment. We can very clearly distinguish the image that has been gradually reconstructed on the basis of the original impression from the image which corresponds directly to it. This is due to the fact that the particular moment is not precisely like those which directly preceded or follow it : the sudden appearance and disappearance of the flash of light marks it off from them ; this distinction makes it less difficult for introspection to neglect or voluntarily to exclude the subsequent filling out of the original ideational image. Here, as everywhere, psychological experiment does not enable us to dispense with introspection, but, on the contrary, renders introspection possible by furnishing the conditions which it requires for exact observation.

Experiments of this kind with momentary impressions may be made in any sense-department. But visual impressions are best suited to the purpose, because they can most easily be selected with a view to their apprehension as independent ideas. The impression itself is, it is true, not entirely momentary ; light-stimulation has a physiological after-effect. However, in impressions which pass very quickly this after-effect is so short that we may neglect it for our present purpose. Visual experiments are made with the apparatus represented in Fig. 37. It is intended for the demonstration of the phenomena to a large audience. If you should merely wish to perform the experiment upon yourselves, the dimensions could, of course, be very much smaller. The apparatus consists of a black screen, held in grooves in front of a black vertical board, some two metres high, and falling when the spring F is pressed. In the screen there is a square opening large enough to enclose a large number of objects that can be separately ideated, such as letters of the alphabet. This opening is so placed that when the screen is raised it shows simply the dark background, but during its fall passes very quickly before the objects to be observed, and instantly covers them again. On the part of the screen below

R

the square opening a small white circle is so placed that before the fall of the screen it exactly covers the centre of the visual surface afterwards briefly exposed to view. This circle serves as a fixation-point to put the eye in the most favourable position for the apprehension of the impressions. *A* in the figure gives a lateral, and *B* a front, view of the apparatus. In *A* the screen is raised, and covers the objects which are to be seen ; in *B* it is represented in the moment of falling before them, so that a number of impressions (letters taken at random) have just become visible. If we imagine the motion continued, we shall have these hidden again the next moment by the upper, closed part of the screen. The size and distribution of the visual objects and the distance of the observer from the apparatus must be so chosen that all the separate letters which are to be seen shall fall within the region of clearest vision. It is of course true that in these experiments every visual object is, strictly speaking, visible not only for a single moment, but for a measurable, although relatively small portion of time ; and, further, that this time is not exactly the same for the different objects. In the apparatus represented in

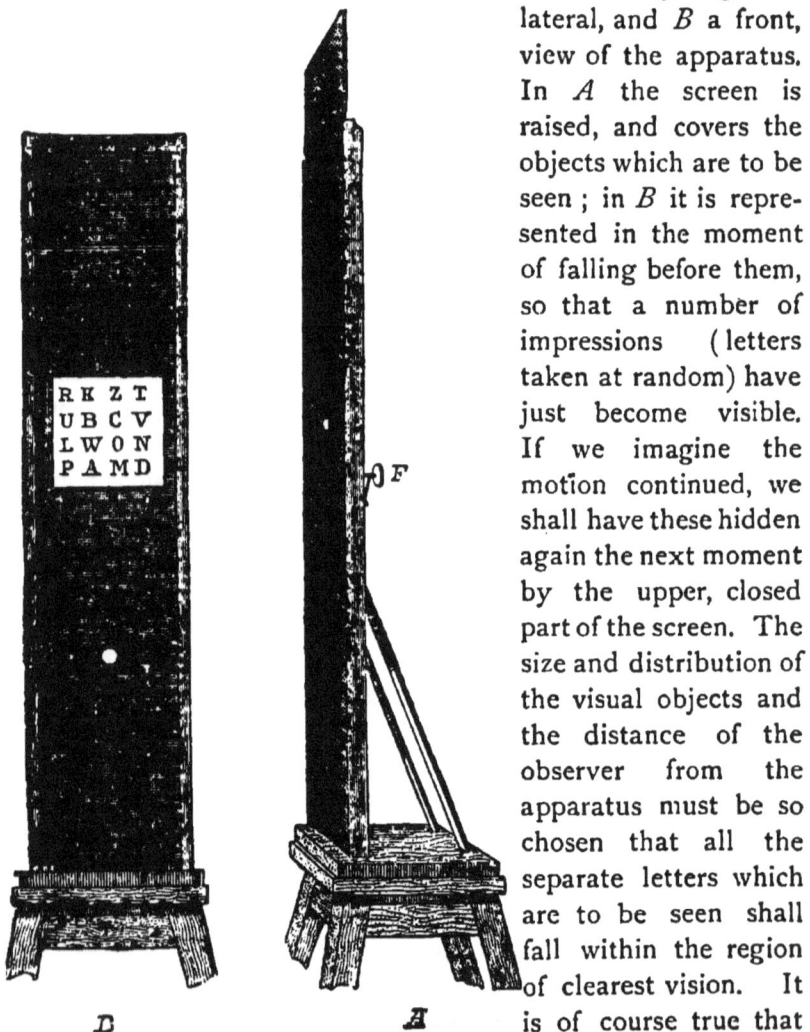

FIG. 37.

Fig. 37, the upper line is visible for 0·09", the lower for 0·07", and the middle for 0·08". These times are, however, so small in comparison with the much greater duration of the after-image that for our present purpose they may be regarded as really momentary.

Experiments of this kind show that four, and sometimes even five, disconnected impressions (letters, numerals, or lines of different direction) may be distinctly perceived. If the separate impressions are so arranged that they enter into combination with one another in idea, the number becomes three times as great. Thus we are able to cognise instantly two dissyllabic words of six letters each.

But the result of such investigations is to call our attention to other phenomena, which render it obvious that we cannot really learn anything in this way regarding the total range of consciousness. We notice that the letters, numerals, words, etc., which we clearly apprehend at the moment when the screen falls, by no means exhaust the conscious content of the moment. Besides these impressions which are clearly apprehended, there are present in consciousness others which are less distinct, or wholly indistinct. In addition to the four or five letters which you were able to read, you would have noticed, *e.g.*, some of which you cognised only the approximate outline, and others about which you had only the quite indefinite idea that they were present and were visual impressions. The experiments show, therefore, that this method can only enable us to determine the number of *clear* and *distinct* ideas present in consciousness, and can give us no information of the entire number which it contains. The number of clear ideas for the sense of sight amounts to 4 or 5 when they are comparatively simple and familiar ; if they are complex, the number varies from 1 to 3, according to the degree of complexity. In the latter case the number of simple ideas present in a clear ideational complex may be as many as 12. You will notice, further, that the impression falling directly upon the yellow spot is usually more distinctly apprehended than any of the others. But this is by no means necessarily the case ; laterally seen objects may take precedence over the objects seen directly, especially if the attention is voluntarily directed upon them.

§ III

So that, even although our first method has told us nothing of the actual range of consciousness, it is worth while to spend a little time in examining the results obtained. Apart from the information they afford us as to the number of ideas clearly apprehensible at a single moment, their indication of different degrees of ideational clearness is particularly worthy of remark. It is true that the distinction of clear and obscure ideas did not escape the keen observation of Leibniz, and since his time has scarcely been disputed. But ordinary introspection does not admit of such definite and direct determination of the relations of the different degrees of clearness as that afforded by the method of instantaneous impressions. The experimental method demonstrates, *e.g.*, the correctness of Leibniz' hypothesis that there is no abrupt transition, but always a continuous gradation from each degree of ideational clearness to the next. In the experiment with momentary visual impressions given above, we distinguished *three* kinds of ideas in consciousness : the clear, the more obscure,—where a partial discrimination is still possible, —and the quite obscure, in which we only cognise the presence of *some* conscious content belonging to a definite sense-department. We must now understand that these ideas only differ in degree of clearness, and that all three degrees are connected by continuous gradations. For the two extremes, however, we may employ the terms which Leibniz introduced. We may term the appearance of an obscure idea in consciousness a *perception*, and the appearance of a clear one an *apperception*. These two names must not be understood to carry with them any presuppositions, either metaphysical or psychological. They merely express a fact, for which (as is usual in science) we choose the name proposed by the investigator who first called attention to it. We leave out of account any assumption, any theory, derived from the observed facts which Leibniz and his successors may have connected with these terms. Notice only that the relation of clear to obscure ideas furnishes an obvious analogy to that of objects distinctly or indistinctly seen in the field of vision ; and that it is therefore natural to refer the distinction of perception and apperception to consciousness itself, just as in external vision we account for the different degrees of distinctness by reference to differences in acuteness of vision in different portions of the visual field. We may say, then, that the *perceived* ideas are those which lie in the field of consciousness, while the *apperceived* are situated at its fixation-point.

Now what are we to understand by the *clearness* of an idea? The word, like all the names of psychological concepts, has been transferred from external objects to the conscious subject. We use the term 'clear' to denote luminous or transparent objects; *i.e.*, those which are themselves easily perceived, or which aid the sense of sight in perceiving others. When the word is applied to consciousness, therefore, it must express some similar characteristic in introspection. An idea is clear when it is apprehended in introspection more perfectly than others which, in contradistinction to it, are called obscure. The only difference between the original and the transferred meaning of the words is this: in the former case the property of clearness may belong to the object without reference to our perception of it, but in the latter the idea is only clear in so far as it is clearly perceived in consciousness; a difference which again has its source in the fact that our perceptions of mental processes and the mental processes themselves are completely identical. Ideas are only ideas in virtue of our perception of them. Internal perception (introspection) is simply the fact of internal experience itself; and we are only looking at this experience from different points of view when we speak of it at one time as idea, feeling, etc., and at another sum it up as internal perception.

We are apt to identify the clearness of ideas with their *distinctness*, and to define one of these concepts by the other, 'distinct' being 'what can be clearly cognised,' or 'clear' 'what we perceive distinctly.' Now it must be admitted that the two properties are generally found together. But they are not at all identical; each of them denotes a different aspect of, or a different reason for, the advantage which a given idea possesses in consciousness. An idea is 'clear' solely in virtue of its own properties, just as in the use of the word in external reference pure water is termed 'clear,' and not 'distinct,' because it is transparent, so that any object that happens to be in it may be seen through it. On the other hand, an idea is called 'distinct' with reference to the definiteness with which it is marked off from other ideas. Thus an object lying in clear water is distinctly seen because it stands out sharply from its surroundings. Similarly a tone is clear when we can fully apprehend its peculiar quality; it is distinct when definitely distinguishable from the other elements of a compound clang, or from other simultaneous sound-impressions.

As applied to our ideas, then, clearness and distinctness denote properties which depend directly upon the activity of ideation, or, what is the same thing, of introspection. One and the same idea under the same objective conditions may be at one time clear, and at another more or less obscure. And for this reason we must be especially careful not to confuse the clearness of an idea with its *intensity*. That is simply dependent upon the intensity of the sensations which constitute it. The intensity of perceptual ideas is determined by the strength of the sense-stimuli, that of memorial ideas by other conditions, which have nothing to do with ideational clearness. At the same time intensity usually promotes clearness and distinctness. Other things equal, the more intensive idea is usually the clearer, and very weak ideas in particular are but seldom clear and distinct. Nevertheless it may happen, if the subjective conditions of perception so determine, that an intensive idea is obscure and indistinct, and a weak one clear and distinct. A very weak over-tone in a clang, *e.g.*, may be heard clearly and distinctly, while the more intensive ground-tone is less clearly perceived, and a loud noise simultaneously given perhaps hardly noticed at all.

It follows from all this that the clearness of ideas necessarily depends upon the condition of consciousness for the time being. Inasmuch as they help to determine this condition, the intensity of impressions and memorial images also exerts an influence upon the clearness and distinctness of ideas. But since the state of consciousness is certainly not entirely dependent upon those conditions, they are by no means the only factors of importance in the matter. Our final definition of clearness must, therefore, be that it is *that property in virtue of which an idea has an advantage over other ideas in introspection.* But it is not difficult to see that this definition is only a description of the word 'clear.' As a matter of fact, it is just as impossible to define the clearness of an idea as to define the intensity or quality of a sensation. We can distinguish these fundamental properties of our mental processes from each other by showing that under definite conditions they vary separately and independently. But the differences could no more be brought home to one who had not experienced them than can the distinctions of colour to the congenitally blind.

§ IV

On the other hand, the becoming clear of an idea is regularly associated with other mental phenomena, which not only assist the introspective discrimination of clear and obscure ideas, but also throw some light upon the subjective conditions of the processes which we have distinguished as perception and apperception. These phenomena are of two kinds,—they consist partly of *sensations*, partly of *feelings*. The sensations which accompany apperception belong to the category of muscle-sensations. They are especially noticeable in cases of external sense-perception. If we are directing our attention upon a particular tone, or a particular visual object, to the exclusion of other impressions of light and sound, we have in ear or eye definitely graduated muscle-sensations, which are probably referable to the *tensor tympani* and to the muscles subserving the accommodation and movement of the eye. The same sensations may be perceived, though less clearly, to accompany memorial ideation, at least when the ideas are vivid. An object seen with the mind's eye is referred to a certain distance from us, and we consequently accommodate the muscular apparatus of our eye to it. The tones of a melody which we recall in memory may give rise to a tension in the ear as clearly perceptible as though they were real. Even the fainter pictorial ideas which constitute abstract thinking are not wholly without this sense-accompaniment. When we are trying to remember a name or are pondering a difficult problem we notice the presence of strain-sensations. These are partly sensed in the eye,—visual ideas being, as you know, predominant in consciousness,—partly in the forehead and temples, where the muscles lying directly beneath the skin, which play a part in mimetic movements, are strained to a degree more or less proportioned to the amount of internal effort.

The connection of these muscular tensions with mimetic expressive movements leads us directly to the second accompaniment of the apperceptive process,—to *feeling*. There could be no emotional expression without feeling. Feelings precede apperception proper, and continue to exist during the course of the process. They are different in the two cases, though their passage in consciousness is continuous, so that those which precede apperception and those which accompany it form an affective totality, which by the fact of its continuity resembles emotion (of which we have to speak later), and which indeed not infrequently becomes actually transformed into emotion. Our perception of these attendant feelings, like our perception of the sensations discussed above, is most distinct when the clearness of their ideational substrates is very great, and especially when this clearness is mainly the result of the trend of consciousness itself, and not of external conditions ; *e.g.*, to put it again concretely, when we are voluntarily recalling an idea which we have previously had, or when we are expecting an impression, etc. Even when the condition is not one of expectation proper, the feeling preceding apperception is very closely related to that of expectation. The feeling accompanying the process, on the other hand, may be compared to that of satisfaction, to the relaxation of a tension, or, again, if the expectation is not realised, to that of disappointment or failure. It is quite true that these feelings are only clearly perceptible under the special conditions which characterise expectancy, recollection, etc. But careful introspection seems to show that feelings of the same kind are never entirely absent where ideas which were formerly obscure become clearer, even although their intensity be much less and their quality exceedingly variable. At least, if there is any essential difference, it is only in the case of the antecedent feelings. Their period of duration, *e.g.*, may be very much shortened (though they hardly ever entirely disappear) when the object of apperception is an external sense-impression, or when memorial images crop up unexpectedly.

§ V

The whole circle of subjective processes connected with apperception we call *attention*. Attention contains three essential constituents : an increased clearness of ideas ; muscle-sensations, which generally belong to the same modality as the ideas ; and feelings, which accompany and precede the ideational change. At the same time the concept of attention proper has no reference to the first of these processes, but only to the last two. Apperception, therefore, denotes the objective change set up in ideational content, attention the subjective sensations and feelings which accompany this change or prepare the way for it. Both processes belong together, as parts of a single psychical event. It may happen in certain instances that the objective effect is distinct, while the subjective aspect of the process does not attain a liminal intensity. Or it may happen, as when an expectation is unfulfilled, that the subjective constituent attains a great intensity, while the objective is entirely overshadowed. But these are only extreme cases of a series which, like all mental series, contains terms arranged without break or interruption. Attention in the wider sense is not,—and this is the important point,—a special activity, existing alongside of its three constituent factors, something not to be sensed or felt, but itself productive of sensations and feelings. No ! in terms of our own psychological analysis at least, it is simply the name of the complex process which includes those three constituents. Their nature makes it plain enough why we regard attention as subjective activity, without our needing to assume any special consciousness of activity independent of the other mental elements. The concept of activity always presupposes two things,—first, a change in the condition of something ; and secondly, a subject whose states vary with that change in such a way that the two can be exactly correlated. We then regard the subject as the active subject, and the changes set up as the effects of its activity. Now the sensations and feelings constituting attention are not accidentally and equivocally associated with the apperceived idea, but stand in very definite relation to it. The attendant strain-sensations and the preceding or accompanying feelings are entirely governed by the nature of the apperceived idea ; if it changes, they change also. So that the phenomena which go to make up the apperceptive process possess all the characteristics required by the notion of an activity proceeding from an acting subject. This acting subject is given us simply and solely in the sensations and feelings accompanying the act of apperception. And since we find among these elements not only a constant alteration, but also a continuous connection of earlier processes with later, we come to regard the acting subject as persisting through all its changes. Language has given an expression to this view, which has been of determining influence upon the further conceptual development of the distinctions in question, in constantly rendering the notion of the permanent subject by the sign for the first person in simple verb judgments.

§ VI

It is in this way that the concept of the self ('I') arises : a concept which, taken of itself, is completely contentless, but which, as a matter of fact, never comes into the field of introspection without the special determinations which give a content to it. Psychologically regarded, therefore, the self is not an idea among other ideas; it is not even a secondary characteristic, common to all or to the great majority of ideas ; it is simply and solely the perception of the interconnection of internal experience which accompanies that experience itself. Now we have already seen that perceptions of this kind,—perceptions which refer to the occurrence of a process, the manner in which it runs its course, and so on,—are sometimes 'transposed back again into ideas. There is a deep-rooted tendency to hypostatise mental events, a tendency evinced by those theories which have regarded ideas themselves as permanent objects (pp. 221, 222). And there is a very special tendency to transpose the 'self' into an idea of this character, though, as a matter of fact, it is nothing more than the way in which ideas and the other mental processes are connected together. Since, further, the manner of this connection at any particular moment is conditioned by preceding mental events, we tend to include under the term 'self' the whole circle of effects which have their causes in former experiences. The 'self' is regarded as a total force, which determines particular events as they happen, unless, of course, they are occasioned by the action of external impressions or of those internal processes which we experience just as passively as we do the external. And since the principal effect of the preconditions of consciousness is the determination of the appearance and degree of clearness of ideas, we further bring the 'self' into the very closest connection with the process of apperception. The self is the subject which we supply for the apperceptive activity. It is plain enough that there is involved here a transference of the relations observed in external perception to the sphere of internal experience. The self, you see, is regarded after the analogy of external objects, which we take to be the same in spite of variation in their properties, because the variation is always continuous both in time and space. But without the continuity of our mental life we should not be able to cognise the continuity of objective things ; so that in this interplay of developments we have the self figuring both as cause and effect. The perception of the interconnection of mental processes, which crystallises in the concept of the 'self,' renders possible the distinction between objects and their changing properties ; and this distinction in its turn inclines us to ascribe an objective value to the concept.

There is another reason for this in the fact that the body, with which all the states of the self are connected, is itself an external object. In the first place, the self is a product of two things,— external perception and internal experience ; it is the body, together with the mental processes connected with it. Later on reflection destroys this unity ; but even then there remains some faint trace of that object-idea which attached to the self of sense. And where the current view of life is the practical one, with its naïve sense-reference, the body takes its place unquestioned as an inexpugnable constituent of the self.

LECTURE XVII

§ I

SIMULTANEOUSLY with the development of self-consciousness, which we described in the previous lecture,
proceeds the development of another complex process,—that of
attention. The two developments are in many respects similar.
States of attention, like those of self-consciousness, present
certain external differences which may be regarded as opposites;
though it is true that, to place the opposition in a clear light,
we must more or less neglect intermediate processes which
would enable us to pass from one to the other. For the
extreme cases, however theoretically possible, never actually
occur in the purity in which they can be obtained by analysis.
However, if we disregard the concrete for a moment, we shall
find evidence enough for the general possibility of the extreme
cases.

We saw that in every act of apperception there are two
principal conditions of the resulting effect,—first, the momentary
condition of consciousness, itself determined partly by external
influences, partly by those of its own earlier states which are
directly related to these influences, and therefore with greater
or less regularity associated with them; and secondly, the
entire previous history of consciousness, which may operate in
the most various ways to alter the effect due to this momentary
state. You must not, of course, suppose that these two conditions are at work in the individual case in the sense of two
opposing forces. That would be impossible. For the earlier

states which are directly connected with any particular
objective impressions,—ideas, feelings, or whatever these states
may be,—themselves form part of the previous mental history.
In other words, we have to do with a difference in degree, and
not with a difference in kind. But this does not prevent the
results in the two cases from appearing as opposites. Suppose
that the direction of the attention is determined merely by
some chance stimulus, and by a 'state of mind' which is permanently associated with that stimulus, or has been brought
about by accidental circumstances. Then the immediate
impression which we have of our internal experience is that of
a passive receptivity of what is going on in our minds. Suppose, on the other hand, that the direction of the attention is
determined by more remote conscious tendencies which have
arisen from previous experience, and which are not directly
related to the particular impressions of the moment. Then we
have always the impression of a productive activity. Apper-
ceptions of this kind we regard as actions of our 'self'; 'self' is
just an expression for the total effect exerted by our previous
mental experiences as a whole, without particular reference to
any special components of those experiences, upon the mental
processes which are running their course at any given moment.
To make this difference clear, we will call the first form of
attention 'passive,' and the second 'active,' attention. Let me,
however, warn you again, even at the risk of repetition, that in
calling attention 'passive' we by no means deny to it any
character of activity, that is, decline to see in it the operation of
previous experiences. The contrary is true : such experiences
are always operative ; it is only that the extent and direction
of their influence are limited and circumscribed. Neither is it,
of course, to be thought that external influences and the states
of mind that follow from them are wholly without effect in the
case of active attention. It is only true that they retire into
the background before the influences of dispositions established
during long periods of time and strengthened by mutual connection ; they are none the less continually at work to modify
these dispositions. To repeat once more, we are dealing with
extreme cases, which can never occur in absolute purity, because
the processes on which they depend are the final terms of a

continuous process-series. In both cases consciousness is functioning in the same way ; the difference is only a difference of more and less, of a greater and a narrower range.

If we could only appeal to the ideational side of mind, then, we should not seldom find it difficult to decide in a particular case whether a particular apperception were active or passive. So that here again we find feeling playing a large part in our immediate apprehension of our own actions. You remember the general characteristic of feeling,—that its peculiar quality gives expression to the total attitude of consciousness. In the present case, the presence of active apperception is invariably and unmistakably indicated by a feeling of activity. We can no more describe this than we can any other feeling ; we can only attempt to determine it by enumerating those of its conditions which belong to the ideational side of consciousness (*cf.* Lecture XIV.). The degrees of intensity of this feeling give us a direct measure of our own activity ; that is, of the preponderance of our whole mental nature over momentary and transitory excitations. There can be no doubt that we must regard it as a total feeling in our previous sense of the word (p. 219). It determines the attitude of consciousness at any moment. But its own peculiar and variable quality is itself determined by the special feelings dependent upon concurrent ideas and their mutual connections. Even passive apperception, therefore, has its attendant feelings ; only that these are associated to form a total feeling with a character of its own, either exclusively conditioned by the quality and intensity of the ideas that happen to be present in consciousness, or (and this is especially the case in apperception of very intensive external sense-impressions) consisting in a feeling of inhibition, which appears to arise from the sudden arrest of existing tendencies in the formation of ideas. In its latter form it may be intensified by sense-feelings of unpleasantness or pain, without, however, being dependent upon these.

. Considered from the point of view of these attendant feelings, the process of apperception and attention appears in a connection which points at once to those elementary mental processes which we have already discussed. Feelings we found to be invariable forerunners and concomitants of volition.

They indicate the direction which an act of will will follow before it has itself become conscious ; and when it has attained its full force, they are still present to colour and explicate its effect. A second characteristic of volition beside feeling is an alteration in the state of consciousness on its ideational side, referable not to external influences, but to past mental dispositions. Both characters attach to the process of apperception, and, since the conditions of each form of this process pass over into one another without interruption, attach to active and passive apperception alike. For the raising of an idea to a higher level of distinctness can only come about in passive apperception when there are present certain positive mental dispositions to favour its preference. So that ideas and the feelings that are connected with them serve as motives to the act of apperception, while apperception itself shows all the characteristics of an act of will. More than this, its two fundamental forms, the active and passive, obviously correspond to the two fundamental forms of conative activity,—the passive form, the impulsive act, and the active, or act of choice. We act impulsively when we apprehend an impression under the constraining influence of external stimuli, and of the ideas immediately and directly aroused by them ; we choose when out of a whole number of concurrent ideas we raise to a higher level of distinctness some particular one which long-established mental tendencies dispose us to regard as the fittest at the time. And the coincidence of the internal with external voluntary acts is proved by this,—that not seldom our decision is prefaced by a clearly perceptible strife between different motives.

Now it is plain that these internal acts of will are not only the analogues of the external, but at the same time their condition. There can never be an external act save as the result of a previous inner selection, and this holds again both of the impulse and of the act of choice. So that apperception is the one original act of will. It can exist without the consequences which follow upon other acts of will, whereas these always presuppose as their condition some internal act.

§ II

There is another property of apperception and attention that demands consideration under the head of its relation to will, and which plays an important part in the sequence of mental processes. We observe an alternation in the internal activity of attention, just as we find in external voluntary acts alternating periods of rest and activity, recurring either at regular intervals, or, as the conditions chance to vary, after pauses of varying length. You know, for instance, how difficult it is to follow a lecture word for word with uniform attention. If it were really necessary for our understanding of the whole that we should apperceive each single word with equal clearness, it would be altogether impossible to follow what is said. But in most cases the context enables us to fill out passages to which we have not been especially attentive. And, to a certain extent, that holds also of the speaker. Language is fortunately of such a nature that a whole number of verbal ideas which are indispensable to the expression of thought associate by frequent repetition to a scheme which comes, so to speak, of itself; so that the attention may be rested while the speaking is following the lines of the customary association. We may assume that these fluctuations of attention are, as a rule, pretty irregular : they vary with external impressions and internal necessities. In other words, since its two conditions change, we shall not expect to find in attention as a whole any periodic function of consciousness. We are, however, able, by special experimental arrangements, to introduce regularity into the conditions, and to keep them practically uniform for a considerable length of time. If this is done, we still find that apperception is by no means constant at a certain intensity ; it still rises and falls, and its periods, owing to the uniformity of conditions, are fairly regular.

For the purposes of such experimentation it is best to employ very weak sense-stimuli, such as can be easily perceived with some strain of the attention, but fall below the limen of consciousness with any relaxation. Under these conditions we find a reciprocal relation to exist between intensity and distinctness of ideas. This is, of course, closely related to the law that intensity favours distinctness in ideation. If we allow a very weak impression which lies just above the limen of stimulus to affect a sense-organ, any momentary relaxation of attention will allow it to fall below the limen. In other words, the previously perceived impression becomes imperceptible. This phenomenon may be regarded in two ways. It may be considered, first, as a decrease of sensation-intensity from the minimal perception-magnitude to zero ; or it may be looked upon, secondly, as the sinking of a previously relatively distinct idea below the limen of consciousness. There is no real contradiction between the two interpretations. The two sets of expressions can only be equivalents if the concepts "stimulus-limen" and "limen of consciousness" mean the same thing, regarded from different points of view : an impression which passes the stimulus-limen crosses at the same time the limen of consciousness. That is, the equivalence of the two expressions is due to the fact that the stimulus-limen is a value depending upon stimulus-intensity on the one hand and upon the state of consciousness,—*i.e.*, of attention,—on the other.

Weak auditory stimuli furnish us with the simplest means of observing the periodic fluctuations of attention under the influence of constant conditions. If you place a watch, say at night-time, when everything is quiet, at such a distance from your ear that its ticking can be just heard with strained attention, you will find that at intervals of three to four seconds the regularly recurring impressions alternately appear and disappear. Very similar fluctuations of sensation may be noticed if the skin is stimulated by a uniform induction current of very slight intensity ; only in this case the periods are somewhat shorter. Sight can be best investigated, not by means of approximately liminal stimuli, but by the aid of stimulus-differences at the limit of noticeability. The difference-limen, as we may call it, takes the place of the stimulus-limen in the other two sense-departments ; the difference is alternately noticed and not noticed. The phenomenon may be very conveniently studied on quickly rotating discs. A small piece of a black sector measuring only

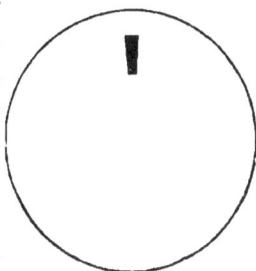

FIG. 38.

S

a few angular degrees is painted upon a white disc. In quick rotation we have a grey ring on a white ground. If the sector is made of the proper breadth, the ring will be just noticeably different from its background. If you fixate it continuously, you will find that it is alternately visible and invisible.

It has often been conjectured that the phenomena which we have been describing are dependent upon purely physiological conditions, lying in the peripheral nerves and organs of sense ; *e.g.*, upon a periodically restored exhaustion of the organs or upon an alternation of strained and relaxed movement. But, so far as the matter has admitted of experimental test, these hypotheses have not been confirmed. Where peripheral changes were found, they proved to be either effects of the fluctuation of attention, or secondary conditions which, though they might affect the temporal course of the phenomena, were not the occasion of them. It has also been noticed, especially in observations upon the concomitant feelings, that so soon as the impression falls below the limen there is a sudden and stronger strain of the attention, immediately attended by the reappearance of the sensation. All these facts make for the assumption that the phenomena in question belong directly in the sphere of the functions of the attention. But we must not suppose, of course, that these functions have no physical attendants, central or peripheral. So that the conditions which serve to vary these will also affect the time-relations of the fluctuations of attention.

§ III

The instantaneous production of transitory sense-ideas, which served in the first place to facilitate the investigation of the varying degree of ideational distinctness, has proved a method of widening our knowledge of a large number of important mental phenomena. It has, however, been found inapplicable to the problem of the range of consciousness for the solution of which it was originally employed, just by reason of this continuous gradation of the distinctness of ideas. At the same time, observation of the effect of sudden visual impressions pointed out the path which investigators of this problem must follow.

Suppose that, at a given moment, a complex impression affects the eye in such a way that only a part of it can be clearly apperceived. It may be a large number of letters, or a complicated geometrical figure (Fig. 37, p. 242). And suppose that the moment after, there is given either a similar or a somewhat different impression. The comparison of the two complex impressions is found to be based not only upon clear apperception, but also upon those constituents of the idea which had been merely obscurely apperceived. It is not infrequently possible to say in such a case that the two impressions are " alike " or "unlike," without our having any account to give of the elements which condition the judgment of unlikeness in the second instance. It follows from this that the more obscure constituents of an impression are taken up into, and are capable of modifying, its total idea. But if the experiment is varied in the way that a complex image is divided into halves and one of these presented at one moment, the other after a small but noticeable interval, it is found that the two successive impressions cannot be at all combined, as were the simultaneously given constituents. If the two half-images, a and b, are compared with the total image, $a + b$, shown in a later experiment, one of two things may happen. Where the complexity of the impression is considerable, it may simply not be seen that $a + b$ is identical with the sum of the successive ideas a and b. Or, if the identity is recognised, it is clearly perceived that reflection and thought have taken the place of direct observation. Suppose, for instance, that the first impression given is that of a uniform duodecagon. If the same object is shown again in a second experiment, it will be at once recognised and clearly distinguished from a decagon, although there can have been no counting of the angles, and although nothing is otherwise known as to their number. Now suppose that in a second series of experiments there presented first one half of the duodecagon, then its other half, and thirdly the complete figure. No one will obtain from the perception of these three objects the idea that the two first together made up the last. That is to say, our subjective perception marks off as totally distinct processes the immediate and perceptual recognition, and the mediate and logical recognition of a compound idea. The former is an instantaneous process of perception, the latter a serial process of comparative judgment. With the former, again, is bound up a characteristic feeling,—an unfailing constituent of perceptual recognitive processes to which we shall recur later ; whereas the latter shows no trace of it.

These invariable and obvious characteristics of immediate recognition, which are, of course, not confined exclusively to the ideas of sight, furnish us with a means of answering the general question as to the range. of consciousness. For immediate recognition it is necessary that the recognised idea has, at some time or another, been present in consciousness as a whole. The problem, then, will now be to determine how many separate ideas may be combined in a total image without doing away with the possibility of an unfailing perceptual recognition when the same impression is repeated. The separate ideas which are thus connected into a complex need not necessarily come from impressions which are objectively simultaneous. Suppose, for instance, that a number of auditory impressions be given in fairly quick succession. The series forms a total impression, of whose constituents there are certainly more than one in consciousness at any moment. Thus we should obviously be quite unable to estimate the rapidity at which one hammer-stroke in a series followed another, if there were not present in consciousness at the instant when a new sound comes one or more of those preceding it ; if, that is, the time-intervals between every two strokes were not directly given in perception. You can see that the same conditions will hold for the perceptual comparison of different series of this kind as hold for other complex impressions. It is only what has been present in consciousness at some time or other as a whole that can be a unity for perception, and as such a unity compared with another similar whole. And there are special reasons for preferring auditory impressions to those of other sense-stimuli in our present investigation. In the first place, it is particularly easy to obtain a relative simplicity and uniformity of sound-impressions. Secondly, the sense of sight, the only possible rival of that of hearing, is liable to disturbance owing to the differences between direct and indirect vision. Lastly, we have had most practice in the apprehension of uniform series of auditory impressions ; so that it is easiest for us in their case to perform the act of recognition at once, and with the necessary certainty. You can see for yourselves how experiments can be carried out. The very simple technique of such experimentation is shown in Fig. 39. It requires a metronome, *m*, of the kind generally used for marking time in music. Metronome-strokes will serve as the

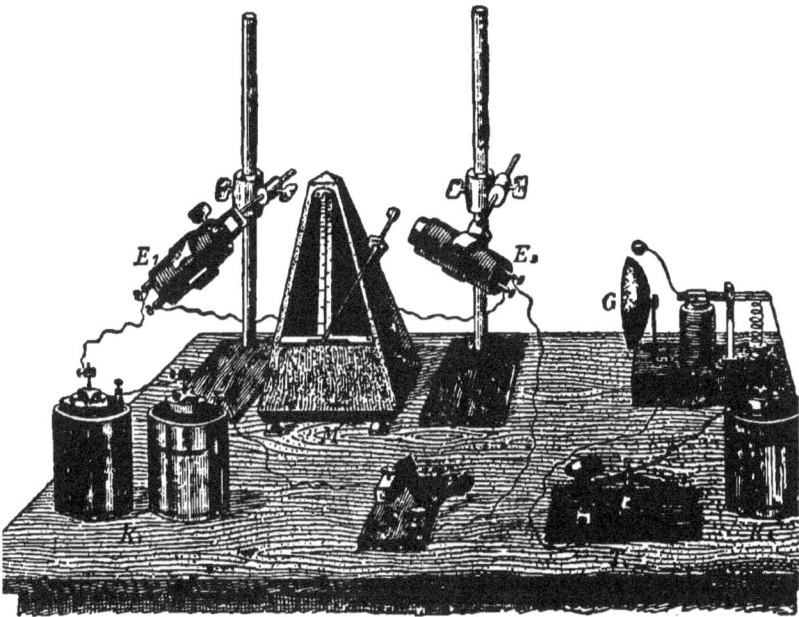

FIG. 39.

simple ideas, the maximal number of which in consciousness we have to determine. Affixed to the pendulum of the metronome is a small iron plate, projecting on either side. This is arranged between two electro-magnets, e_1 and e_2, in such a way that the pendulum can be arrested or set swinging at any moment by the closing or opening of a current passing through them from the battery k_1. The current is made by simply closing the key *s* with the left hand. In order to mark off for perception the separate series of metronome-beats, we make use of a small electric bell, *g*, supplied by a second current, k_2. This current is made for a moment, and then broken again by an instantaneous pressure upon the button of the telegraph key *t*. The experiments are carried out as follows : a signal is given to the observer that all is ready ; and then the experimenter opens *s*, and lets the pendulum swing. Simultaneously with the first stroke, he presses *t*, and the bell sounds. After the right number of strokes has been given, a second series is at once begun, its first stroke being again marked by a simultaneous bell-stroke. As soon as the second series is concluded, the experiment is brought to an end by closing *s*; *i.e.*, by the arrest of the pendulum by one of the two electro-magnets. If we denote the metronome-strokes by quavers, and the bell-signal by an accent placed above them, an experiment consisting of two successive series may be represented in this way :—

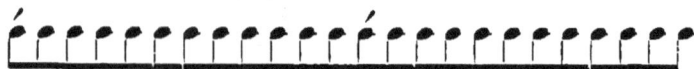

In this instance, the two series are of equal length. In actual experimentation the second series is made to contain one stroke more or less than the first ; while the length of the series and the rapidity of the pendular oscillations will also be varied. The pendulum can, of course, be slowed or quickened within limits wide enough for the present purpose by moving the running weight up or down. The problem, then, is to ascertain how long a series may be, at a given rapidity of vibration, for the immediately following series to be cognised as equal when equal, and as unequal when unequal, without there being any counting of the strokes. And a further question is how the length of the just recognisable series varies with variation of the rapidity of vibration.

A circumstance which facilitates these experiments, and tells at the same time very strongly for the correctness of the interpretation which we ascribe to them, is this : that the point at which immediate recognition ceases to be possible can be very accurately indicated by the observer. This point is really settled at the conclusion of the first series ; for the series is either apprehended directly as a coherent whole, or, if the limit is past, appears as a discrete and indeterminate impression. Peculiar feelings are connected with both phenomena, feelings of distinctly opposite character, which make the observer fairly certain at the beginning of the second series whether he will be able to institute any comparison at all between the coming group and that which has just ended. The most favourable

objective conditions for the apprehension of the largest possible number of impressions are obtained when the interval
between every two strokes is 0·2″–0·3″. The number decreases
if the pendulum moves more quickly or more slowly; and
grouping ceases to be possible at a lower limit of 0·1″, and at
an upper of about 4″. Particularly interesting are the subjective impressions in the neighbourhood of the upper limit.
You recall in this case the previous strokes as each new one
comes, but your recollection is accompanied very plainly by the
same feeling of cognition that you have when you recall previous ideas which have disappeared from consciousness. Each
single stroke, that is, stands to the foregoing one as (where
grouping is possible) each whole series stands to the preceding
series.

Within the limits of the possibility of conscious grouping
there comes to light a further phenomenon which is so variable
that it exercises a decisive influence on the result. If we give
ourselves quietly up to the apprehension of the impressions,
we observe that the separate strokes are not all alike, even
though they are really and objectively perfectly equal. We
alternately accentuate and slur them, just as we do in marking
time in speech, by a voluntary and regularly alternating intensity of accentuation. If we denote the accented impressions
by points placed above them, we have the two series of our
former figure as they usually occur in reality :—

The series of twelve strokes, that is, consists not of twelve
equivalent ideas, but of six ideas, each of which has two parts.
With this simplest form of accentuation we are able to group
into a recognisable series at most sixteen single strokes; *i.e.*,
eight bimembral ideas.

But the same phenomenon may show itself in more complicated form. The series need not be divided on this simplest
scheme in 2 : 8 time. There may be different degrees of
accentuation regularly alternating with one another and with
slurs; and so there may arise a more complicated rhythm.
There need not be any intention to form these secondary

groups : a certain degree of complexity may result simply from the effort to hold as many impressions as possible together in consciousness. You may quite easily obtain the following system, *e.g.*, in which the different degrees of accentuation are again denoted by points, the strongest by three, the next by two, and the weakest by one,—

and so on. By employing this graduated accentuation, we divide up the total idea of the current series into component ideas, each of eight single impressions.

Detailed and varied observations of this nature have shown that we are able with strained attention to hold in mind, and to compare with an immediately following group of similar extent, a series consisting of five of these compound impressions ; that is, of forty separate impressions. If the ideas are made as little complex as possible, therefore, eight is the maximum of grouping; if they are made as complex as possible, five. But, on the other hand, the number of ideational elements simultaneously present in consciousness may be raised through progressive complexity from sixteen to forty.

We never find more than three accents employed to divide a series, for the obvious reason that more than three cannot be distinguished with certainty. This reminds us of the fact that in other cases of purely quantitative discrimination in immediate sensation we cannot pass beyond the three-limit without imperilling the accuracy of recognition. We can easily interpolate a sound of mean intensity between a stronger and a weaker, but hardly more. Ordinary language designates grey as an intermediate quality between black and white ; and grey itself is further distinguished as dark grey, grey, and bright grey. This limitation of our capacity to graduate quantitatively may well be referred to that principle of relativity which underlies mental measurement in general (*cf.* pp. 62, 63). The principle tells us that any estimation of intensive magnitude must be made in terms of other magnitudes simultaneously ideated. So that we can easily apprehend a given sensation in its relation to a stronger and a weaker, but are hopelessly at sea when required to hold in mind any larger number of sensible relations.

If we look at the metrical forms employed in music and poetry, we find again that the limit of three degrees of accent is never exceeded. The absolute amount of accentuation may, of course, be very different in different cases. But in immediate perception these different degrees are always arranged in three principal classes, which alone are of any real importance in metrical division as a basis of classification for rhythmical forms. As a matter of fact, however, music and poetry never push their use of this aid in the formation of easily comprehended ideational series to the extreme limit of conscious grouping. Each member in a rhythmical series must be referred to its predecessors ; and for this to be done with pleasure and without effort, it is necessary that the grasp of consciousness be not too heavily taxed. So that a time like the 6 : 4 is one of the most complex of the rhythms employed in music. Its scheme is the following :—

It contains, you see, only twelve simple impressions. We must, of course, remember that in this case there is present, beside the intensive, a qualitative tone-variation, capable of far wider variation and, therefore, setting all the narrower bounds to intensive change.

§ I

WE have solved the problem of determining the ideational content of consciousness at any given moment. The next question that arises is that of ideational succession. · This falls into two parts: we must investigate, first, the time-relations of ideational change, and, secondly, the qualitative relations obtaining between the changing ideas. The actual train of ideas must always be regarded under both its temporal and qualitative aspects. A quantitative consideration of its time-relations cannot, therefore, neglect the qualitative relations of the single ideas, just because mental time-relations in general must be essentially dependent upon the quality of conscious content. At the same time, we shall find it best to separate the two sides of the problem so far, at least, as to deal predominantly with temporal properties, and to attend only to those more general and fundamental qualitative relations which exercise a decisive influence upon them.

The coming and going, the rise and fall, of ideas have been often enough described, though a guarantee for the absolute correctness of the description cannot always be found. The alleged facts rest partly upon all manner of speculative assumptions, partly on the uncertain ground of introspection. Introspection unsupported by experiment can just as little lead to any certain result here as it could in the inquiry regarding the range of consciousness. And, unfortunately, the phenomena of ideational succession with which introspection was specially concerned happened to be just those which are least of all accessible to exact investigation,—the internal train of fancy and memory-images which we find running its course in the absence of external sense-perceptions. There was total disregard of the ideas directly excited by sense-impressions, or directly and uniformly connected with sense-perceptions. In their case there seemed to be no need of question, inasmuch as in sense-perception the course of objective impressions and the train of subjective ideas were entirely congruent.

§ II

The first indication of the wrongness of this opinion, and the first sign that the shortest and indeed the only road to the investigation of the temporal course of conscious processes lay in observing the ideas directly aroused by external stimuli, came to psychology from outside, from a science in which observational methods had come in the course of time to a high development,—astronomy. Astronomers had noticed certain sources of error in the temporal determination of movements of the heavenly bodies which, while they tended to invalidate the objective value of an observation, cast at the same time a most instructive light upon the subjective peculiarities of the observer.

Suppose that we have to determine the time of the passage of a star at some distance from the pole across the meridian. We may employ an old astronomical method, which is still sometimes used for temporal determinations, and which is called the 'eye and ear method.' A little before the time of the expected passage, the astronomer sets his telescope, in the eyepiece of which there have been fixed a number of clearly visible vertical threads, in such a way that the middle thread exactly coincides with the meridian of the part of the sky under observation. Before looking through the instrument, he notes the time by the astronomical clock at his side, and then goes on counting the pendulum-beats while he follows the movement of the star. Now the time-determination would be very simple, if a pendulum-beat came at the precise moment at which the star crosses the middle thread. But that, of course, happens only occasionally and by chance : as a rule, the passage occurs in the interval between two beats. To ascertain the exact time of the passage, therefore, it is necessary to determine how much time has elapsed between the last beat before the passage and the passage itself, and to add this time,—some fraction of a second, —to the time of the last beat. The observer notes, therefore, the position of the star at the beat directly before its passage across the middle thread, and also its position at the beat which comes immediately after the passage, and then divides the time according to the length of space traversed. If f (Fig. 40) is the middle thread of the telescope, a the position of the star at the first beat and b at the second, and if $a\ f$ is, *e.g.*, twice as long as $f\ b$, there must be added $\frac{2}{3}''$ to the last counted second.

FIG. 40.

When the errors dependent upon accidental circumstances have all been eliminated, these measurements still show differences between different observers. They persist even when there is no external reason discoverable. The fact was first noticed in the annals of the Greenwich Observatory for 1795. The astronomer writes that he dismissed his assistant as unreliable because he had acquired the habit of seeing all stellar transits half a second too late. Not till many decades later was the scientific honour of the assistant vindicated. It was the celebrated German astronomer Bessel who proved that this difference between two observers is only a special case of a phenomenon of universal occurrence. Bessel compared his own results with those of other astronomers, and came to the surprising conclusion that it is hardly possible to find two observers who put the passage of a star at precisely the same time, and that the personal differences may amount to a whole second. These observations were confirmed at all observatories, and in the course of the experiments many other interesting facts came to light. It was found, for instance, that the personal difference between two observers is a variable quantity, fluctuating, as a rule, but little in short periods of time, but showing larger variations in the course of months and years.

It is plain that these differences could not possibly occur if the idea of an impression and the impression itself came simultaneously. It is true that differences might appear between the determinations of various observers owing to uneliminable errors of measurement, but these would. disappear if a sufficient number of observations were taken. A constant and regular difference, such as this actually is, is only explicable on the assumption that the objective times of the auditory and visual impression and the times of their subjective perception are not identical, and that these times show a further difference from one another according to the individual observer. Now *attention* will obviously exercise a decisive influence upon the direction and magnitude of such individual variations. Suppose that one observer is attending more closely to the visual impression of the star. A relatively longer time will elapse before the apperception of the sound of the pendulum-beat. If, therefore, the real position of the star is a at the first beat and b at the second (Fig. 41), the sound will possibly not be apperceived till c and d, so that

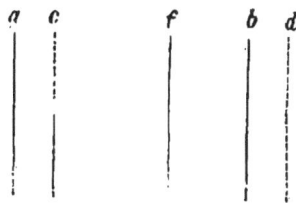

FIG. 41.

these appear to be the two positions of the star. If $a\,c$ and $b\,d$ are each of them $\frac{1}{8}''$, the passage of the star is plainly put $\frac{1}{8}''$ later than it really should be. On the other hand, if the attention is concentrated principally on the pendulum-beats, it will be fully ready and properly adjusted for these, coming as they do in regular succession, before they actually enter consciousness. Hence it may happen that the beat of the pendulum is associated with some point of time earlier than the exact moment of the star's passage across the meridian. In this case you hear too early, so to speak, just as in the other case you heard too late. The positions c and d (Fig. 42) are now inversely related to a and b. If $c\,a$ and $d\,b$ are again $\frac{1}{8}''$, the passage is put $\frac{1}{8}''$ earlier than it really occurs. If we imagine that one of two astronomers

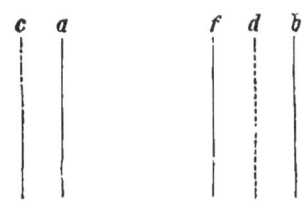

FIG. 42.

observes on the scheme of Fig. 41, the other on that of Fig. 42,—in other words, that the attention of the one is predominantly visual, that of the other predominantly auditory,—there will be a constant personal difference between them of $\frac{2}{8}'' = \frac{1}{4}''$. You can also see that the smaller differences will appear where the manner of observing is the same in both cases, but there are differences in the degree of the strain of the attention ; while the larger differences must point to differences like those just described, in the direction of the attention.

It is, unfortunately, not possible in these astronomical observations to eliminate the errors introduced by the mental tendencies of the observer. We do not know the time of the actual passage of the star, and we can only infer from the personal differences that the observed time of passage is not the real time. But the exact deviation of the individual observer from the true time remains undetermined. Hence the explanation which we have offered for personal differences in general and the larger ones in particular is so far no more than a hypothesis. To prove that it is right, we should have to determine the actual position of the star at some point in its passage and compare this with the estimated position given by different observers. This is, of course, impossible ; the heavenly bodies are beyond our control. But there is nothing to prevent the repetition of the phenomenon by artificial means under circumstances which readily allow of a comparison of actual and estimated times. A very simple apparatus of this description is represented in Fig.

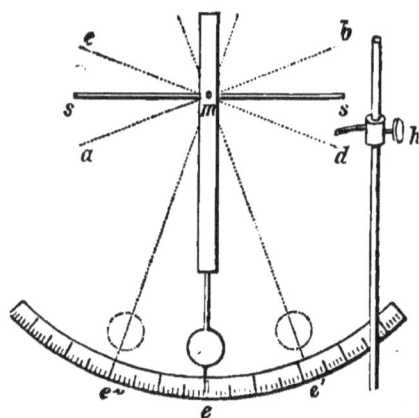

FIG. 43.

43. It is the one by the aid of which I carried out my first experiments on the time-relations of mental processes in 1861. It consists of a large, heavy wooden pendulum. The bulb carries a pointer, which, as the pendulum swings, passes over a circular scale. Near the point of rotation *m* there is fixed to the stem a horizontal metal bar: *s s*. A movable, vertical standard, *h*, has attached to it a small metal spring, also in the horizontal line. The spring is fixed in such a way that the bar *s s* in passing by produces a single short click, the end of the bar and the point of the spring just touching each other, while the shock is so slight that the course of the heavy pendulum is not noticeably affected. By watching the course of the pointer attached

to the bulb of the pendulum, while the upper part of the instrument remains concealed, we can determine at what point of its passage to or fro the click of the spring takes place. For example, if the pointer appears to be at e' at the moment of the sound, the bar $s\,s$ will be in the position $a\,b$, and this will mean that the passage is put too early. If the pointer seemed to be at e'', the bar would be in the position $c\,d$, and this would mean that the passage was put too late. If we know the duration and amplitude of the pendular vibration, and measure the angular difference between e' or e'' and the actual point at which the bar $s\,s$ comes in contact with the spring, we can easily calculate the interval between the giving of the sound and its apperception. To obviate the influence of preconceived judgments, the spring is given a slightly different position in each experiment, so that the observer never knows when the sound is really coming. It was found by this method of investigation that a slow vibration-rate gave a time-displacement averaging $\frac{1}{8}''$. The time of the click was put that much too early ; the sound-impression was associated with the position of the pointer, which in actual fact preceded it by $\frac{1}{8}''$. Later experiments with a more adequate technique [1] have shown that the magnitude and direction of this time-displacement are conditioned in the most various ways. Of especial importance is the rapidity at which the sound-impressions succeed one another. In a slow series we tend to antedate the passage, in a rapid series to postdate it. Moreover, the temporal localisation of the sound becomes later if other impressions,—*e.g.*, electrical cutaneous stimuli,—are given simultaneously with it. The nature of these influences confirms the explanation of the varying time-displacement in astronomical observations offered above. For all the conditions which delay our apprehension of the passage are conditions which either prevent or retard the completion of a preparatory adjustment of the attention. To the former class belongs a high rapidity of the succession of sound-impressions ; to the latter, the simultaneous excitation of other senses.

---

[1] I have described and figured the apparatus, under the name of *Pendelapparat für Komplikationsversuche*, in my *Physiol. Psych.*, 3rd ed., ii., p. 344.

§ III

But however interesting these astronomical observations and the psychological experiments by the eye and ear method modelled upon them may be for a theory of attention, they give us no direct information with regard to the duration of mental processes. It would be altogether wrong to take the absolute difference between the actual and estimated time of the passage for a time-period corresponding to the duration of any particular mental act. For we have seen that this difference simply depends on the interrelation of the objective change of impressions and the variation in attentional adjustment. It will be positive, negative, or zero, according to the experimental conditions ; the latter, of course, when the rate of succession is found at which the actual and apparent times of passage are approximately coincident.

But there is another method which has brought us nearer the desired result. Like the first, it has come to psychology from astronomy. To avoid the considerable personal differences of the eye and ear method, and at the same time to obtain greater accuracy in the estimated times of stellar transits, the astronomers have recently been led to prefer a different mode of procedure, which is called the registration-method. The passage of the star across the field of the telescope is observed in precisely the same way as before, except that there is no counting of pendulum-beats. At the moment when the star passes the meridian thread, a movement of the hand is made, which records the transit upon a chronometrical instrument. The apparatus employed is usually as follows:—an endless sheet of paper is transferred by clockwork from one cylinder to another, so as to move at a constant velocity before a twofold registration instrument. One half of this consists in a writing lever, which is moved by an electro-magnet every time that the pendulum of the clock swings through its position of rest. If the pendulum makes one complete to-and-fro movement in the second, the lever moves every half-second, making a momentary elevation in the line it is describing upon the moving sheet of paper (*u u'*, Fig. 44). The other half of the registration instrument consists in a similar writing lever, which is connected with a key of the kind used in telegraphy. The observer keeps this key closed by pressure of the hand, and opens it at the moment when the star crosses the middle thread of the telescope. A movement of the lever follows, the beginning of which can be exactly determined from the simultaneously recorded half-seconds of the pendulum. Thus, if $U\,U'$ is the half-second line upon the moving paper, and $R\,R'$ the line recording the reaction-movement of the observer's hand, we can ascertain the time at which the second lever began to rise, c, by drawing a perpendicular, $c\,b$, to the line $u\,u'$, and measuring the time $a\,b$ which has elapsed between b and the beginning of the last half-second. This is done, again, by putting space for time: if $a\,b = \frac{1}{4}\,a\,a'$, $\frac{1}{4}''$ must be added to the time-value of a.

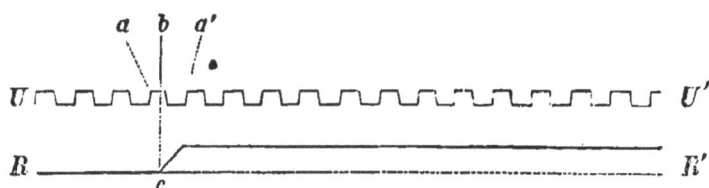

FIG. 44.

Astronomical observations of stellar transits by the registration-method showed, as had been expected, smaller personal differences than those of the eye and ear method. But the differences by no means disappeared. They may still amount to hundredths or even tenths of a second. And this is not difficult to understand. We cannot suppose that the reacting hand-movement takes place simultaneously with the actual passage of the star: a certain time will elapse between the transit and its perception, and again between the perception and the execution of the movement, which may possibly be different in different individuals, and so condition 'personal differences.' Indeed, the composition of these lesser time-values is plainly a matter of more complexity than that of the times found by the eye and ear method. In the first place, physiological processes occupying a certain period of time enter into the total movement-process under consideration. The impression of the star upon the meridian thread must be conducted to the brain, must arouse an excitation there ; and then, before the hand-movement can take place, the impulse of will must be conveyed to the muscles, and

T

these stimulated to contract. To these two purely physiological must be added the psychological or psychophysical processes of apperception of the impression and impulse of will. Even though in actions like this, where the movement answering to the impression is so accurately anticipated, the two acts of apprehension and corresponding movement may possibly be exactly coincident in time, still it seems necessary to suppose that this whole psychophysical process will occupy no inconsiderable, perhaps the major, part of the total time elapsing between sense-impression and reaction-movement. The supposition becomes probability when we take into account the magnitude of the personal differences which are still found to occur. Differences so large as these may be expected where the processes involved are mental, but hardly where they are physiological and similarly conditioned. But neither does the registration-method tell us anything of the actual time-values of the various processes. We do not know the real time of the stellar transit, and so are

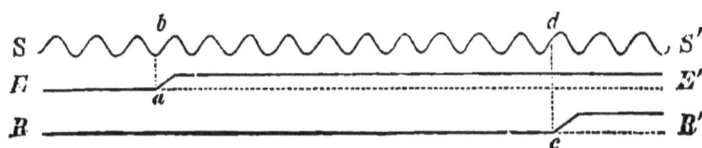

FIG. 45.

still restricted to the inference that, since the differences between separate observers are so considerable, the times whose differences they are must themselves be comparatively large.

But here, again, it is not difficult to introduce artificial experimental conditions which allow of the measurement of the absolute times in question. We can use for this purpose the same astronomical method, with the single difference that, in addition to the instruments for registering time and movement, there is introduced a third lever to mark upon the moving paper the moment at which the external sense-stimulus is given. It is also advantageous, since the times and time-differences to be determined may possibly be very small, to substitute for the vibrations of a clock-pendulum in the record of the time-curve some other and more accurate chronometrical instrument. The best is a vibrating tuning-fork ; and the technique is so far simplified that a tuning-fork with a bristle attached to it can very well trace its own movements. For instance, if $S\,S$ in Fig. 45 denotes the line written by the vibrating fork, and $R\,R'$ that of the hand-reaction, a third line, $E\,E'$, between the two, will represent the self-registration by the stimulus of the moment of its objective occurrence. From the beginning of the elevation indicating the giving of the stimulus and from the beginning of the reaction-movement answering to it, perpendiculars, $a\,b$ and $c\,d$, are drawn to the tuning-fork curve, and the time between b and d measured by reference to the known duration of its vibrations. If, *e.g.*, the fork makes 100 vibrations in 1″,—a hundred full vibrations, each comprising one hill and one valley of the record,—every tenth part of a to-and-fro movement corresponds to $\frac{1}{1000}$″, a space-value which is not too small to admit of fairly accurate measurement. The distance $b\,d$ would then correspond to some $\frac{10\cdot4}{100} = \frac{104}{1000}$ or $0\cdot104$″. The time thus measured between impression and reaction-movement is called the *reaction-time*. It is made up, as we have seen, of purely physiological and of psychological processes ; and we cannot separate the two, or ascertain with even conjectural probability the time-value of the mental component. But although the mere determination of the reaction-time possesses scarcely any importance for psychology, it, nevertheless, is the necessary first step in all possible mental time-measurements. Recognising this, we are bound to consider it in some little detail.

Since the reaction-time may in certain circumstances amount to $0\cdot1$—$0\cdot2$″, but the time-values of the mental processes connected with it and approached by means of it often be considerably larger, this method of counting the vibrations of the tuning-fork becomes too cumbrous and tedious for experimental use. There is generally substituted for it an apparatus of more recent construction, which works as follows : the vibrating body does not record its movement upon paper, but regulates the course of a very rapidly running clockwork. A vibrating spring, which takes the place of the less convenient tuning-fork, interlocks with a toothed clock-wheel in such a way that at each vibration the wheel can only turn by the width of one tooth. The same wheel is connected with a clock-face, on which the times that have elapsed can be directly read off. To allow of the measurement of longer periods, connections are introduced in the wheel-work of the clock similar to those which in an ordinary seconds watch join the wheel that carries the seconds hand with that regulating the large hand. There is further introduced a system of electro-magnets, which enables us to arrest or start the movement of the hand at any moment by the make or break of a current. It is now easy to arrange the experiment in such a way that the giving of the external sense-impression sets the clock in movement, and the reaction of the observer stops it. The difference between the position of the hand before and after gives us the reaction-time.

§ IV

In making experiments of this kind with chronometrical instruments, or drawing inferences from experiments, we must remember that the chronoscope, whose unit is a thousandth of a second, cannot be regarded as a simple watch. It would be quite wrong to read off the times from it without paying regard to the numerous sources of error which minute chronometry always involves. Unless the apparatus is continually and accurately tested, and the observer thoroughly practised in the technique of such experimentation, there can be no hope of obtaining reliable results. And you will find during the course of practice that there are individuals who are entirely incapable of any steady concentration of the attention, and who will, therefore, never make trustworthy subjects. That should not be surprising. It is not every one who has the capacity for astronomical or physical observation ; and it is not to be expected either that every one is endowed with the gifts requisite for psychological experimentation. This is, unfortunately, not seldom forgotten. And the consequence is, that the literature of psychological chronometry, which has assumed such imposing dimensions in the course of the last few years, gives but a scanty store of sifted grain to the inquirer who would turn it to psychological account.

At the same time, the simple reactions to impressions of sound, light, and touch are happily possessed of certain characteristics which render it an easy matter to separate the useful from the useless, provided only that the experiments are described in sufficient detail to allow of their being definitely known. The first consists in the mean value of the reaction-time, the second in the relative constancy of this value. In opposition to all the earlier statements of large individual differences in reaction-time, it has been shown with ever-increasing certainty since the introduction of the more accurate observational methods that, other conditions equal, there is great uniformity in reaction-time,—a uniformity which is independent of all individual differences when once practice has been carried sufficiently far. Equality of conditions means, first, likeness of quality and intensity of the sense-impression, and, secondly, likeness in the condition of the sensory and motor apparatus concerned in the reaction-process. As regards the former point, it is noteworthy that the different sense-departments show certain constant differences, and that very weak stimuli lengthen the times, though these are absolutely constant for impressions of moderate intensity ; as regards the second, that the condition of the organs of sense and movement, however uniform their functioning, never fails to present one difference, which is determined by differences in the direction of the attention, and therefore so far psychophysical in nature. The attention may be principally directed upon the expected sense-impression. In this case the specific muscular apparatus of the sense-organ (*e.g.*, the *tensor tympani*, or the muscles of accommodation) are strongly innervated, the muscles concerned in the reaction-movement only weakly. Or the attention may be mainly turned to the movement which is to be made. In this case the energy of innervation is distributed in the converse way. We may, therefore, designate the first kind of reaction, where the sense-organ is attended to, the *sensorial* form ; the second, where the attention is directed upon the organs of movement, the *muscular*. No one, then, can be regarded as capable of experimenting upon the time-relations of mental processes until he is able to change at will from either of these forms of reaction to the other. The question as to which form we have in a particular case can be easily answered by reference to the magnitude of the measured times and the degree of their average constancy. If we take as our time-unit, for simplicity's sake, not the second, but the thousandth part of a second, and denote this unit by the Greek letter σ (sigma) written above the line, we may put it that the sensorial reaction amounts to 210—290$^\sigma$, the muscular to 110—180$^\sigma$. The smaller number in each case gives the time for touch and hearing, the larger for that of sight. The mean variation of the separate experiments in an experimental series of at least 25 observations amounts in the first case to 20—40$^\sigma$ (where again the larger number refers to the sense of sight), in the second to 10—20$^\sigma$. Whenever, owing to insufficient practice or general inability to concentrate the attention, there is oscillation between the two kinds of reaction, or neither is attained in its extreme form, we find differences in the duration of the average values and (what is a still less mistakable indication) in the magnitude of the mean variation.

§ V

We may designate this kind of reaction to an expected impression of sound, touch, or light the *simple reaction*. In each of its forms it may be regarded, for the reasons given above, as a composite process, containing both physiological and psychological constituents. And the association of these constituents is so difficult of analysis that no conclusion can be drawn regarding the duration of the mental terms of the series. But in our consciousness these mental terms are separated off from the others : and we may evidently insert other mental acts in the same process, and so lengthen the total time of reaction by the precise interval which they require for their completion. Such reactions we may call *compound*. And we shall obtain the duration of the interpolated mental process by subtracting the simple from the compound time. For we may surely assume that the purely physiological processes are the same in both cases ; and that the apprehension of the impression and the impulse of will, implied in the simple reaction, recur in similar form in the compound. There is always one necessary condition, it is true : that the sensorial form be taken as the basis of comparison, and not the muscular ; the muscular is so automatic in character that the interpolation of new mental processes is impossible. For instance, in a first experimental series the observer may react to some light-impression without regarding its qualitative character, the reaction-movement simply following the impression upon the eye. In a second series qualitatively different light-impressions may be presented irregularly and at random, and the observer required only to react after he has distinguished the quality of the particular impression. By subtracting the previously determined simple time from this longer time we get a *discrimination-time*; that is, the time required to complete an act of discrimination.

But now we can easily go a step farther. We may require the observer after discrimination to choose between different reaction-movements, and not to react until he has chosen. Thus two light-impressions, say a red and a blue, may be given in irregular order, the rubric being that red shall be reacted to with the right hand and blue with the left. Here, you see, there are two mental acts over and beyond the simple reaction,— first, the act of discrimination as before; and secondly, a new process, that of choice between two actions. If we subtract the compound time of the first order,—the discrimination-time,— from the compound time of the second order,—the time of discrimination with choice,—we obtain a *simple choice time*. Besides this, we may obtain *compound* choice times; *i.e.*, the duration of the act of choice between more than two movements. Since we have ten fingers at our disposal for experimental purposes, we can prescribe choice between as many as ten movements. In this case, of course, the association between the ten-finger movements and ten qualitatively different associations must have been made stable by practice, if the conditions of the experiment are to resemble those of simple choice in everything except the number of impressions. Impressions which are to be discriminated only may, naturally, be varied to a very much greater degree. We can determine the visual discrimination-times, for instance, not only of colours and brightnesses, but also of letters, words, geometrical figures, and other more or less well-known objects.

Yet another method of obtaining reaction-times of the second order is to set out from the time of discrimination or cognition, and to require that the reaction-movement shall follow only when some idea has been associated to the idea aroused by the impression. By subtracting again the cognition reaction from the total association reaction, we obtain an *association-time*; that is, the time required for the appearance in consciousness of an associated idea excited by a perception. You will see at once that here again the conditions can be varied at will, whether by limiting the association to definite groups of ideas, by requiring the completion of trains of logical thought suggested by the sense-perception, or what not.

We cannot, of course, enter in this place into the details of these measurements. Here is a brief table of figures which give the average time-value of some of the above-mentioned mental processes in thousandths of a second [1] :—

Cognition of a colour	30
„ „ a short word . . .	50
Choice between 2 movements . .	80
„ „ 10 „ . . .	400
Association	300—800

A simple geometrical figure (triangle, square, etc.), or any other equally simple visual object, seems to be cognised almost as quickly as a simple colour. A single letter requires about the same time as a short word. Both these facts show the immense influence of habitual practice. The total impression of a well-known object is so familiar to us, that the need of analysing it into its constituents in order to distinguish it from other objects is as remote as is the possibility of such analysis in the case of a simple colour. In the same way, when once we have learned to read, we do not divide up a word into its component letters, but apprehend it as a single total impression. And there belongs here a further observation of interest. A letter printed in the ordinary German type requires 10—20$^\sigma$ more for its cogni-

---

[1] The two first lines of this table are taken from the results of an investigation recently carried out in my Institute by E. B. Titchener (*Phil. Studien*, vol. viii., part 1). They are very much smaller than the values ascribed to the same acts in earlier researches, and published in the 3rd edition of my *Physiol. Psychologie*. The reason for the difference is the more careful observation of a uniformly sensorial form of reaction, both in the discrimination-experiments and the simple times with which they were compared. The earlier numbers were gained in experiments made before the discovery and consequent utilising of the difference between the two simple forms, and by observers who inclined to react muscularly,—a circumstance which increased their cognition-times by something like the difference between the sensorial and muscular forms ; *i.e.*, by about 80—100$^\sigma$.

tion than a letter of the same size in the Roman character. But there is no such difference between words printed in the two types : the German word can be read in just about the same time as the Roman. The single German letter is harder to cognise, because of all its fine strokes and flourishes. You can see this very easily if you take the capitals ; now and again there occurs a word printed throughout in large letters. It is true that cognition is also retarded in this case by the un-accustomed nature of the whole impression. And the same factor is operative to make us slower in reading substantives printed or written in Grimm's way with small first letters,—a fact which tells against the advisability of riding this Germanistic hobby.

We can easily understand why the times of the cognition of complex objects, of association and of the formation of judg-ments, should be, not only longer, but at the same time more variable, than the others. The more complicated· the processes become, the more dependent is each particular result upon the individual conditions of observation, and especially upon the disposition of the observer, itself determined by numberless past experiences and incalculable chances. A further general conclusion to be drawn from the numbers in the table is, that the duration of mental processes is by no means so brief as has often been assumed. The phrase ' quick as thought' does not refer so much to the actual rapidity with which idea succeeds idea in consciousness, as to our undoubted ability to drop out the intermediate terms in a train of thought, and so pass at one bound from the first to the last link in the ideational chain. Apart from this, it is obvious that the absolute time-values of the various mental processes are of no importance whatsoever in themselves ; they only become important when they·help to throw light upon the nature and interconnection of our ' states of mind.' And for this reason the quantitative examination of the temporal course of ideas must always go hand in hand with the qualitative investigation of their mutual relations. If it pays heed to these facts, the psychological chronometry of the future may be looked to for the solution of many an important problem.

§ I

IF we look for a moment at the coming and going of ideas in our minds, we cannot fail to see that the plot of the play is determined by two influences,—accidental external sense-impressions and previous experiences. Which of the two preponderates at a particular time depends upon circumstances. Cast your eye over a landscape, or follow attentively the rendering of a musical composition, and you will find yourselves seemingly wholly given up to the external impressions; subjective tendencies coming in only secondarily, and rather as feelings than as ideas. Now try to recall the events of the past few days. External sense-impressions are hardly noticed; and the train of ideas, so far as it is clear and distinct, consists solely of reproductions of previous mental experiences. These are both extreme cases; ordinarily we find ourselves in some intermediate frame of mind. Memory-ideas are aroused by sense-perceptions, and again interrupted by new impressions. Wherever the influence of past experience is traceable, we find the memory-ideas aroused evidencing a definite relation to the condition of consciousness at the time. Sense-perception varies with every variation of the environment; but the memory-image is always suggested, whether by a sense-perception or a previous memory-image. You will object that now and again a recollection crops up suddenly and for no apparent reason. But attentive introspection will in most of these cases enable you to discover the thread of connection with your present state of mind. However little obvious this connection, then, we may safely assume that it is there. If it escapes our observation altogether, as it may do, that is only because the conditions are not favourable for its apprehension.

The interconnections of memory-ideas and sense-perceptions, or of memory-ideas with other memory-ideas, are called *associations of ideas*. The term belongs to the English 'association' psychology. It was first employed to cover the phenomena of memory only, but afterwards extended to all possible connections of ideas originating in the preconditions of consciousness. As customarily used, it is at once too narrow and too wide : too narrow because it leaves out of account a whole number of connections for the sole reason that in them the ideas do not come to consciousness in succession as in ordinary recollection, but, owing to special conditions, appear simultaneously as a complex totality ; too wide because it embraces all successive ideational connections, the act of recollection aroused by a simple sense-impression and the most involved process of logical thought. Now, true as it may be that in both these cases the ideational connection is determined by mutual relations implied in past conscious experience, it is equally true that they are in all other respects so different that to treat of them without further discrimination cannot but obscure the analysis of their constituent processes and hinder the understanding of their inter-relation. We shall ourselves mean by *associations* simply those ideational connections which do not exhibit the characteristics of the activity of logical thought. What these are we shall discuss later.

§ II

The starting-point of the doctrine of association, in the usual sense of the term, was observation of the reproduction of earlier ideas. It has hardly done more than put into modern form what had been taught as the psychology of memory from Aristotle down. But conscious recollection implies a distinction between the inducing and the induced ideas ; if the two are not discriminated, the process cannot be that of conscious recollection. Now it is evident that this recognition of an idea as having been previously experienced is a character which may possibly attach to a revived idea, but need not by any means necessarily do so. The simpler case of association we must rather admit to be that in which ideas are connected simply by reason of their mutual relations in consciousness, and without there being any direct apprehension of the connection as an act of recollection. Certainly memory presupposes association,—on the assumption, that is, that no idea comes into our minds without cause,—but not every association involves an act of memory. That is, we must obviously set out first of all from the phenomena of association proper, and then go on to determine what new conditions are necessary for the association to become recollection.

In this wider sense association embraces a whole number of connective processes in which the associated ideas do not succeed one another, but come to consciousness as a simultaneous ideational complex. There can be no question here of an act of recollection, for the simple reason that the induced idea associated with the inducing is not in any sense separate from it,—in other words, cannot be independently compared either with it, or with any other idea. We may call connections of this kind, in which the primary inducing and the associatively induced idea form a simultaneous conscious complex, 'associations in simultaneous form,' or, for brevity's sake, *simultaneous associations*. There belong here, in the first place, all those products of the fusion of simple sensations which compose our complex sense-perceptions. These latter always consist of a connection of several sensations forming a simultaneous complex idea, such as a compound clang or some spatial idea of sight or touch. One difference there is between these connections and other associations,—that the sense-impressions which arouse the constituent sensations are themselves interconnected; so that the re-excitation of earlier ideas, though of course not entirely inhibited, is completely overshadowed by the connective tendencies obtaining among the sensations. The sensations composing a clang or a visual perception depend upon a simultaneous activity of sensory stimuli. At the same time this difference does not imply any essential difference in the psychological character of the general process; that is, if we regard as the chief characteristic of association this property of mind to connect certain ideas or ideational elements automatically into a complex idea. And there is every reason for doing this, since, surely, certain types of clang, certain spatial arrangements of sensations, are every whit as familiar as, nay even more familiar than is, for instance, the connection of a perception with a similar memory-image. But this means that in these simultaneous associations of perception also there is nothing to prevent a sensation-element which is not actually given in the sense-impression being supplied by immediate reproduction. We have seen, for instance, that eye-movement influences the perception of visual space even when the organ is at rest ; thus we may be deceived as to the position or movement of external objects in consequence of having intended, but not actually executed, an ocular movement, and this just because of the intimate association of movement-sensation with impulse of will.[1]

These fusions of uniformly connected sensations, which constitute sense-perception in general, are not very obviously related to the 'associations' of current psychology. More akin to these are the interconnections of the perceptions of different senses. We see a musical instrument, and hear a clang from it. Our eye apprehends the white, crystalline nature of a lump of sugar at the same time that our tongue is experiencing a taste-sensation of the quality sweet. There arises in this way a connection between sensations and perceptions of different senses so intimate, that if but one sense-impression chances to be actually presented, or the memory-image of an impression aroused in the domain of one sense only, the other sensation is at once mentally associated with it. We hear the piano, and no sooner do we hear it than a vague visual image of the instrument crops up in our minds. Or we taste sugar in the dark, and there is at once associated with its taste a general notion of what it looks like. These connections of ideas of disparate senses which are referred to the same objects, and so belong closely together, we may term with Herbart *complications*. There can be no doubt that they are simultaneous associations. One sense-impression is so inti-

---

[1] For similar instances *cf.* Lecture V. (Associations of Tone-sensations), and Lectures IX. and X. (Associations of Spatial Perception).

286 *Lectures on Human and Animal Psychology*

mately associated with another, or, at least, the origin of the
two is so little distinguishable in time, that the disparate
constituents show themselves in consciousness only as the interrelated parts of a single idea.

Most frequent and most important among the complications
are verbal ideas. They stand, as a rule, in a twofold connection : the acoustical impression is associated, first, with a
sensation of movement, and then, secondly,—at least, in many
cases,—with the visual impression of the printed or written
characters. Movement-sensations are evidently complicated
with other ideas as well. They acquire an especial significance
from the fact that the memory-image of a movement is apt
at once to arouse the movement itself. The consequence of
this is, that movement-sensations often act vicariously for the
sensations of certain senses whose memory-images are so faint,
that we either do not perceive them at all or only imperfectly
and by the aid of the muscular sensations customarily associated with them. It is, for instance, in most cases illusion
when you think that you can recall the scent of a rose.
Observe the working of your minds in the act of recollection
more carefully, and you will find something like this. First,
you have a more or less distinct visual picture of the flower,
and then, secondly, a movement-sensation in the nose corresponding to the inhalation of air, and then again, thirdly,
a sensation of touch and témperature, arising from the ˌair
actually inhaled, your movement-sensation having been attended by an actual movement. The sensation of smell proper
is either entirely absent, or, at least, so faint that it is altogether
overshadowed by the other components of the complication.
In the same way the complications consisting in memory-
images of impressions of taste contain hardly anything of the
sensation of taste, which is, however, quite adequately represented by the movement-sensation, which varies for different
gustatory substances with variation of the accompanying
mimetic expression.

§ III

Apart from these cases of the confusion of definite sensation-
qualities with muscular and tactual sensations, the separate

constituents of a complication are in general clearly distinguishable, belonging as they do to disparate modalities and occurring in other connections under other conditions. The same cannot be said of a second important class of simultaneous associations,—the connections of an externally excited sense-perception with its related memory-images. This type of association we will call *assimilation*, and speak of the memory-image as the assimilating element, the sensations following from the sense-impression as the assimilated. These expressions imply that the memorial constituents are the determining factors in the result, while the incoming sense-impressions are determined by them. This is so far true. An impression may be apprehended in the most different ways, according to the disposition in which the mind has been left by previous experiences. The resultant complex idea is, therefore, a mixed product of the impressions given in perception and of an unknown number of memory-images. But, just because the idea is a single complex, there can be no question of analysis into these two constituents. Hence the reproductive elements are invariably referred to the sense-perception, which now contains constituents not to be found in the impression which aroused it. On the other hand, real constituents of the sense-impression may be wanting in the resultant idea, owing to their conflict with reproductive elements of greater intensity.

The process of assimilation then, unlike that of complication, is not one to be discovered by casual introspection. To examine it, we must carefully compare the impression with the idea aroused by it. The comparison shows the incongruence of the two, and so leads us to look for the ground of their difference where alone it can be found,—in the activity of previously experienced ideas. When once our attention has been called to their influence, we have the key to a whole number of phenomena of ordinary life and of experimental practice, which, though striking enough, are generally left unnoticed or unexplained. We ordinarily read over a printer's error without seeing it; that is, we read the familiar word-picture into the impressions presented to us. Or we fill out a sentence in a lecture which we have heard indistinctly without remarking that we have not heard it clearly. On the other hand, we are equally liable to hear wrongly by supplementing the indistinct sounds by a wrong set of memory-ideas. The crude outlines which serve to represent a landscape in stage scenery look by artificial light and at the proper distance to be a perfect reproduction of a real scene. Here the ready assistance rendered by the appropriate memorial elements is made still more effective by the confused outlines of the retinal image. Outline drawings of tridimensional objects, if purely schematic and unshaded, can be seen at will as tridimensional or areal, and if the former, as extending in this direction or in that ; it simply depends upon which of our familiar space-ideas we employ. For instance, the outline drawing of the head on a coin can be seen pretty much at pleasure as cameo or intaglio. You are all familiar with the puzzle pictures which have, *e.g.*, the head of some well-known man outlined in the foliage of a tree. At first it is quite difficult to find the head. But when once you have it, it stays ; and you are hardly able to get rid of it again, however much you try. The same thing may often be noticed in stereoscopic observations. For a time it will be impossible to get the idea of depth, and then on a sudden it comes with even plastic clearness. What all these cases mean is, that the assimilating memorial elements have required some little time to be called into activity by the appropriate constituents of the external sense-impression.

But, of course, it is not usually a single memory-image which unites with the given impression in the process of assimilation. A stereoscopic object which we may never have seen in the exact form in which it is presented will arouse the co-operation of a whole number of memorial elements, taken from a whole number of originally separate perceptions ; and may be able in this way to call up the idea of three dimensions. But, for this very reason, it would be wrong to suppose that sense-impressions are first of all present in assimilation as independently co-existent ideas, and then fuse to an ideational unity. The assumed stage of independent co-existence of the components is neither discoverable by introspection nor actually possible, since, as a general rule, the assimilating effect proceeds from a large number of ideational elements, originally distributed through quite different ideational series. We can only imagine that every sense-impression acts as an excitant to numerous tendencies remaining from previous impressions ; and that such of these as are appropriate to the impression, and, at the same time, more easily excited than the others, help to form the resultant idea. Lastly, in all these processes of assimilation, which follow directly upon sense-impressions, the peripherally excited sensations are so far of influence upon the memorial elements that they increase the intensity of the reproduced sensations. That is the only possible explanation of the fact that even in the normal assimilation it is impossible to distinguish between the ideational elements aroused by external stimulus and those excited by association. The impossibility becomes still clearer when the elements of the latter kind obtain so exclusive a predominance that the resultant idea is wholly inadequate to the sense-perception. Assimilations of this class we term *illusions*. In the illusion, we imagine that we perceive something which is not there ; that is to say, we confuse memorial elements with sense-impressions. And that again is only possible when there is no noticeable difference in the intensity of the two constituents.

The occurrence of a process of assimilation can be proved with absolute certainty, as these instances show, when the resultant assimilation-product is a sense-perception, whether actual or more or less illusory. In either case the new idea is so different from the sense-impression that the activity of assimilation is a matter of direct inference. But you will see that it is at least extremely probable that assimilations take place in terms of memory-image pure and simple ; and we have an obvious indication of this in the fact that a particular perception is not as a rule assimilated by a particular memorial idea, but by an indefinite number of such ideas. Suppose, then, that there is no sense-perception present, but that some memory-image crops up of itself. It will undergo continual variation by assimilating other ideas which refer to similar objects. So that we cannot draw any hard and fast line between a memory-image proper and what is called a fancy-image. Psychologists are accustomed to define memory-images as ideas which exactly reproduce some previous perception, and fancy-images as ideas consisting of a combination of elements taken from a whole number of perceptions. Now memory-images, in the sense of this definition, simply do not exist. The ordinary memorial representation is determined by several perceptions of the same object. Thus, if we try to recall a person whom we have often seen, we never represent him exactly as he was on any particular occasion ; our idea of him is the resultant of many perceptions, whose constituents, mutually supplementing or inhibiting one another, combine partly to deepen, partly to soften, the general outline. This explains the indefiniteness of most memory-images. Even when we are recalling an object only once seen, the idea does not coincide with the original perception ; some elements are wanting, others which do not belong to the object are wrongly transferred to it from similar ones. Try, for instance, to draw from memory some landscape picture which you have only once seen ; and then compare your copy with the original. You will expect to find plenty of mistakes and omissions ; but you will also invariably find that you have put a great deal in which was not in the original, but which comes from landscape pictures which you have seen somewhere else. So that, according to the ordinary definition, every memory-image would be a fancy image, and ideational reproduction a concept with no corresponding reality. For there is no memory-image that reproduces either the primary perception-image, or any other memory-image of that same perception. And that is not hard to understand when we remember that our ideas are not permanent objects, but processes which can never exactly recur, because the conditions of their occurrence are never twice alike.

§ IV

The process of assimilation is, therefore, always a compound process, set up in any particular case by an incalculable number of elementary connective processes. We may now ask as to the character of these last,—the indecomposable and fundamental processes of connection. In answering that question, we must again set out from those cases of assimilation which begin with a sense-impression, since they furnish the best material for the determination of the conditions of the phenomena. There can be no doubt that there are always two connective processes running side by side in every case of assimilation, whether this be normal and initiated by a sense-perception, or illusory and implying a misinterpretation of sensory impressions. First, the sense-impression calls up previous similar sensations, and, secondly, by the mediation of these sensations arouses other ideational elements not contained in the given impression, but which were connected with it on other occasions. What first happens when you look at a stereoscopic object is, that certain of the outlines correspond to those of some material object known from previous perceptions. But these coincident elements would in themselves be wholly inadequate to suggest the actual picture of a tridimensional object. There must be further aroused elements not present in the actual figure, but associated in previous ideas with the coincident parts, and now necessary to complete the image as that of some definite thing. When we read a wrongly printed word correctly, the primary suggestion proceeds from the rightly printed letters in it : they arouse the corresponding memory-images of the same letters, and these, again, recall to mind the letters which were visually connected with them in earlier perceptions, and which, taken together with them, give the correct picture of the word. So the disturbing elements in the impression are overlaid by the reproduced ideas.

The older doctrine of association,—*i.e.*, of successive association,—distinguished associations as those of similarity and contiguity. In the first form an idea is excited which in certain of its characteristics resembles the exciting idea ; in the second, an idea which at some time or other was in temporal or spatial connection with it. If we apply these terms to simultaneous associations, we may obviously call what was above denominated the second act in assimilation a contiguity-association. On the other hand, we cannot reduce the first act in the same way to a similarity-association. Two objects are similar when certain of their characteristics correspond, while others are different. Now it does not seem possible that an impression should directly call up the memory of another, if this differs from it more or less. It can, surely, only arouse a memory-image like itself. There may, of course, follow upon the excitation of these like elements the reproduction of others which are unlike, provided that these have been connected with the like in previous ideas. In other words, a similarity-association always points to the union of a likeness-association with a contiguity-association. The wrongly

printed word calls up the right image through the likeness-association of the coincident letters, and a contiguity-association,
which takes the right elements, not given in the actual impression, from previously seen word-pictures. The result of this
compound process is a so-called similarity-association, for the
wrongly printed and rightly ideated words are similar, but not
like. It is just the same with the stereoscopic idea. The outlines which originate the suggestion arouse a likeness-association,
which is at once supplemented by one of contiguity, which partly
fills out the resultant idea, and partly corrects it by suppressing
any disturbing elements in the impression. Since no two ideas
are absolutely like, it might be objected that our likeness-association is itself nothing better than a similarity-association. But,
as a matter of fact, we are not concerned here with associations
between complete ideas, but with connections between constituents of ideas. Absolute likeness between two ideas is impossible, for the very reason that every likeness-association gets at
once attached to it a number of contiguity-associations, the
final result being either a 'similarity' or 'contiguity' association,
according as one or other of the elementary processes predominates. We cannot discover any other elementary processes than
these two,—the connection of the like and that of the contigu-
ously associated in time and space. Each of them must of
necessity be present in every concrete association. An idea can
only call up an earlier idea, if it has some elements in common
with that ; and since the reproduced idea contains unlike as well
as like constituents, the likeness-connection implies the formation
of a contiguity-connection. Likeness of the elements, you see,
is directly effective. If a new impression contains elements like
those of an earlier impression, these will separate from the rest,
having become so much more familiar by repetition, and will
preponderate in consciousness. Contiguity is only mediately
effective. It works by way of reviving other elements externally connected with these like constituents of previously
excited ideas.

In view of this essential difference between the two processes,
is it right to speak of a likeness-connection in the same sense as
a contiguity-connection ? When a given impression arouses an
idea consisting in part of elements belonging to the impression,

in part of elements not actually present in it, but connected with it in previous perceptions, there can be no doubt that we are justified in speaking of a connective process in referring to these unlike constituents ; the corresponding excitations must be set up by an impulse proceeding from the impression. But the elements which pass directly from the impression into the assimilation-product hardly seem to need anything which might be reckoned among the association-processes: they are directly given by the external sensory stimuli, and would appear rather to be the condition of the origin of a connective process than its result. In other words, is not the assimilation-process entirely and exclusively referable to association by contiguity ?

However tempted we may be at first sight to answer this question in the affirmative, a little more consideration will convince us that such an answer would be incorrect. As a matter of fact, the elements which pass from impression into assimilation-product are not in their second connection what they were in the first ; so that the word 'pass' is only in place if we are comparing the result with its constituents, and not as referred to the actual process. The 'passage' involves the action of intermediary processes in two different directions. First, the passage of elements of an impression into the resultant idea is favoured, as we know from experience, by the frequency with which they have presented themselves in previous impressions. The only explanation of that is, that the corresponding excitations are intensified by the dispositions left by the action of previous impressions. This intensification will, of course, be directly connected with the present impression ; it will not do to assume, as those psychologists do who make ideas into permanent entities, that constituents of the new idea connect with constituents of some previous one. It will simply happen that the greater frequency of certain kinds of excitation implies the ascription of a higher intensity-value to any one of that kind which affects us. And it follows from this that the final result is due not simply to the impression, but,—and this holds of all association-processes,—to the connection of the impression with the after-effects of previous excitations. Secondly, this passage of elements of an impression into the assimilation-product implies the presence of another process, the direction of which is just the opposite of that of the former,—a process of disappearance of elements which are contained in the impression, but supplanted in the idea by other new and incompatible elements called up by contiguity. That is to say, the like elements are not by any means after the 'passage' what they were before it. They are partly intensified by previous practice, partly weakened or, at least, severed from their original connections by inhibitory influences. All of which goes to show that the likeness-connection is as truly the result of many and different processes as is the connection by contiguity. At the same time, we must not forget that these determining processes are essentially different in the two cases. The best terms to indicate this difference in brief are perhaps those employed above : the likeness - connection is immediate, the contiguity - connection mediate.

LECTURE XX

§ I

IN close connection with assimilation stands, as we saw above, the *successive association* of ideas. This is the process to which the general name of 'association' was originally confined ; and it is still customary, even at the present day, to speak of *laws* of association, and to distinguish in that way connection by similarity from that by co-existence in space or succession in time, or sometimes from that by contrast. It is scarcely necessary to say that these are really simply forms, and not laws, of association ; they are not universally valid conditions of its origin. They merely serve to furnish classificatory concepts under which the ready-made products of association can be subsumed. But, curiously enough, the authority of Aristotle and the constant inclination of the human mind towards logical schematisation have worked no less harm in this department of psychology than they worked in the sciences of nature. Aristotle had distinguished four kinds of memory, in terms of the logical opposites 'similarity' and 'contrast,' 'simultaneity' and 'succession': just as he had arranged the fundamental qualities of all natural bodies under the rubrics of the contraries 'hot' and 'cold,' 'moist' and 'dry.' And these four forms have held the field, despite the evidence of observation, down to our own day. It is now pretty generally agreed that 'contrast' may be omitted, or, where anything corresponding to it occurs, referred to 'similarity'; while spatial co-existence and temporal succession are brought under the general head of external contiguity. This means a reduction of the four forms to two,— association by similarity and association by contiguity. And the reduction is so far good that the different cases of successive connection may, as a rule, be arranged in one or other of the two classes. At the same time, the terms still tend to suggest the wrong idea: that they are the distinguishing marks of elementary processes, instead of the classificatory headings of association-products, each of which is constituted by a whole number of simple processes. In the matter of constituents there is, of course, no essential distinction to be drawn between the two. forms. For it is obvious that just the same processes must be operative in successive association as in assimilation,—the only difference between them being that the successively associated ideas are not combined into one simultaneous idea, but remain temporally separate, in obedience to conditions which we have still to discuss.

Apart from this, however, we shall expect to find every successive association composed of two processes : a direct connection of like elements of different ideas, and a connection, attaching itself immediately to this, of such elements of previous ideas as have been externally contiguous to those like constituents. If, as we look at the total result, the connections of the like elements are predominant, we speak of a similarity-association ; if the external connections are the stronger, of a contiguity-association. Thus it is an association by similarity when the picture of a landscape reminds us of the reality as we have actually looked upon it. Different as the picture and the retinal image may actually be, there are certain outlines that correspond. These call up the memory-images of earlier perceptions, and cause us to transfer to the picture many of the elements of the original which it does not really possess. Now this revival of elements which are not given in the picture is plainly an association by contiguity. Certain of these contiguity-elements work by way of assimilation ; they make the similarity of the picture to the original appear greater than it is. Others make against assimilation ; it is through these that we are able to distinguish picture and reality at all,—that the result of the whole process is not a simultaneous assimilation, but a successive similarity-association. On the other hand, if we have read the letters *a b c d*, we are inclined to continue *e f g h*. This is a case of association by contiguity. But here, too, the original process is the direct connection of like with like. The letters when read call up the like letters previously read or heard. And it is at this stage that contiguity-effects must intervene, if the visual idea is to be apprehended as agreeing with previous perceptions. Then, by a further operation of contiguity, the absent letters are supplied to complete the usual series.

Association by similarity and association by contiguity, then, differ in two points as regards the nature of their constituent processes. First, there is a general predominance of the elementary connection by likeness in the former, of that by contiguity in the latter, form ; secondly, in the similarity-association our attention is directed upon the common properties of the ideas, in the contiguity-association upon their divergences. The association of the picture with the landscape is an association by similarity, because the resemblance of the two ideas makes us neglect not only their differences, but also contiguity-connections which are essential for the institution of a comparison. The association of the letters of the alphabet is an association by contiguity, because we attend only to the added letters, and not to the cognition of the first letters and the likeness-connections which it almost invariably involves.

§ II

The result of all this is, that there are two fundamental forms of connection between ideational elements : connection by likeness and connection by contiguity ; and that both of them are concerned in every case of actual association. For the proof of this fact our best recourse will be to the simplest cases of association. They possess the further advantage of exhibiting with especial clearness the conditions which differentiate successive from simultaneous association, and particularly from assimilation. The constituent elementary processes are, of course, the same in both forms.

The simplest case of assimilation is the cognition of an object ; the simplest case of successive association, its recognition. We cognise a picture as a picture even when we are perfectly sure that we have never seen it before ; we recognise it in remembering that we have seen it, just this particular picture, on some previous occasion. The simple act of cognition is a process of assimilation. The present impression calls up earlier ideas: there are set up connections by likeness and contiguity, but there is no analysis into a succession of ideas ; the presented and the memorial elements combine at once to a single idea, referred to the actual impression. The fact that the resultant idea is, however, no new one, but one which, upon the whole, is familiar, expresses itself in the character of the accompanying feeling. We will call this the feeling of cognition. Since feelings always have some ideational basis, we may suppose that the indefinite memory-images in the background of consciousness, whose function is the assimilating of the given impression, serve as the intellectual substrate of this particular feeling.

From this process of cognition is developed that of recognition. The steps are three in number.

Most closely related to the act of cognition is the process of immediate recognition. In this we are either unconscious or but obscurely conscious of the connecting links by whose aid recognition is effected. And here again two alternatives are possible : first, the idea is merely accompanied by the consciousness that it has been before our minds before, at some time or other, once or oftener,—that is, the recognition takes place without there being any recollection of attendant circumstances. Secondly, though the recognition is immediate, it involves the recollection of attendant circumstances. We recall the temporal relations and spatial surroundings in which we previously made the acquaintance of the recognised object. In both cases the act of recognition is accompanied by a feeling. Where the first form of the process occurs, this is generally indefinite, and may be connected with the emotion of doubt. But it becomes distinct and vivid as soon as there is conscious localisation of the recognised idea in time and space. We may call this feeling the feeling of recognition. Now the recollection of attendant circumstances consists merely in the excitation of secondary ideas externally contiguous with the recognised object in previous experiences. In other words, the act of recognition requires these contiguity-connections for its completion.

The second form of immediate recognition furnishes the transition to a third form of the general process,—to mediate recognition. In this we are clearly conscious from the outset that recognition is brought about by the mediation of secondary ideas. Think how often you meet a person whom at first sight you take to be an absolute stranger. But he tells you his name, and on a sudden the face that was so unfamiliar shows you the features of an old acquaintance. Or there may be other mediating circumstances. You see a third person whom you have often noticed in his company, and your eyes chance to fall on a coat or a travelling-bag that awaken your memory. Here again there is a special feeling regularly associated with the act of recognition. This feeling comes later and arises more gradually than the immediate recognition-feeling. At the same time, you will find that it may be very vivid, even when the apprehension of the agreement between the present idea and a previous one is still quite indefinite.

There can be no doubt that instances of mediate recognition (in this sense) occur which are wrongly taken for cases of immediate recognition, the reason being that we are not clearly enough conscious of the auxiliary ideas which mediate the recognition. Thus it has been shown by experiment that it is quite easy to hold in mind three shades of grey between the extremes of black and white, and to recognise each of them correctly, and apparently immediately, after the lapse of some time ; while, if one more shade is interposed, their recognition is uncertain, and mistakes are many. Now there are in common use in language just three designations of shades of grey,—dark grey, grey, bright grey. So that it is not hard to see a reason for the definite limitation of recognition. We have only to assume that one of these three verbal ideas was involuntarily associated with each of the three impressions, and that it mediated the recognition. A musical ear can recognise a particular musical clang after a long time, if this possesses a definite tonal quality, and has its definite place in the musical scale marked by some note-name. But recognition becomes impossible very soon after the giving of the impression, if some other tone is taken which cannot be definitely associated with a name like c, $c\sharp$, d, etc.

We have seen that these different forms of recognition pass over into one another by degrees. It might appear doubtful whether they should be regarded as different processes at all, and not rather as modifications of one and the same process, differing only as regards their secondary conditions, *i.e.*, in the clearness or temporal course of the various conscious elements. Thus mediate recognition, and immediate recognition with attendant circumstances obviously differ only in this,—that while in the former the secondary ideas are first apperceived, and there then arises the consciousness of the agreement of the principal idea with that experienced before, in the latter these secondary ideas are only clearly apprehended at the same time with the agreement of the two principals, and maybe even later. Now, the apperception of an idea is not the same thing as its appearance in consciousness. Our discussion of the phenomena of time-displacement (pp. 270, 271) showed us that when two ideas, *a* and *b*, follow each other in rapid succession, the second, *b*, may be apperceived before the first, *a*, which certainly anticipated it in consciousness. It is possible, *i.e.*, that an apparently immediate recognition with secondary ideas is in reality a mediate recognition as well. The secondary ideas might exert just the same influence in this latter case, although they were longer in coming to clear consciousness. The difference between the two forms will then depend essentially upon the rapidity with which the feeling of recognition arises. If it is excited by the bare entrance of the secondary ideas into consciousness, we call the recognition immediate. If there are required a longer operation and a greater degree of clearness of the secondary ideas, the act of recognition becomes mediate.

Now, if the difference between these two forms reduces itself to a difference of degree in the efficacy of the secondary ideas, it is plain that there can be no sufficient reason for regarding simple recognition without attendant circumstances as a process *sui generis*. If the secondary ideas, the whole scope of whose activity is that of auxiliary forces, may in cases be raised to clear consciousness only after the recognition has taken place, it will surely be possible that they may entirely disappear from consciousness as soon as the result of that recognition begins to take effect. As a matter of fact, a closer glance at the conditions serves to make the possibility a probability. Immediate recognition occurs, first, where the objects are completely fami-

liar to us from repeated experiences, and, secondly, where we
have come to know them but a short while before, or under
circumstances in which they made an especially deep impression
upon our feelings. Now these are conditions which render intelligible a quick apperception of the object, with an accompanying feeling of recognition, but which by no means indicate as
probable the absence of the otherwise universally present secondary ideas. When we see a person with whom we are daily
associating, there are so many reproductions of the most various
situations in which we were in his company, that it is very difficult for any particular one among them to come to clear consciousness. But at the same time there may always be operative
a certain number of these obscure secondary ideas, which will
explain the presence and vividness of the feeling of recognition.
It is a little different when we see some one for the second time
whom we met but a short time before. The recognition-feeling
in this case is certainly not without its foundation in attendant
secondary ideas. But these are fewer in number, and not in
opposition to one another. They therefore possess a more definite character, and so are, as a rule, easily perceived, if the attention is directed upon them. In other words, the process in this
instance appears to assume the character of an immediate recognition, for the sole reason that the connection of the secondary
ideas with the object is still so close, that no perceptible time is
required for the excitation of the recognition-feeling.

It would seem, then, as though the feeling of recognition,—
which, as introspection shows, takes on very different shades in
the different cases we have been discussing,—depended in every
instance upon the excitation of auxiliary ideas. But the time
which these secondary ideas take to appear is not always the
same ; and that points to another difference between the process
of assimilation and the act of recognition. If a given impression
calls up an earlier idea without exciting secondary ideas,
whether clear or obscure, and without reviving the feeling which
is dependent upon them, the result is an assimilation. Impression and idea are combined to form a simultaneous whole ; the
conditions of recognition are wanting. We perceive the object
as one of a class with which we are familiar without referring it
to anything definite of which we have had previous experience.

And it is for this reason that we speak of an act of cognition rather than of recognition. In other words, we may oppose the act of recognition in general, as a simple case of successive association, to the act of cognition, which is a simultaneous association.

We have already seen that contiguity-connections are involved even in the act of cognition. We should not be able to bring our visual idea of an object under the head of a familiar class, if the likeness-connections which arise first did not immediately call up contiguity-connections from earlier perceptions. But since these last remain wholly indefinite,—they may possibly belong to quite different and unrelated ideas,—the result is only a feeling of cognition : the object is regarded as a new presentation, but one belonging to a class of known ideas. So that, although the recognition-feeling is certainly akin to the cognition-feeling, there is a greater qualitative difference between these two than exists between the different shades of the recognition-feeling mentioned above. And the feelings are not only different in quality, as is to be expected when we consider the different conditions under which they appear, but also in intensity ; the feeling of recognition is generally far more intensive. Parallel with these differences, again, run differences in time-relations : the recognition-feeling comes later, and its gradual intensification can usually be clearly followed in introspection ; while the cognition-feeling generally seems to appear simultaneously with the impression. These are differences which are at once explicable when we remember the different nature of the underlying association-processes.

The two feelings correspond most nearly in all their attributes in the case of simple recognition of persons or objects familiar to our every-day experience. Here the total process of recognition is very closely related to that of assimilation. On the other hand, the feeling of recognition is most characteristically itself in the case of mediate recognition.

(There can be no doubt whatever, in the case of mediate recognition, that the secondary ideas upon which the recognition-feeling depends are actually present in consciousness.) Indeed, we are in this process not only conscious of the presence of the secondary ideas : we see quite clearly that the attendant feeling is bound up with them. But what shall we say of immediate recognition, when these auxiliary ideas are either not noticeable at all, or are only found by introspection after the act has taken place? We may assume either that they are below the limen of consciousness, and rise above it, if at all, only at a later stage; or that they are in consciousness throughout, but in so obscure a fashion as not to be perceived at first. The experiments made upon the different degrees of clearness of ideas with momentary transitory impressions, which we have already referred to (pp. 241 ff.), declare for the latter hypothesis; it can hardly be doubted that the auxiliary ideas are in consciousness, however dimly and obscurely. The different ways in which those experiments showed us that obscure ideas could make their presence known in consciousness correspond exactly to the various phenomena of immediate recognition. Sometimes it is possible, after the completion of the act, to represent its circumstances in detail; sometimes there is left only the indefinite feeling that the object was seen. So that the not infrequent impossibility of localising the recognised object in space and time is no proof at all against the presence of obscure auxiliary ideas. On the other hand, it is at least a very improbable supposition that ideas which have disappeared from consciousness can nevertheless exert an influence upon it in the form of a definite feeling. For if an idea that has disappeared can still excite a feeling in consciousness, it must possess positive attributes in its unconscious condition which completely resemble those attaching to it as a conscious process. The recognition-feeling, you see, is essentially the same whether the recognition is immediate or mediate, *i.e.*, mediated by clearly conscious secondary ideas. That supposition, in other words, would plainly commit us to the adoption of the untenable position that ideas which have disappeared from consciousness still persist in an unconscious condition, possessed of the same attributes as attached to them in consciousness: or, to put it a little differently, the vanished ideas would be indestructible objects, and not those dispositions towards the repetition of previous processes which the facts tell us they really are.

§ III

|The process of recognition was described above in general terms as a simple form of successive association. |To this must now be added that its various forms present a continuous serial transition from simultaneous to successive association. Immediate recognition, which approximates most nearly to the simple act of cognition, comes in all respects very near that of assimilation. The only indication of an ideational basis lying outside the recognised object is the characteristic feeling of recognition, which even here usually requires a certain time to arise. If these extraneous ideas come subsequently to clear consciousness, the simultaneous association passes over of its own accord into one of succession. And just the same may take place with the act of cognition. When an object has been assimilated by previous ideas of the same kind, one of two things may happen : certain particular secondary ideas, contiguous with the assimilating ideas, may enter into consciousness, or from the whole number of assimilating ideas there may be singled out particular ones, which subsequently attain to a greater clearness.

(If in a series like this an associatively excited idea is apprehended as having been previously experienced in its own special quality, the process becomes a successive act of memory.) Such a memorial act results directly from the different forms of cognition and recognition when the ideational acts which are in them given simultaneously or almost simultaneously are divided up into a clearly conscious temporal series. And just in these transition-cases we may perceive with especial clearness the condition of such a temporal analysis. This condition is given in the fact that the separate constituents of a total associative complex require periods of various duration to attain to clear consciousness. In immediate recognition and in the simple act of cognition there is no clearly noticeable succession, because no sooner is the impression given than the elements of the assimilating ideas which cohere with it are also apprehended. But even in mediate recognition there is not only a distinction between the secondary ideas and the principal impression, but a temporal dissociation of them in introspection : the secondary ideas come first. This time-relation may vary further in the most diverse ways. The principal idea may be assimilated first, the secondary ideas coming later as revivals of earlier experiences; that is a case of 'association by contiguity.' Or an assimilation-process of the usual form, involving an indefinite number of assimilating ideas, may run its course, and then certain of these ideas be held in consciousness by themselves; that is an ordinary 'association by similarity.' If this consists in an act of recognition of the kind described above, and if there are further associated with that other secondary ideas previously contiguous with the cognised object in space and time, the process is one of recollection.

The analysis of associations into temporal series depends in all these instances upon two conditions. First, one of two revived ideas may enter consciousness later than the other. This is realised in the case of mediate recognition, and in the simple memorial processes developed from it. Secondly, while several revived ideas may appear simultaneously in consciousness, and perhaps exert each its own influence upon the state of feeling, they may nevertheless be successively apperceived, coming one after the other to the conscious fixation-point. This is, of course, the case in all acts of recognition with temporal and spatial localisation. At the same time, this condition is probably quite often crossed by the other; it is in consequence of this temporal and spatial localisation that individual ideas become conscious at all.

The result of these paragraphs is to show that the connections operative in successive associations are the same as those which constitute simultaneous associations. The first thing always is, that certain elements of our ideas call up the like elements of other ideas. To these attach others which at some time or another were connected with them. And the whole process is perpetually shaped and modified by two influences,—the mutual intensification of the like and the mutual repression of the opposing elements. So that all our mental experiences are continuous and interconnected; the sum of ideational elements which consciousness has at its disposal forms an unbroken, interlaced and intertwined whole, within which each separate point can be excited from any other point by the mediation of

objects $a, b, c, d, e, f. \ . \ . \ .$ In a second and immediately following experimental series, each of these objects is connected with another, and in such a way that certain members of the series have attached to them the same secondary ideas ; *e.g.*, $a\,a$, $b\,\beta, c\,\gamma, d\,\delta, e\,a, f\,\gamma, g\,\beta. \ . \ . \ .$ After some time the principal terms of the series, $a, b, c, \ . \ . \ .$ are presented again in a different order, and without the secondary objects $a, \beta, \gamma. \ . \ . :$ *e.g.*, $f, b, a, g, e. \ . \ . \ .$ If sufficient time is allowed after each impression for the formation of an association, it is found that in a relatively large number of cases there are associated ideas from the same series, which were connected with identical secondary ideas ; *e.g.*, e will be associated to a, g to b, etc. This result is most striking when the principal ideas, $a, b, c, d. \ . \ . \ .$ are familiar objects (a house, a tree, etc.), and the secondary ideas, $a, \beta, \gamma, \delta. \ . \ . \ .$ arbitrary signs (*e.g.*, letters from a language with which the observer is unfamiliar). In this case the secondary ideas which are such effective aids to association are but seldom clearly remembered. And so the observer, when he is asked why he associates the definite idea e to another, a, replies that he cannot tell. If you recall what we have said above as to the efficacy of such secondary ideas in the acts of cognition and recognition, you will see that we must assume in this present case that the secondary idea a was obscurely present in consciousness and excited e, with which it had previously been connected, whereupon e, which had frequency and familiarity in its favour, appeared by itself in the foreground of consciousness. That is, the only difference between this process and an ordinary association is, that here certain members of the associative series remain unknown, with the result that the connection appears to be broken at the places where it is mediated by them. Occasions to this indirect form of association cannot be rare. We shall, therefore, be justified in always referring to it purporting instances of the spontaneous origin of ideas in consciousness, although, from the nature of the case, it is only exceptionally that we can prove the efficacy of the unapperceived middle terms.

LECTURE XXI

§ I

ASSOCIATIONS are due to the interconnections obtaining within the whole circle of our ideational consciousness. And it is a necessary corollary from this that all the relations into which ideas can enter with one another take their origin from those connections by likeness and contiguity which lie at the root of the association-process in general. But it is equally plain that the inference so often drawn, ' all ideational connections are associations,' is wholly unjustifiable. This inference has its source in an error with which we are already familiar, —that which transformed the forms of association into ' laws of association.' It rests upon the supposition that these forms are themselves elementary processes, whereas they are really, as we have seen, complex products resulting from the elementary connections by likeness and contiguity. But while we grant that all the possible interrelations of ideas are reducible to these two elementary types, (we do not mean to assert that the association-products can be exhaustively and without exception classified under the heads of simultaneous and successive association. There is one limitation which must not be disregarded. We never speak of association except where the elements which mediate the connection belong to a restricted circle of ideas. Thus assimilation is confined to perceptions of so homogeneous a character that they can be connected to form one single idea, complication to disparate impressions, which are inseparable concomitants in perception. The same thing holds of successive associations by similarity and contiguity, which, you remember, only differ from simultaneous associations in the (specially conditioned) temporal separation of the individual acts of ideation.

(Now there can be no question that we find processes in consciousness which are inexplicable in terms of these associations between similar or frequently connected perceptions, although certainly presupposing the existence of the association-products.) Let us consider for a moment one particular class of ideas, which is of all the processes of the same generic kind that which most resembles association as regards conditions of origin, but which nevertheless is quite characteristically different from it. I mean the ideas which we call *concepts*. If our eye lights suddenly upon the picture of a man, the first thing that occurs is an effect of assimilation : we cognise the picture as that of a man in virtue of its likeness- and contiguity-relations to previous perceptions. If these relations are of so individual a nature, that they suggest a similarity-association with some definite person, the originally indefinite act of cognition passes over into an act of recognition. There may then further attach to this a number of successive contiguity-associations ; we remember, it may be, the circumstances in which we saw the recognised face for the last time, or upon some special occasion, and so on. All these processes arise under obvious associational conditions ; but not one of them gives us the concept of man. It is, indeed, perfectly true that if the concept happens to be a very familiar one it may be more or less clearly present in the association. In the nature of the case, however, that is not necessary : the simple cognition of an object as known does not in any way imply a concept, although simple acts of cognition and recognition must inevitably precede the formation of concepts in general. How, then, do we distinguish a concept from an ordinary idea, which is cognised as agreeing with some other or with several others ? It is obvious that the conceptual quality cannot be a specific attribute of the idea which stands for the concept in consciousness. There is nothing in this, considered apart from its relations, to distinguish it in any way from any other particular idea. No ! the only distinguishing mark of the conceptual idea consists in the accompanying consciousness that the particular individual idea has only a *vicarious* value, and that therefore any other particular idea which belonged under the same concept, or was in any way to be thought of as an arbitrary sign of it, might be just as well put in its place. This accompanying consciousness is also attended by a characteristic feeling, a *conceptual feeling*, which is wholly different from the feelings of cognition and recognition, and points to a divergent ideational substrate. And this feeling, again, can only consist in concomitant ideational processes, running their course very likely in the more obscure regions of consciousness. In the present case, these processes must evidently be those which give to the concept-idea the peculiarities distinguishing it from other ideas. Processes of this kind are processes of *judgment*, as may be seen from the fact that concepts do not exist from the first in isolation, but obtain their conceptual significance only as elements in judgment. Whenever, therefore, we think of the concept in isolation, we are thinking of it as a constituent of an indefinite number of judgments ; no other case is possible. In this instance, then, the secondary ideas will be obscurely conscious judgments, in which the concept occurs ; and they will tend in particular to be judgments which somehow contribute to a definition. If, *e.g.*, we think of the concept 'man' in isolation, we have before us at the fixation-point of consciousness either the image of some individual man, or the word 'man' (as vicarious sign), or perhaps a complication of the optic and acoustic images. In its more obscure and outlying regions, and probably moving restlessly from this part to that, are a number of judgments in which the concept is involved, and of which only an occasional one may rise here and there to clearer ideation. But, obscure as they may be, they serve to invest the concept-idea with the consciousness of its vicarious significance, and with the resultant concept-feeling. You see, this significance and the feeling which accompanies it attach to the immediate introspective perception that in all these judgments the idea might have been a different one.

Now we have already seen that acts of cognition and recognition are also attended by secondary ideas, which on the one hand give rise to the peculiar feelings accompanying these processes, and on the other, if they happen to come clearly before consciousness, arrange themselves to a temporal series of interconnected ideas. But if cognition, recognition, and concept are so far alike, it is just at this point that we are able also to lay our finger upon the essential difference between the first two processes and the third. This temporal series is always a similarity- or contiguity-association, in which (as the name implies) each idea persists as an independent unity. For objects which resemble one another, or which are contiguous in space and time, may certainly be combined to form more complex ideas; but every part of the resulting compound process is still independent, so that if it becomes dissociated from its companions, it continues unimpaired in consciousness. But with concept-ideas, as with all conscious content which belongs to a logically coherent thought-process, the case is very different. (The significance of the individual is now entirely dependent upon the whole of which it forms a part.) Dissociated from this, it no longer possesses any significance of its own. Or if it seems to, the explanation is the same as for the concepts which we think of in isolation; we can confer a significance upon it in so far as, for this special purpose, we leave the logical connection in which it belongs undetermined. Thus the concept 'man,' when we think of it without reference to any context of judgment, can only have this significance: that it may be subject or predicate of a large number of judgments. Only as such an indefinite element of a logical thought is it a 'concept' at all. In all other cases the corresponding idea would be simply and solely a concrete particular idea.

An objection might be raised to this train of argument. 'It may be quite true,' you will say, 'that concepts and their connections are in many respects different from ordinary associations arising between particular ideas. But that is no reason for refusing to reduce them, and all logical thought-processes with them, to associations in the wider sense, perhaps associations of a peculiar and more complex nature.' The objection does not hold. The differences between the two kinds of conscious process are characteristic and fundamental, as evident to subjective perception as they are manifest in the uniformities objectively discoverable. To employ one name for both, and so to suggest the view that the processes are of the same kind, would serve not to clear, but seriously to embarrass, the path of investigation.

§ II

The most obvious *subjective* mark of the intellectual conscious process, as distinguished from association, is the accompanying feeling of *activity*. The very best means of arousing a purely associative,—*i.e.*, not logical,—train of ideas is to assume the most passive attitude possible, to get free of that activity of thought which requires volition to initiate it, and is attended by the activity-feeling. The question what this activity is, and in what the feeling of activity consists, has been already answered in our discussion of voluntary action (Lecture XV.). In virtue of its subjective characteristics, intellectual activity falls at once under the concept of internal voluntary action, or *active appercep-tion*. In this sense, then, we may distinguish intellectual pro-cesses from associations, on the purely psychological basis, as *apperceptive connections of ideas*. By a 'free' or 'voluntary' action, we do not, of course, here, any more than in our previous discussion, understand an unconditioned action. The phrase simply means that changes in consciousness are involved which are not explicable by reference to individual ideational connections, but only as resulting from the general tendency of all our conscious content at any given time,—in the last resort, that is, from the whole of the previous development of consciousness. If we term the result of this total development our 'self,' we must regard this self as the cause of all in-tellectual processes.

These considerations make it self-evident that the range of the subjective relations to which these internal processes owe their origin may vary enormously. No excitation, of course, can affect at once the whole number of our acquired disposi-tions. For the action to take rank as a voluntary intellectual activity, it is only necessary that groups of ideas be impli-cated which do not stand in any obvious associational relation to those directly preceding them. And, in the same way, it is inevitable that associations intrude upon the intellectual functions. And with reference to this it is especially signifi-cant that the intellectual ideational connections, once formed, themselves enter into contiguity-relations, and can therefore be revived in the form of external association. In this case, of course, there is no trace of the feeling of activity which else-where accompanies the intellectual processes. ꜱSuch a transition from apperceptive trains of thought to association is of the very greatest importance : it facilitates constructive mental work in a high degree. In this sense it forms one of the principal constituents of those various practice-processes which gradually enable us to perform voluntary actions, at first matters of intention and reflection, as appropriate mechanical reactions to definite external stimuli. So universal is it, that for the accomplishment of external acts of will the interposition of a voluntary decision is only necessary at certain critical moments ; their detailed execution is relegated to the mechanism which practice has perfected. And, in the same way, the active work of thought in the intellectual processes becomes more and more confined to the essential moments in the flow of thought, while our thinking on all subordinate points goes on with no other aid than that of logical association. The more practised in thinking we are, the more numerous become the middle terms which suggest themselves, and the more real force and energy has thought to bestow upon decisive issues.

With these subjective characteristics are united not less im-portant *objective* peculiarities, distinguishing the intellectual processes from association. They are given in the totally different character of the temporal succession in the two cases. In successive association one idea follows on the other as the various likeness- and contiguity-connections operative in the case determine. Each particular idea retains its independence. ꜱAnd since in a long associational series a new idea is regularly associated with only one of its predecessors, generally the most immediate, the series is subject to abrupt changes of the most varied character. Beginning and end in particular may be wholly unrelated, however complete the chain that connects term with following term. In contradistinction to this, the intellectual processes always begin with *general ideas*. These differ from the complex ideas resulting from simultaneous association in that they do not consist of connections which (like position in time and space) appear as direct objective attributes of the idea ; but the relations existing among their constituents are regarded as conceptual determinations, into which the complex object is analysed by the activity of ⟩ thought. The basis of such a general idea, however, is always a complex idea associatively produced. Thus the impression of a red house gives rise, by means of associative fusions and assimilations, to a complex visual idea. This only becomes a general idea when the red colour is separated from the idea of the house as such. For then attribute and object are conceptually thought, and brought into mutual relation in the general idea.

§ III

The first products of intellectual activity, then, are *simultaneous general ideas*. · Their only difference from ordinary, associatively formed ideas consists in this,—that the ideated object is regarded as analysable in terms of arbitrarily selected thought-relations. As soon as one or more such analyses have been carried out, their simultaneous connection gives rise to a train of thought. This process is best seen in the acts of logical thought in the narrower sense ; *i.e.*, in the processes of judgment which are expressed in language. These are completely different, even in outward form, from the associative series. In the latter one idea joins on to another indeterminately ; but logical thought is governed by a *dichotomic* law, admitting of no exception, except when associations intrude upon the apperceptive train of ideas in the manner specified above. The clearest expression of this law is to be found in the grammatical distinction of the parts of the sentence. The division is either simple, as in the simple sentence, where subject and predicate each consist of one single idea ; or it is multiple, as in all kinds of compound sentences, where each of the principal constituents can again be subdivided on the same plan, the subject into substantive and adjective, the predicate into verb and object, verb and adverb, etc.

But this external law or uniformity is the result of internal conditions. Thought is always a discriminating and relating activity. That its analysis follows the above rule is due to the fact that it separates the constituents of a general idea only to bring them at once into some mutual relation,—a relation determined after the comparison of numerous ideas, partly alike and partly different. The general idea of the red house and the resultant judgment, 'the house is red,' can plainly not arise until many ideas of houses have been formed, with varying colour-attributes. Then, and then only, will it be possible in a particular case either to abstract the attribute from the object of experience, or to refer it to an object thought as existing independently of it.

It might perhaps be objected to this discrimination of the general idea from the judgment that when once we have separated the former from the ordinary complex idea we have *ipso facto* constituted it a judgment. It is impossible, *e.g.*, to think of an object as logically connected with any of its attributes without this connection finding immediate expression in a judgment. But certain as it is that the simplest judgments (like 'the house is red') can hardly be distinguished in fact from their corresponding general ideas ('red house'), the possible confusion ceases with them. For consider those acts of thought which presuppose a continued process of subdivision. When we are about to express a complicated thought, what is first of all in consciousness is the total thought in the form of a general idea. But it is quite impossible to say of this that it is identical with the judgments into which it is analysable. No! we can perceive well enough that while the whole thought is already there as a general idea before its expression, its separate constituents are only raised to clear consciousness in proportion as their analysis is actually carried out. General ideas are, therefore, the more indefinite the more comprehensive they are, the more numerous the acts of judgment they require for their complete determination.

We must, however, remember that the logical judgment is not the only,—more than that, it is not the original,—form which apperceptive ideational processes take in consciousness. The commonest case is, that general ideas of a more or less comprehensive character are consciously analysed in the form of sense-perceptions. The unitary character of the complex process means even here that when a general idea is held in mind for a considerable time each act of division attaches closely to its neighbour. But the purely perceptual nature of the content and the absence of any formulation of it in logical terms render the dichotomic law subordinate to the vague impression of an analytic process, standing in a definite context and setting out from a single general idea. To this must be added that it is a more obvious fact in such instances than it is in conceptual logical thought that ideational analysis implies an ideational *explication.* The products of analysis, at first but obscurely apprehended, oftentimes fail to obtain a more clear and distinct content until they have entered into new associations. This perceptual form of intellectual elaboration is the *activity of imagination.* In other words, imagination is in reality a thinking in particular sense-ideas. As such it is the source of all logical or conceptual thought. But it continues to exist alongside and independently of the latter both in the unguided play of fancy of our every-day life, and in the finished creations of the artistic imagination.

§ IV

To give a complete account of the intellectual functions would extend beyond the limits of our present undertaking. Part of that account,—the description of the conceptual or logical forms of thought,—lies within the province of logic ; part of it,—the consideration of imagination as a form of intellectual activity,—within that of æsthetics. But it is of psychological interest, and desirable for the right understanding of the relation of association to intelligence, that we should pause here to cast a glance at the alterations produced in the train of ideas and the mental processes to which it gives expression by the different forms of *mental disturbance.*

The most clearly marked and most permanent of these disturbances are the various kinds of insanity. The particular forms of insanity, as you know, are so many and so different, that pathological psychology has as good a claim to rank as an independent discipline, beside normal psychology, as has the pathology of the body to be separated from its physiology. This latter separation was long ago effected ; the two disciplines are independent sciences, quite apart from the fact that pathology has also practical applications. Add to this that every mental disturbance involves, besides alterations of the intellectual processes and associations, other and equally

fundamental modifications of the mental life : especially derangements of sense-perception and of the emotions, as
compared with which the ideational disturbances often appear
as merely secondary consequences, though it is true,—and you
will understand it when you consider the inextricable inter-
dependencies of all these mental processes,—that these alterations of ideational content will react again upon the affective
and conative side of consciousness. Here, however, we shall
leave all this out of account, and simply look at mental disturbance from the single point of view of alteration in the train
of ideas. And in this respect we shall find it in its fundamental character, despite differences of detail, uniform and
homogeneous throughout.

This much may, however, be said with regard to the alterations produced by mental disturbance in the spheres of sensation
and emotion : that, regarded purely psychologically, they
embrace the most various divergences from the norm in all
directions,—from the apathy of idiocy, which is only to be
moved by the most intensive sense-impressions, to the enormous
excitability of delirium, when the slightest external or internal
stimulation suffices to call up hallucinations and misleading
illusions ; or from the deep depression of melancholy, which
clouds the present and the past alike, to the bursts of maniacal
passion and the immovable cheerfulness of paralysis. The
deviations from the normal train of ideas will, of course, be
correspondingly different ; its course will be too quick or too
slow, too crowded or too sparse, as the case may be. And
these deviations stand in an intimate relation to emotional
changes ; they are not really disturbances of which these latter
are independent, but the two derangements are both implied in
any mental disturbance. In melancholy and paralysis alike,
the train of ideas moves haltingly, arrested by definite impressions and memories, only that the affective colouring is wholly
different ; while in mental exaltation and mania it is accelerated,
springing from topic to topic without order or control.

All the more noteworthy is it, then, that, despite all these
differences, the departure from the norm is constant in one
particular respect,—in regard to the relation of associations to
the intellectual processes. If there is any single criterion of

mental disturbance, it is this,—that logical thought and the voluntary activity of the constructive imagination give way to an incoherent play of multifarious associations. If the disturbance has not gone very far,—*e.g.*, in the first period of a slowly developing disease,—this change in conscious content may be hardly noticeable, either because long intervals of sanity interrupt the progress of the derangement, or because the latter appears to confine itself to some particular ideational and affective connection. But even in these cases, borderland-cases, midway between the normal and the abnormal, where a good natural constitution may more than hold in check the encroachments of a disease,—even in such cases there can be no doubt that from the moment at which consciousness is overcome by the disturbance the normal equilibrium of association and active apperception is once for all destroyed.

The most general way in which this disturbance of equilibrium manifests itself is by a *defective concentration of the attention.* It arises from the liability of the intellectual processes to be continually interrupted by sudden associations. And the states of mind in which the patient is always concerned about particular impressions or feelings are only an apparent exception to this rule. When the melancholiac broods incessantly over the crushing sorrow which he supposes himself to be experiencing, it is not that he voluntarily directs his attention upon it, and so controls the direction of his thoughts ; his mind is dominated by an ever-present group of intensively toned ideas, against which the will struggles often enough, but struggles in vain. In the condition of mental exaltation we have an unnaturally strong excitation of the sense-centres, giving to the associational contents the character of external sense-impressions ; so that these associations, which normally make against the influence of the active attention, are rendered unusually powerful. Ideational assimilation plays an especially large part in this case. /In normal mental life the assimilating elements are just strong enough to render acts of cognition and recognition possible ; in hallucination they become so potent as to throw the sense-impression into the position of a mere external accident, which sets in motion ideational tendencies that have not the remotest resemblance to itself.

Notice the way in which the insane express their thoughts. Their language is abrupt ; it will vacillate in a purposeless way between the most heterogeneous subjects, or it will come back again and again to the same topic, for no assignable reason. All this is not hard of explanation, if we suppose a lack of voluntary control over the unruly crowd of associations. Of course, this lack of control may exist in very different degrees, from the eccentricity of thought which just oversteps the norm to the wildest flights of fancy, in which thought follows thought without the possibility of the mind's dwelling for even a short time upon any. The last stage is a total incapacity to frame any judgment that is at all complex. The patient begins to utter some sentence. But his attention is taken captive by new sense-perceptions, or by some extraneous circumstance, perhaps by associations aroused by the sound of his own words. So another and heterogeneous train of thought is set going, only to be interrupted in its turn by still other associations. And so the mad hurry goes on, until mental exhaustion puts a temporary end to the tangled drift of ideas.

When we consider the vast importance of language for the development of thought, we can understand that associations of words and sounds play the leading part among the thousand forms of ideational involvement in the fancy-flights of the insane. Words of like sound are heaped together in meaningless confusion in the middle of a sentence ; or a word will suggest some totally heterogeneous thought, in which it also happens to occur. So that the speech of the insane is our best opportunity to observe 'similarity'- and 'contiguity-associations' in simultaneous and successive form, and with the utmost variety of content. But it not seldom happens that when the flight of ideas is covering a large range of topics some particular word will call up not another word, but perhaps simply a series of articulate sounds, belonging, it may be, to like-sounding words, or contained in others which have accidentally become connected with the first. When this is the case, the language of the insane becomes an unintelligible jargon, composed of articulate sounds that occur in real speech, but putting them into quite new connections. There may then arise in the patient's mind the delusion that he is speaking a previously unfamiliar language, an idea which in its turn may give rise to other delusions. But if we consider with some attention the bewildering confusion of sound-conglomerations whose genesis we have traced, we shall see that even here the influence of practice, so important for association in general, is distinctly traceable. The more frequently a particular sound-complex has been repeated, the greater is the inclination to utter it again. More especially does it tend to enter into fresh associations, whether with other sounds or with external objects ; and so there may arise a dialect of insanity which in certain of its constituents possesses all the characteristics of a new-formed language : certain sounds or sound-complexes become determinate symbols of definite concepts. At the same time we can hardly say that this language is the invention of the insane. It owes its origin to the blind chance of associational activity ; and this continues to modify it in the most haphazard way.

There is scarcely anything more interesting to the psychologist than the observation of the gradual decay of the intellectual functions as manifested in the language of the insane. Written records are even better for this purpose than oral speech, for the torrent of words, which must flow with incredible rapidity to keep pace with the flights of fancy, cannot readily be followed. There are printed works extant in all literatures that declare themselves as products of a deranged mind. In them we have most beautifully demonstrated the separate trains of association, their intrusion upon the course of logical thought, the gradual disintegration of this latter, and not infrequently the influence of hallucinations and fantastic misinterpretations of sense-impressions. The final stages of this mental decay are, of course, usually lacking ; but all the rest are there. But I chanced once to pick up a book,—'privately printed for the author,' as you can imagine,—in which nearly every stage of the process of disintegration, from one end to the other, could be clearly traced. Its opening sentences are correct in form and expression, although their content shows from the first the beginnings of abnormal thought. Then follow, with increasing frequency, descriptions of unmistakable hallucinations and linguistic solecisms, while the intrusion of disconnected associations makes itself more and more evident ;

till finally, on the last few pages, there is not a single sentence that is brought to its correct grammatical conclusion.

When we subject these phenomena to a careful analysis, especially as they are exhibited in such permanent form in productions of the diseased mind which are more accessible to investigation than the spoken word, we see at once how inexact and superficial it is to speak of them as due to 'a lowering of the mental functions.' As regards the flights of fancy, for instance, the mental functions are rather raised in a particular direction than lowered. The normal mind has not at its disposal anything like such an abundant supply of associations as is not seldom met with in mental derangement. At the same time, it is in this very motility of association that the germ of decay is to be looked for. It is the unfailing symptom of the weakening of voluntary control over all those connections of the ideational elements which the manifold ramifications of the associational network render possible in consciousness. We may safely say that no intellectual function is possible until these relations and connections have been collected from previous impressions. But, nevertheless, mental activity only becomes intellectual when the total force resulting from the whole sum of these previous experiences, the will, controls and gives definite shape to the associative material lying to its hand. In relation to these associations, the will is at once an active and an inhibitory force,—it furthers the connections concerned in the predominant interest of the moment, while it inhibits all that might draw the attention some other way. You see, then, that a sane man may voluntarily call up experimentally, as it were, a train of ideas which very nearly corresponds to that of the insane. He has simply to repress the regulative and inhibitory function of the will with regard to the associations which crowd into consciousness. Put yourselves in this condition, and write out the thoughts and ideas which come to you 'of themselves'; you will have an inextricable tangle of fragments of half-completed thoughts, of chance impressions, with here and there a new-formed association,—a picture that you might easily take to be that of a deranged mind.

It is true that this luxuriant growth of associations, which

Y

dwarfs the intellectual functions as creepers cramp and stunt the trees that they entwine, is not a permanent condition. If the process of disturbance continues, the associations become increasingly restricted to fewer and more stable ideas, which repeat themselves over and over again. These 'fixed' ideas, called up in the first place by the particular tendency of the diseased mind, become more and more insistent as the process of associative practice keeps even pace with the general disintegration. When these fixed associations become exclusively dominant in consciousness, the influence of the will is destroyed once and for all. At the same time, the patient becomes less responsive to external stimuli. The emotional depression consequent on disquieting hallucinations, the affective disturbances caused by painful impressions, disappear as the sensibility becomes less, and the general mental dulness greater ; and give way to a mood of uniform cheerfulness or indifference. And with that is reached the final stage of mental derangement, the saddest for the observer, the most happy for the patient himself.

§ I. Dreams. § II. Sleep-walking. § III. Hypnotism and Suggestion. § IV. Autosuggestion and Post-hypnotic Influence.
§ V. Errors of the 'Hypnotism-psychology.' § VI. Theory
of Hypnotism and Suggestion.

§ I

WE have seen that a person of sound mind is able of his
own will to give himself up to the play of association,
and so induce a state of mind which more or less resembles
the ideational condition of the insane. That is not all, however.
We are all of us normally subject to experiences which bring
us still closer to a realisation of mental disturbance. One such
condition of what we may call normal temporary insanity is
that of *dreaming*.

In every relation of life we find the *omne ignotum pro mag-
nifico* borne witness to. Mankind tends always to regard the
unaccustomed as more wonderful than the usual and normal.
The glamour of mystery surrounds the unfamiliar, just because
it is unfamiliar; while the commonest phenomena, which so
often present the really most difficult problems, are looked
upon as matters of course. Former ages regarded the insane
as favoured of Heaven and illuminated above their fellows, or
as possessed of devils,—according as the pendulum of circumstance swung. And even to-day the subjective ideas of these
unfortunates are at times affected by such thoughts : thoughts
which arose in the first place from the contemplation of mental
derangement in its various forms. Even after this view had
died out as regards insanity, dreams were still invested with
something of the miraculous. The popular belief in premoni-
tion by dreams we need not notice. But there are still philosophers who incline to think that when we dream the mind

has burst the fetters of the body'; and that dream-fancies transcend the activity of the waking consciousness, with its close confinement to the limits of space and time.

An unprejudiced observation of the phenomena of dreaming must convince us that, beautiful as these theories are, they are pretty nearly at the opposite pole from truth. When we are awake, we are generally able by an effort of will to overcome petty bodily distractions without much difficulty. (The dreamer is absolutely at their mercy; the train of his ideas is diverted by every chance impression that affects his senses, by every accidental association.) The commonest causes of the most vivid dreams are indigestion, palpitation of the heart, difficulty of breathing, and troubles of that sort. It is a matter of dispute whether there is any such thing as dreamless sleep; and it will probably always remain so, seeing how easily we forget what we have dreamed. But it is certain that if such a state does occur, it will be most easily found in cases where all bodily stimuli are wanting, or where they are at least too weak to call up ideas.

The physiological nature of sleep we need not stop to discuss. But little is known about it, if we except the general fact that sleep is one of those periodic vital phenomena which originate without exception in the central nervous system. There is also one fact that is psychologically important in a teleological consideration of life: that during sleep there takes place a reparation of the forces expended in the waking state by the functioning of the various organs. Sleep is not seldom hindered in the performance of this important duty by its attendant, the dream. Vivid and unrestful dreams will detract from the refreshing effect of sleep. (The actual phenomena of dreaming, however, make it probable that its greater or less interference with sleep is due to a greater or less degree of abnormal irritability in the sense-centres of the brain or in particular parts of them; this in its turn being caused, perhaps, by a disturbance of the intracranial circulation) · A' confirmation of this view is found in the fact that pathological alterations in blood and blood-supply (such as occur, *e.g.*, in fever) may considerably intensify dream-phenomena, and even occasion similar mental conditions, those of febrile delirium, in the waking state.

We have already, then, in outline the essential characters of the dream-idea. It is a hallucination ; its intensity is as great as that of a sensation given in direct perception, and it is as such that it is regarded by the dreamer. The principal dream-constituents are memorial images, but memories which are interwoven altogether at random,—whether they refer to the immediate past or to some more remote experience, whether they belong together or are wholly unrelated,—by the unregulated play of association. Dreaming has, therefore, at first sight, some resemblance to the normal activity of imagination ; it tends to combine memory-ideas in new and unaccustomed ways. But it entirely lacks that purposive arrangement and grouping of ideas which is the one criterion for the discrimination of imagination from the activity of memory.

The world of memory and the world of dreams are alike dominated by ideas of sight. Auditory ideas are also found to occur. No other sense appears to furnish dream-material to any considerable extent except when directly stimulated from without. Of course, there may be direct external excitation in the cases of sight and hearing also; and indeed it is probable that dream-ideas are aroused in this way far oftener than is generally assumed. It may even be that the predominance of the visual idea in the world of dreams is to be accounted for not solely by the very great importance of sight for memory, but also by the peculiar nature of the eye, which is more exposed than any other sense-organ to the continual operation of weak external stimuli. If we look attentively at the darkened field of vision of the closed eye, we notice an unceasing appearance and disappearance of light-phenomena : now single points of light will shoot like meteors from side to side ; now a veil of twilight is drawn across the blacker background ; now again parts of this appear in the most brilliant colours. There can be no doubt that these phenomena persist during sleep to call up the memory-images that resemble them, and of which consciousness possesses so large a store.

Dreaming, then, is related to the train of ideas in the waking consciousness in that its proximate cause is usually some external sensory excitation, to which memorial images readily attach themselves. But in two respects the processes are very different. The ideas called up by the sense-impressions are more or less fantastic illusions; and the consequent successive associations do not possess the character of ordinary memory-images, but of hallucinations: and, like these, they are taken for actual experiences. It is, therefore, very seldom,—if at all, only at the moment when sleep is passing over into waking,— that we dream of remembering anything. Dreaming is as immediate as any experience can be. It shows no trace of the usual discriminatory marks of imagination and reality.

If dreaming reminds us, from this point of view, of mental disturbance, it has one characteristic which does not appear in the same degree in any of the manifold forms of insanity,— its complete restriction to the ideas of the immediate present. The illusions or hypostatised memorial images of the deranged mind can never do more than partially prejudice the normal apprehension of an object; while there is a clear distinction drawn between them and the customary activity of imagination and memory. And dreaming occupies a peculiar position in yet another respect. If we look at the part played in it by hallucination, we shall be inclined to parallel it with the initial stages of certain mental disturbances, which bring with them an abnormal excitability. But in the incoherence of dream-ideas, in the clouding of judgment and the lapse of self-consciousness, we have a series of phenomena which only find a parallel in the most extreme forms of mental disorder. Probably the greater number of dreams come and go without involving any really intellectual process at all. The dreamer acts, or looks on at action, without ever making his experience the subject of reflection. It is generally when dreaming is, as it were, dovetailed in with waking,—just before we fall asleep or shortly before waking,—that a real activity of intelligence is noticeable: we make dream-speeches, or carry on dream-conversations. But the language used is of a curiously mixed kind. Sometimes there is an almost normal capacity of connected expression, though when we analyse we find that the dream-speech consisted entirely of familiar phrases and current turns of language. Sometimes there is no normal connection in the thoughts uttered ; the whole is a strange medley, the judgments meaningless, the conclusions wrong. It may be that the confusion of thought extends even to the sound of the words employed ; so that we have new formations of articulate syllables, like those which occur in the talk of the insane. And these are connected, by the way, with the same set of subjective ideas ; the dreamer, like the maniac, thinks that he is speaking fluently a new and unknown language, or perhaps some real language, which he has actually studied, but only very imperfectly mastered.

All these phenomena tend to show that the relation of the intellectual functions to associations in dreaming is altered very much as it is in the more advanced stages of mental derangement. The control of the will over the mob of ideas and feelings has been abrogated. The dreamer is completely at the mercy of associations instituted by accidental external impressions. And, in addition to that, the hallucinatory character of dream-ideas gives them their peculiar ability to pass for real experiential events.

§ II

A special kind of dreaming, which is generally a symptom of an abnormal excitability of the nervous system, is *sleep-walking.* It is a dream carried one step beyond the hallucinatory conversations mentioned above. For sleep-walking consists simply in this : that the connections between conscious states and external voluntary acts which hold in our waking life are realised to the same extent during sleep. And since, of all these connections, that of idea with the muscles subserving language has become the most customary and automatic, we can understand that these will be most often and most readily exercised. Sleep-walking, then, like dreaming, has nothing mysterious about it ; it is simply an event of infrequent occurrence, which for that very reason has sometimes been looked upon as inexplicable. Indeed, when we consider the character of the connections obtaining between sensations and the movements which they stimulate, it becomes rather a matter for surprise that dream-walking is not a far commoner phenomenon than it is. We may explain the facts in one of two ways. We can assume either that the sensory centres are more exposed during sleep to the operation of the various causes of excitation than the motor centres are, or that the latter are in general subject to certain inhibitory influences. However this may be, there can be no doubt of the very great utility of this separation of the world of dream-ideas from the sphere of external action. Think what would happen if we actually did everything that in our dreams we imagine we are doing!

But sleep-walking presents other differences from ordinary sleep. To mention in particular is an increased excitability of the sense-organs in presence of external stimuli. The sleep-walker sees and cognises external objects up to a certain point. But his dream-perceptions are of an illusory character, and so he misinterprets them : he may take the window for the door, or the ridge of the roof for a boulevard. While, therefore, he can perform simple acts, and especially such as have become more or less automatic by practice, he will hardly go beyond these. The tales that are told of wonderful dream-performances,—the sleep-walking mathematician who solves a difficult problem or the schoolboy who regularly does his work in this very convenient way,—we may consign without more ado to the limbo of the fabulous. No reliable observer has ever confirmed reports like these ; and they conflict with all that we know of the nature of dream-ideas in general.

§ III

It is but a short step from sleep-walking to phenomena which have lately formed the topic of much discussion,—the phenomena of *hypnotism.* The principal condition for the induction of the hypnotic state is a vivid idea of a passive surrender of the will to that of some other person, who is able to influence his subject by words, acts, or gestures. An abnormal excitability of the nervous system favours this influence. But the experiment will, as a rule, succeed in the long run even with persons who at the outset were proof against it ; or, in other words, the frequent repetition of the state facilitates its induction, and furthers the passage of the initial stages into the higher.⌉ Other conditions which have often been regarded as auxiliary or as the sole causes of hypnosis,— especially weak and uniform stimulation : the steady gazing at an object, the 'magnetic' stroking of the skin, and so on,—are, obviously, only secondary and indirect means to the end. They serve in part to weaken the active attention, in part to arouse the idea of an influence militating against the independence of the subject's volition. It was proved by the cures worked by Mesmer and his successors in terms of 'animal magnetism' (which is in essentials just the same thing as hypnotism) that it was necessary for the success of the experiment that the patient should *believe* in the efficacy of the strokings and other manipulations, but that this belief was sufficient ; so that the passes and all the rest of it might be left out, if only the subject thought that they were there.

The symptoms of hypnosis vary according to the stage which the condition has reached, and the susceptibility of the subject. We can distinguish three degrees of it, which, from their resemblance to the corresponding stages of normal sleep, have been termed *drowsiness, light sleep*, and *deep sleep*. The similarity is, however, confined to merely external characteristics, and in particular obtains only for the appearance and behaviour of the subject before definite influences are allowed to play upon him and assume control of his perception and volition. It is this possession, as we may term it, which really differentiates the two states. ⎸Even in the light hypnotic sleep 'suggestion,' —the influencing of ideation and volition from without,—begins to play a part.⎞ The hypnotised subject cannot open his eyes of his own accord, cannot perform any voluntary movement whatever, though he recovers this power the instant that an action is suggested to him by a word of command from the hypnotiser. The skin is anæsthetic, which is never the case in sleep ; so that needle-pricks are often not sensed at all, or, if they are, only as pressures from blunt points. Conjoined with anæsthesia we find all the various phenomena of 'automatic reaction to command.' The subject executes movements that are suggested to him, puts his limbs in the most uncomfortable positions, and keeps them there until another command suggests relaxation. In many persons a rigid or tetanic state of muscles appears even in the absence of suggestion. On waking,—the hypnosis can be dispelled instantaneously by a word from the operator,—the subject generally retains some sort of cloudy recollection of what occurred to him during his sleep.

This power of recollection serves to distinguish the lighter from the deeper hypnotic sleep,—*somnambulism*, as it is also called,—after which the memory is a simple blank. All the other symptoms are also much exaggerated. The automatic reaction in particular extends beyond movement to sense-perception. The somnambulist will objectify any ideas that are suggested to him. There are two proofs of the intensification of ordinary fancy-images, arbitrarily aroused by the suggestion of the hypnotiser, to hallucinations: first, this confusion between imagination and reality, and secondly, the production of complementary after-images of the suggested perceptions. Here is a particular observation to illustrate the second fact. The operator calls out to the subject, ' Look at that red cross on the wall!' When the latter has found it, he is told to look on the ground, and asked, ' What do you see there?' The answer is, ' A green cross.' That is, the after-effect of the hallucination is just what that of the actual impression of a red cross would have been (*cf.* p. 109). Illusions of taste are also very easily induced. The subject will take a glass of water for champagne, and drink it with every sign of satisfaction ; while, if he is told a moment after that he has been drinking ink, he will spit it out with equally evident marks of disgust. At the same time I am not sure that all these are cases of real hallucination. When we consider how very rarely hallucinations of smell and taste occur in dreaming, we shall be inclined to think that in these instances of suggestion too the sensations composing the aroused ideas may be confined to those of mimetic movement.

There are many other phenomena, manifested especially in the somnambulistic stage, which have often been employed to shroud the hypnotic sleep in the veil of mystery and wonder. Thus a suggestion may very readily lead to the formation of the idea that the subject is to obey the suggestions and commands of one person only, generally of the operator, while remaining

indifferent to any attempt at influence on the part of others.)
There then arises what the animal-magnetism school term the
rapport of the medium with the magnetiser. This is really, as
we have indicated above, only the result of a secondary suggestion, which is favoured by the special circumstances of the
induction. The exclusive direction of the attention of the subject upon the operator is itself at times sufficient to produce
this *rapport*, even without any express command, especially if
he is always put to sleep by the same person, as is of course
the case in the instances of 'magnetic' cure. There is no
intrinsic reason, that is, why the hypnotic subject should not be
accessible to other influences. Without them we should be unable
to explain the fact of *autosuggestion*. (Autosuggestion implies
an abnormally strong tendency towards hypnosis.) Frequent
hypnotising may lead in the long run to an irresistible passion
for the hypnotic sleep, in which case the impulse to obtain it
acts like the morphine-habit or habituation to any particular
stimulant or sedative. The confirmed hypnotic will try in every
possible way to procure the enjoyment which he craves. And
he has in autosuggestion, when he has once discovered how to
use it, a means lying always ready to his hand. By voluntarily
arousing and fixing the idea that he will fall into the hypnotic
sleep, he can induce all the phenomena that ordinarily accompany it. It appears, moreover, that in the condition as brought
on both by autosuggestion and by suggestion from without
there may be a continuous memory from sleep to sleep, such
as is sometimes observed in dreaming and in certain forms of
periodical mental derangement. The psychology of the act
of recollection gives us the key to the explanation of this phenomenon. It is wholly unnecessary to assume the existence of a
mysterious mental double, the 'other self' or second personality,
or to set up any other of the fanciful hypotheses so plentiful in
this field. There is, as you know, one invariable condition of
the occurrence of an act of recollection :—if we wish to bring
about a complete renewal of a past experience, we must reproduce the whole of the ideational and affective content of consciousness which characterised that experience. Now, since there
is a great difference between the waking consciousness and that
of the hypnotic sleep and since the difference increases with

the progress of that sleep, we can readily understand that its suggestions will be forgotten on waking; while we also see how the recollection of those suggestions will be possible when the hypnotic state is renewed.

Many cases of the *post-hypnotic effects* of suggestion appear explicable on these two principles of the conditions of memorial functioning and of autosuggestion. When, *e.g.*, it is suggested to a somnambulist that he will perform a definite action at some particular time of another day,—take a certain walk, execute a given order, mix a special draught and offer it to some third person who is described in detail,—the idea recedes into the background of consciousness at the moment of waking; but, as the appointed time draws near, comes up again in obscure form, as the 'feeling' that something particular is going to happen. This idea of the time of the performance of the action, usually furthered by special insistence upon it in the primary suggestion, is still more intensified, in accordance with the general laws of association, when the time actually arrives; and from this moment the indefinite idea of a commission undertaken,—*what* commission is still wholly unknown,—exerts an autohypnotic effect. There follows a partial revival of the somnambulistic state, sufficient to re-excite the memory of the suggested ideas, and at the same time to exclude any consideration of the motive or purpose of the action. This is then performed in an automatic, lethargic condition, which, however, as an imperfect repetition of the previous complete somnambulism, does not preclude the memory of it in the normal waking state. Asked why he did so-and-so, the awakened subject is unable to give any explanation; or simply says that he could not help it, or that it was suggested to him in sleep, the latter answer giving clear evidence of a persistence of the suggested ideas in an obscure form into the waking state. (Where the post-hypnotic effect takes place immediately after the awaking from somnambulism, it is probably the direct result of this persistent operation on the part of the suggested ideas.) It looks sometimes, under these circumstances, as though the subject were not really fully awake; his behaviour is more like that of a person just aroused from ordinary sleep. Indeed, those of you who are subject to very vivid dreams may now and again observe quite analogous phenomena when you are suddenly awaked from one. You think and act for a short while in terms of the preceding dream-ideas; but waking impressions keep continually mixing with them, until at last they gain the upper hand, and you are fully awake.

But it is plain enough that autosuggestion cannot be the exclusive cause of post-hypnotic effects. When it is a question of executing some simple order directly after waking, or after the lapse of a definite time, the actions very frequently follow without any symptom of a partial relapse into hypnosis. We must therefore suppose that the suggested idea, with its corresponding motor impulse, may be latently operative in consciousness; so that it will come to the conscious fixation-point either immediately after waking or at the time suggested. It will then, like every other impulse, continue to exert its influence until the action results, or it is inhibited by opposing forces of the waking consciousness, themselves impulses, sensory or intellectual. This view is confirmed by the frequent occurrence of phenomena of resistance, often successful, to the performance of the suggested action. At the same time, there is positive proof that even here consciousness does not at once return to its normal condition,—proof furnished by post-hypnotic hallucination. The awakened somnambulist, in obedience to suggestion, may see the operator in fanciful costume, perhaps with a red cloak over his shoulders and horns upon his head; he will find a flower in his buttonhole which is not there; or he will overlook a doorway, and declare that the room has no outlet. It is obvious that such hallucinations and illusions require us to assume an extreme excitability of the sensory centres, such as certainly does not exist after waking from normal sleep.

You are familiar with the very astonishing cures now and again worked by physicians who employ hypnotism therapeutically. These must be referred to the post-hypnotic effects of suggestion. It cannot be disputed that a cautious and intelligent use of suggestion may be of avail for the temporary, perhaps even for the permanent, removal of diseases due to functional derangement of the nervous system, or to harmful practices, like alcoholism or the morphine-habit. But it is an equally undeniable fact that suggestion is in the long run just as ineffective for the cure of diseases arising from some palpable pathological cause as would be any other form of command to the patient to grow well again.

There is, indeed, one exception to this rule,—an exception which is explicable from well-known physiological facts. Mental influences may, of course, affect the functioning of the bodily organs, and especially the excitations of the vasomotor and secretory nerves. And suggestion takes rank with other mental influences in this regard, becoming increasingly efficacious as the subject surrenders himself more and more permanently to the power of the suggested ideas. Thus an arrest of any particular secretion,—provided always, again, that the derangement is not due to pathological conditions seated in the organ itself, —may be overcome under the influence of suggestion. Dilatation of the blood-vessels, with all its consequences, may be suggestively induced, particularly if actual external stimuli are present to help produce the effect. An innocent scrap of postage-stamp paper, stuck upon the skin, has been known to answer all the purposes of a blister, if the idea were suggested that it was really a blister which was being applied. It is true that these phenomena cannot be obtained in the case of every hypnotised subject, or even of every somnambulist: the right disposition is requisite for the manifestation of such intensive organic effects. As for the physiological results of suggestion in themselves, they are simply more intensive and permanent forms of familiar and universal relations existing between mental states and bodily processes. If the transient emotion of shame can normally bring about a temporary distension of the blood-vessels of the face, it is after all not surprising that an abnormal excitability of the vasomotor and secretory nerve-system, combined with a domination of consciousness by ideas and feelings tending definitely to oppose the accompanying mental disturbance, should condition a more intensive and permanent physiological reaction to mental stimulus. And in saying this we are stating that hypnotism as a therapeutic agency is a two-edged instrument. If its effects are strongest when the patient is predisposed to it in body and mind, or when suggestion has become a settled mode of treatment, it may obviously be employed to intensify or actually induce a pathological disposition. It must be looked upon, not as a remedy of universal serviceability, but as a poison whose effect may be beneficial under certain circumstances. We find, of course, not only the dabbler in hypnotism,—who has no claim to a judgment on the question, and in whose hands the practice of suggestion becomes a public nuisance,—but also the physician,—to whom thinking men will no more deny the right to employ this dangerous remedy in certain circumstances than that of using any other,—asserting that the hypnotic sleep is not injurious, because it is not in itself a pathological condition. But surely the facts of post-hypnotic hallucination and the diminution of the power of resistance to suggestive influences furnish a refutation of this statement which no counter-arguments can shake. It is a phenomenon of common observation that frequently hypnotised individuals can when fully awake be persuaded of the wildest fables, and thenceforth regard them as passages from their own experience.

§ IV

But we are not concerned in this place with the physiological effects of suggestion, and its consequent significance for therapeutics. (For psychology the questions of special interest are that of the alterations of consciousness occurring during the· hypnotic sleep, and, connected directly with this, that of the nature of suggestion.)'The difficulty of the former lies in the impossibility of actual introspection on the part of the hypnotic subject. When aroused from the lighter form of hypnotic sleep, he has no clear recollection of what has taken place ; while after the somnambulistic state the memorial activity is in complete abeyance. Observation is therefore more difficult here, if that be possible, than it is in the case of dreaming. So there is all the more opportunity for fanciful hypotheses, to which the lay mind is tempted by the unusual and apparently mysterious character of the phenomena. And the lay mind is unfortunately an all too frequent possession of those who have desired to make hypnosis the object of psychological observation. Most hypnotic investigators are either physicians, who employ suggestion for therapeutic purposes, or philosophers, who think that they have discovered in hypnotism a basis for new metaphysical systems, and who, instead of examining the phenomena in the light of well-established psychological laws, reverse the matter and erect their psychological superstructure upon hypnotic foundations. So that it is hardly to be wondered at that the modern hypnotism-psychology has time and again manifested its descent from spiritism. Clairvoyance and the magic of telepathy play a suspiciously important part in it; and though there are found observers who have remained sane enough to hold aloof from all these absurdities, many even of them evince the fatal effect of the influence under which they have fallen by declaring all these superstitions to be after all 'open questions,' which deserve, if they do not demand, a closer examination. Like the scientific superstition of all times, this modern one arrays itself in garments borrowed from real science. It determines the credibility of clairvoyant somnambulists, or the occurrence of a telepathic miracle, by the rules of mathematical probability. It terms this whole field of hypnotic mysticism,— and here again it follows in the footsteps of the spiritism that preceded it,—'experimental psychology.' It organises 'societies for psychical research,' which are devoted to the cult of hypnotic experimentation. The chief danger of all this, it seems to me, does not lie in the abuse of post-hypnotic suggestion for criminal purposes which may happen once in a while. Crimes have hardly as yet been committed by 'mediums' as a result of suggestion. No! the great danger is, that persons of insufficient medical training, working not for therapeutic ends, but 'in the interests of science,'—though there is absolutely no guarantee of the real existence of their scientific devotion,—may exert an influence upon the mental and bodily life of their fellow-men such as, if continued for any length of time together, cannot fail to be injurious.

It must, moreover, be plain to you all that there can be no question of an experimental psychological method, in the exact sense of those words, in this matter of hypnotising. The condition of hypnosis is such as absolutely to preclude the possibility of a psychological experiment in the real sense. The psychological experiment demands from its subject concentration of the attention, practice, skilled introspection, in short,

the fulfilment of all manner of conditions, which, if not altogether and normally out of the reach of the hypnotic subject,
is at least wholly impossible during the course of the induced
sleep. If we compose ourselves to sleep, with the intention of
observing our dreams so far as that is possible, or even if we
take morphine for the same purpose, we are not making an
experiment, not doing anything that in execution or result is
essentially different from simple observation. The conditions
of dream-observation are not altered in the slightest degree by
the fact that we have brought on sleep intentionally. The characteristics of the experimental method are variation and gradation of the phenomena, and elimination of certain conditions.
Such a mode of procedure can be followed out in artificially
induced sleep as little—or, let us say, as imperfectly—as in
natural sleep: we shall gain no more by investigating the former
than by collecting casual observations of normal dreams. And
all this holds in still greater measure of hypnotism, since just
in the cases which present the most interesting phenomena
there is a total absence of any subsequent recollection. We
can only infer what goes on in the mind of the somnambulist
from his words and actions; if we wish to subject him to special
influences, we are hampered by the same conditions as hinder
the investigation of sleep and dreams.

§ V

Any explanation of the phenomena of hypnosis must, obviously, begin with the cognate facts of our normal mental life.
We have not to explain the usual functions of consciousness in
terms of hypnotism, but conversely. The established facts of
normal life, and especially those which best admit of introspective
control, must be employed to throw light on these phenomena,
which, if not pathological, are at least due to unusual conditions.
Now there is one state which you will see at once evinces a very
great similarity to the hypnotic sleep,—*heaviness after waking.*
In this condition we may perform actions, obey orders, answer
questions; but when we have fully awakened we realise that
all this was done half automatically, without any assistance from
the will. In other words, there may be developed an 'automatic

z

reaction to command,' very like that of the hypnotic subject. The sleepy soldier acts automatically at the word of command; the half-roused servant executes a commission; but it is only when fully awake that either remarks what he has done: indeed, if the sleepiness continues, he may entirely forget it. Years ago, when I myself went through several such experiences, I noticed particularly that I felt myself entirely at the mercy of external impressions, and acted, under their influence and that of the first dreamlike associations that they aroused, without any consideration, and so without any consciousness of the occasional wrongness of what I was doing. This self-surrender to external impressions approximates the dream-state; while the capacity of action and the general absence of hallucinations and illusions suggest, on the contrary, the waking condition of consciousness. But this exemption is not altogether constant. Illusions in particular frequently occur in the state of drowsiness.

Suppose now that the condition just described is brought one stage nearer that of sleep,—suppose that the self-surrender to external impressions results in exclusive conscious control by definite ideas and feelings suggested by the operator,—and you have the hypnotic sleep. One of the chief effects of suggestion is to increase the duration of this state. The subject, who fell asleep at the word of command or under the influence of ideas equivalent to it, remains dominated by the idea that he cannot wake except by a new command. So that the most marked characteristics of hypnosis are loss of volitional initiative, re-striction of the receptivity of consciousness to external impressions and the first associations that they call up, and usually a deter-minate direction of the attention induced by the influence of the operator's suggestion. Its effects are intensified by the tendency to hallucination, which, combined with the absolute surrender to external impressions, results in a transformation of the suggestions of the hypnotic consciousness into real objects.

You see, then, that the hypnotic sleep is akin to normal sleep and dreaming, occupying an intermediate position between these and drowsiness, but characterised by the surrender of our will to that of another person, and by the consequent efficacy of suggestion. Especially to be noticed is the inhibition of volun-tary activity. This is not only one of the chief diagnostic symptoms, but an important condition of the origin of the other phenomena. You must not, however, imagine that will-power is altogether in abeyance. Hypnotic actions are always voluntary actions in the wider sense of the word. But they are not self-initiated, resulting from a consideration of motives and a decision of the agent's own mind ; they are impulsive, univocally determined by a suggested idea and by the associations which it directly excites.

Regarded from the point of view of will and voluntary action, then, the hypnotic sleep is an abnormal condition. But it is a condition which follows all the psychological laws of our waking life. And the same is true of *suggestion*, that other so significant factor in the origin and progress of hypnosis. Ideas are continually aroused in our minds by the words which we hear and the actions which we see. Word and act are intimately associated with ideas, and affect our mind and will with all the greater force the smaller the number of conflicting associations or inhibitory intellectual motives that opposes them. Looked at in this way, suggestion reduces itself simply to an external impression, followed by associations characterised less by definite and positive attributes than by the negative quality of the absence of inhibitory influences. This impression and the consequent ideas will, therefore, continue to be the exclusive determinants of volition until other suggestions (*i.e.*, other similarly excited associations) give a contrary direction to the hypnotic consciousness. How these facts enable us to explain certain special forms of suggestion,—autosuggestion, post-hypnotic suggestive influences, etc.,—I attempted to show in describing those phenomena themselves.[1]

[1] For the physiological substrate of the hypnotic condition, and for the discussion of many other points which can only be briefly touched on here, *cf.* my detailed account of hypnotism in the *Philosophische Studien*, vol. viii., pp. 1 ff.

LECTURE XXIII

§ I. Problems of Animal Psychology; Deficiencies of the Science. § II. Methodological Rules. § III. Acts of Cognition and Recognition among Animals. § IV. Association among the Lower Animals.

§ I

IN the preceding lectures we have considered the associative and intellectual processes of consciousness, first in their general and normal features and then under the various aspects which they present in mental disturbance, dreaming, and certain conditions related to that of sleep and dreaming. There now remains one last question, the answer to which is important if we are to understand the nature of these processes and their relation to the other functions of the mind, the question of *animal intelligence*, or, to express it more exactly, of the nature and significance of those animal actions the conditions of whose origin lead us to refer them to mental processes similar to our own associations, and possibly even to our own processes of judgment and inference.

The study of animal psychology may be approached from two different points of view. We may set out from the notion of a kind of comparative physiology of mind, a universal history of the development of mental life in the organic world. Then the observation of animals is the more important matter ; man is only considered as one, though, of course, the highest, of the developmental stages to be examined. Or we may make human psychology the principal object of investigation. Then the expressions of mental life in animals will be taken into account only so far as they throw light upon the evolution of consciousness in man. You will remember that we decided at the outset of these lectures to deal with animal psychology in this second sense, and for the more limited purpose.

If we compare these two ways of treating psychology with comparative and human physiology, we cannot fail to see that the two spheres of investigation are very different as regards methods and appliances. The bodily organs and functions of animals are just as accessible to objective examination as are those of man : indeed, in the living condition for obvious reasons far more completely so. So that there is no need to remind human physiology that it must never omit the comparative reference to animals. It follows this rule of its own accord, oftentimes more than it cares to, or than might be wished in the interests of physiological theory, because it must necessarily have recourse to animals where observation on man is impossible. Human psychology, on the contrary, may confine itself altogether to man, and generally has done so to far too great an extent. There are plenty of psychological text-books from which you would hardly gather that there was any other conscious life than the human. It is entirely different with comparative physiology and comparative psychology. It would be possible, if needs were, to write a monograph upon the physiology of an order or a species, say the infusoria or the frog, without paying any regard to the analogous functions in man. But not the least advance can be made, either in the psychology of a particular animal or in that of the animal kingdom, without starting out from the facts of the human consciousness. And here again it is psychology that has been at fault. Following the analogy of comparative anatomy or physiology, psychologists have attempted without more ado to schematise the evolution of mental life in animals, and then to apply their results directly to that of man. The outcome could be nothing else than that which always follows from the premature use in one connection of concepts found in another. So that Bacon's comparison of the insufficient observation of nature by the Aristotelians of his day to the report of an ambassador who based his knowledge of the measures of a Government upon town gossip, and not upon accurate examination, applies fairly enough to the animal psychology of our own time. It is permeated through and through by the concepts of the every-day psychology, which is thought to suffice for the requirements of ordinary life, and too often also for the sciences which cannot do without psychological reference. The one great defect of this popular psychology is, that it does not take mental processes for what they show themselves to be to a direct and unprejudiced view, but imports into them the reflections of the observer about them. The necessary consequence for animal psychology is, that the mental actions of animals, from the lowest to the highest, are interpreted as acts of the understanding. If any vital manifestation of the organism is capable of possible derivation from a series of reflections and inferences, that is taken as sufficient proof that these reflections and inferences actually led up to it. And, indeed, in the absence of a careful analysis of our subjective perceptions, we can hardly avoid this conclusion. Logical reflection is the mental process most familiar to us, because we discover its presence whenever we think about any object whatsoever. So that for popular psychology mental life in general is dissolved in the medium of logical reflection. The question whether there are not perhaps other mental processes of a simpler nature is not asked at all, for the one reason that, whenever self-observation is required, it discovers this reflective process in the human consciousness. The same idea is applied to feelings, impulses, and voluntary actions, which are regarded, if not as acts of intelligence, still as affective states which belong to the intellectual sphere.

This mistake, then, springs from ignorance of exact psychological methods. It is, unfortunately, often rendered worse by the inclination of animal psychologists to see the intellectual achievements of animals in the most brilliant light. This, of course, is due to the natural pleasure which the objects of our observation always give us, and which is the most effective spur to continuous devotion to a particular subject. In the present case it is transformed into the unintentional endeavour to observe things which shall be as interesting as possible. Unbridled by scientific criticism, the imagination of the observer ascribes the phenomena in perfectly good faith to motives which are entirely of its own invention. The facts reported may be wholly true ; the interpretation of the psychologist, innocently woven in with his account of them, puts them from first to last in a totally wrong light. You will find a proof of this on nearly every page of the works on animal psychology. Take a few

instances, chosen at random, from Romanes' industriously com-
piled volume on *Animal Intelligence*. [1] While we admire the
diligence with which the author has observed and collected the
observations of others, we cannot but notice the unfortunate
absence of the critical attitude in a field where it is especially
desirable. Turn to the chapter on ants. An English clergy-
man writes apropos of the 'funereal habits' of these animals:
'I have noticed in one of my formicaria a subterranean
cemetery, where I have seen some ants burying their dead by
placing earth above them. One ant was evidently much
affected, and tried to exhume the bodies; but the united exer-
tions of the yellow sextons were more than sufficient to
neutralise the effort of the disconsolate mourner' (p. 92). How
much is fact, and how much imagination? It is a fact that the
ants carry out of the nest, deposit near by, and cover up dead
bodies, just as they do anything else that is in their way. They
can then pass to and fro over them without hindrance. In the
observed case they were evidently interrupted in this occupation
by another ant, and resisted its interference. The cemetery, the
sextons, the feelings of the disconsolate mourner, which impelled
her to exhume the body of the departed,—all this is a fiction of
the sympathetic imagination of the observer. Another friend
of the ants gives this account: 'At one formicary half a dozen
or more young queens were out at the same time. They would
climb up a large pebble near the gate, face the wind, and assume
a rampant posture. Several having ascended the stone at one
time, there ensued a little playful passage-at-arms as to position.
They nipped each other gently with the mandibles, and chased
one another from favourite spots. They, however, never nipped
the workers. These latter evidently kept a watch upon the
sportive princesses, occasionally saluted them with their antennæ
in the usual way, or touched them at the abdomen, but
apparently allowed them full liberty of action' (p. 88). The
correctness of this observation need not be questioned. Why
should not a number of young queens have been crowded
together upon a pebble, and some workers have been with them,

---

[1] *Animal Intelligence*, by G. J. Romanes, Int. Sci. Series, vol. xli., 4th ed.
(1886). *Cf.* the same author's *Mental Evolution in Animals* (1885).

and occasiönally touched them with their antennæ, as ants do everywhere? But that they 'sported' and played, that the others 'kept a watch upon them' like chaperones, and now and again did homage to them by 'saluting,'—that, again, is all due to the imagination of the observer. He would hardly have told the story as it stands had not zoology introduced the misleadingly suggestive term 'queens' for the mature female insects. If the adults are 'queens,' the young females must, of course, be 'princesses.' And since no princess ever went out without an attendant or a chaperone, the rest of the narrative follows as a matter of course. Written in just the same style is the following account of the education of ants, taken from the same work. It may serve at the same time as an instance of the more general remarks current in the literature of animal psychology. 'The young ant,' the author tells us, 'does not appear to come into the world with a full instinctive knowledge of all its duties as a member of a social community. It is led about the nest, and "trained to a knowledge of domestic duties, especially in the case of the larvæ." Later on the young ants are taught to distinguish between friends and foes' (p. 59). These illustrations will enable you to judge how much of similar descriptions is actual fact, and how much is due to the imagination of the observer.

How easy it is to misinterpret an observation if the very greatest care is not taken in recording it, and if it is impossible to vary the circumstances by experiment, and so obtain accurate knowledge of its conditions, is well shown by the following facts. Pierre Huber, one of the most reliable students of the habits of ants, stated that he had assured himself that an ant, if taken from the nest and returned after an interval of four months, was recognised by its former companions; for they received it friendlily, while members of a different nest, even though they belonged to the same species, were driven away. Huber regards this as evidence of the extraordinary accuracy of memory in these insects. Now the correctness of his observation cannot be doubted; and, besides, it ¯has been confirmed by another experienced investigator,—Sir John Lubbock. At first sight, therefore, the conclusion seems perfectly justifiable. But if a single individual were really recognised after so long an interval, think what the general mental capacity of the ants must be! Fortunately, Lubbock made the matter a subject of experiment. He took ant larvæ from the nest, and did not put them back again till they were fully developed. The result was, that they too were quite friendlily received. Plainly, then, there can be no question of an act of individual recognition. There must be some characteristic peculiar to all the members of a particular nest, possibly a specific odour, which determines the instinctive expression of 'friendship.'

It is the same with those facts upon which the conclusion has been based that ants and other insects that live together in 'states,' as they are called, possess a fairly well-developed language. The animals are undoubtedly able to call in the assistance of others for the accomplishing of tasks too great for their own strength. But this purpose can be served by quite simple movements, which are common to very many species of social animals. Since these movements are manifestations of impulse, they exert a reflex influence upon the corresponding impulses of all individuals of the same kind. It is true that action must in every case be preceded by certain ideational connections. The ant that finds a load too heavy will connect this impression of weight with the often repeated perception of its mastery by united effort, and that again with the idea of assistance by other ants. But all this is a matter of very obvious association. To assume a supra-associational intellectual activity for processes which consist in manifestations of animal instinct of every-day occurrence, and repeated through countless generations, seems altogether unjustifiable. It would be entirely contrary to everything that introspection tells us of our own impulsive and automatic actions.

§ II

These considerations lead up to a question which it is important to raise with regard to the study of animal psychology in general. We have no other means of estimating the mental processes of animals than in the light of those of our own consciousness. We must employ these in such a way as to gain the best and surest knowledge possible of the animal mind.

How is this to be done? The current animal psychology does not trouble itself to give any very careful answer; indeed, a general answer is not given by it at all. In any particular case, however, as we have seen, it selects from the totality of mental processes the intellectual functions, and explains the mental life of animals in terms of them. Its implicit principle, that is, is precisely the opposite of the approved maxim of the exact natural sciences that we should always have recourse to the *simplest* explanation possible. It attempts to derive all the facts of its objective observation from the most complicated mental processes of human subjective experience. Where there are two alternatives,—derivation from logical reflection and explanation in terms of simple association,—it invariably chooses the former. And the fact that logical reflection is everywhere a possible explanation is taken as sufficient proof that it is the right one. But there are two reasons why this particular inference from effect to cause is not permissible. In the first place, mental activities are so complex and multifarious, that practically every objective action is capable of more than one interpretation; and it needs very careful consideration of all the secondary circumstances to decide the actual psychological conditions of a given result. Secondly, logical reflection, just because everything can be subsumed to it subjectively, may itself be translated into the objective condition of *anything*. So that the psychologist who interprets a fact of consciousness as a logical reflection, when it is not so given, is acting in principle as a student of natural science would who ascribed the properties of certain natural objects to the usefulness which they happened to possess for ourselves or for other organisms. But since logical reflection is itself a mental process, and may therefore take its place with other processes as a possible cause of some psychical effect, the danger of the error becomes much greater, and the proof of it much more difficult. All the more reason is there to emphasise the necessity, when we are attempting a psychological explanation of phenomena which can only be inferred from their objective results, of always inquiring for those special characteristics which are of determining value for one mode of explanation as compared with another.

§ III

If, with this in mind, we ask : what are the most *simple* manifestations of animal life which cannot be explained without the
introduction of the mental factor? we come first of all upon
voluntary actions, referable to acts of *cognition* and *recognition*.
You see at once that the very names employed to denote these
simplest ideational connections show 'traces of the fatal
tendency to dissolve all the facts of consciousness in the medium
of logical reflection. That the processes in question consist of
nothing more than simple associations was shown when we
were analysing the corresponding phenomena of the human
consciousness. We used the word 'cognition' because there was
no other term current to denote the process ; but it is sufficiently
plain from the language of our description that any idea of an
act of logical cognition was very far from our minds. Now
the analogous phenomena of the animal consciousness must, of
course, be treated from the same psychological standpoint. The
acts of cognition and recognition are processes which represent
the simplest modes of connection between present impressions
and past ideas ; in other words, they are mental phenomena,
which in a large number of cases furnish us with the only
reliable evidence of the existence of a mental life. For to prove
mentality we must be able to prove a persistence of the sense-
impression,—some form of memory, however elementary.
Memory of some kind is presupposed in the cognition of an
impression ; memory of a definite kind, in its recognition. In
the former the impression is assimilated by numerous previous
impressions of the same character; in the latter it is associated
with a single or with a strictly limited number of past impressions, the terms of the association frequently arranging
themselves in a temporal series.

In the very lowest animals it is only the simple association
process that we term 'cognition' of an object which plays any
important part ; 'recognitions' are hardly demonstrable. That
the creature knows its proper food, and may be determined in
its knowledge of it by previous impressions, is regarded as the
first and primal indication of the presence of animal,—*i.e.*, mental
—life. But that an animal recognises a certain object,—*e.g.*, the

place where it finds its food,—presupposes a separation and differentiation of ideas which in all probability is not realised at the lowest stages of organic existence. The simple acts of cognition develope from the primitive animal impulses which are at the root of the earliest manifestations of life that can be called psychophysically definite. The origin of the selection of food, for instance, can only be explained on the assumption that inherited organisation determines the correlation of pleasurably toned sensations with certain sense-impressions, and that these sensations are connected with the movements subserving nutrition. The fact, often confirmed by experiment, that even the protozoa congregate in light of one quality and avoid spots illuminated by that of another, must depend upon some such original sensation-character. The discrimination is made in terms of sensation and the affective tone attaching to it, as in all the higher animals. Ants try to escape from a violet light, but crowd together on a blue surface. Lizards and blind-worms avoid blue and all the more refrangible colours, but are fond of red. Again, there may be connected with these instinctive manifestations of mental life others which indicate an intensification of the original affective distinctions by previous impressions. The more frequently experiments are repeated with the different illumination of different portions of space, the more quickly and certainly does the animal decide for its favourite colour. And the fact that all animals prefer colours and brightnesses corresponding to those of the medium in which they live, —the winged insects, *e.g.*, blue or white,—may be referred to the influence of previous impressions. At the same time it is not improbable that these psychophysical influences have in the course of generations modified the sensitive elements of the sense-organs ; so that the affective quality which determines the preference or avoidance of certain stimuli is so far connate that it connects at once with the sense-impressions. Thus the excitability of the eye of the owl and eagle is entirely different from the very first. Nevertheless, there is every reason to suppose that this difference has been developed in the course of generic evolution, simultaneously with the differentiation in mode of life and habit. And the relation of animals to their proper food-supply is an altogether analogous case. The actions prompted by the nutritive impulse will be first excited by the affective colouring of certain definite sensations. But when the enjoyment of food has been once experienced, the new and the old impression become associated. We therefore find that in all animals the discovery and right discrimination of food-stuffs are perfected in the experience of the individual life. This 'experience,' psychologically analysed, consists wholly in simple acts of 'cognition,' *i.e.*, in the assimilative effect which previous impressions have upon new ones of the same character.

It is generally more difficult to say whether an act of *recognition* has taken place in a given instance. In recognition a perception is apprehended as agreeing either with a previous perception referring to some individual object, or with a strictly limited group of such previous perceptions. The process obviously presupposes a more highly developed discrimination of ideas, *i.e.*, a greater store of determinate associations. Often enough actions of the lower animals which are interpreted as recognitions belong to the sphere of indefinite cognitions. A good instance of this is the ant experiment quoted above. The insect returned to its nest by Pierre Huber, after a separation of four months, was really known not as an individual, but in virtue of some generic or family characteristic. This does not, of course, mean that individual recognition is not possible among the same insects, though it would probably only extend over a much shorter time. The ant which brings up its companions to assist it in carrying a load must, naturally, recognise the object to be carried and the road it has passed over,— possibly also some of the helping ants. It is well known that in the higher animals the memory is able to retain particular facts and objects for a comparatively long period. The dog recognises its master after an absence of months ; the domestic cat is thoroughly familiar with the rooms of the house in which it is kept, and at once makes itself at home after quite a long absence. The dog, the elephant, and many other animals have an accurate memory for any previously unknown individual who teases or strikes them, and show it by very evident tokens of displeasure. When it is assisted by instinct and specific sense-development, this faculty of recognition among the animals may far exceed that of man in its own particular sphere. The dog can find its way over a long and complicated road by means of smell ; the carrier pigeon recognises the direction in which it flew some long time previously, thanks to its trained space-memory and far-reaching vision. All these phenomena, which have sometimes been referred to a mysterious ' sense of locality,' are explicable as dependent upon recollection, highly developed in one particular direction and assisted by sense-peculiarities. The carrier pigeon taken from its home in a closed cage to some unknown and remote spot will either not return at all, or only by a long and roundabout way.

§ IV

Wherever recognition is possible, other associations may also be observed. The animal, like ourselves, naturally associates the similar and the contiguous in time and space. And we frequently find in the lower animals what we have observed in man,—that associations give rise to actions whose result is equivalent to that due to the operation of the intellectual functions. It is here, therefore, in the various forms of successive association proper, that the act of interpretation which resolves the mental life of animals into concepts, judgments, and inferences, according to all the rules of logic, finds freest play. But if the whole body of reliable observation is carefully tested, and due regard paid to the *lex parsimoniæ*, which only allows recourse to be had to complex principles of explanation when the simpler ones have proved inadequate, it seems that the entire intellectual life of animals can be accounted for on the simple laws of association. Nowhere do we find the characteristic marks of a true reflection, of any active functioning of imagination or understanding. In saying this, we are, of course, regarding only well-authenticated facts, and not those ' travellers' tales ' of which animal psychology has as many as it has wrong explanations of actual observation. You may remember the story in Pliny's *Natural History* of the elephant who was punished during a performance for his bad dancing, and who secretly set to work to practise in the night, so as to do better the next time. We may be pretty confident in supposing that this tale and many others like it, which are still current in the animal psychologies, are either pleasant inventions or, if they contain a grain of real observation, have received so much imaginative embellishment, that its discovery is practically impossible.

As a matter of fact, the mental life of animals is rich enough to be able to do without all this. The life of the more highly organised invertebrates,—even apart from the phenomena connected with the social instincts, to which we shall return later,— has many features which would astonish an observer untrained in psychological analysis. Instead of giving you a whole number of illustrations which are all variations of the same theme, I will narrate an observation of my own. I remember it very distinctly, because it was the first time that I had brought home to me the mental life of a lower animal. I had made myself, as a boy, a fly-trap, like a pigeon-cote. The flies were attracted by scattered sugar, and caught as soon as they had entered the cage. Behind the trap was a second box, separated from it by a sliding-door, which could be opened or shut at pleasure. In this I had put a large garden spider. Cage and box were provided with glass windows on the top, so that I could quite well observe anything that was going on inside. At first nothing particular happened. When some flies had been caught, and the slide was drawn out, the spider, of course, rushed upon her prey and devoured them, leaving only the legs, head, and wings. That went on for some time. The spider was sometimes let into the cage, sometimes confined to her own box. But one day I made a notable discovery. During an absence the slide had been accidentally left open for some little while. When I came to shut it, I found that there was an unusual resistance. As I looked more closely, I saw that the spider had drawn a large number of thick threads directly under the lifted door, and that these were preventing my closing it, as though they had been so many cords tied across it.

What was going on in the spider's mind before she took this step towards self-preservation—a step, mark you, which but for the *vis major* of the boy-master would have been perfectly adequate to effect the desired result? The animal psychologist will possibly say: 'the spider must first of all have come to understand the mechanism of the sliding-door, and must have said to herself that a force operating in a definite direction could be compensated by another in the opposite direction. Then she set to work, relying upon the perfectly correct inference that if she could only make movement of the door impossible, she would always have access to the victims of her murderous desires. There you have a consideration of general issues, an accurate prevision, and a cautious balancing of cause and effect, end and means.' Well, I am rather inclined to explain the matter otherwise. I imagine that as the days went by there had been formed in the mind of the spider a determinate association on the one hand between free entry into the cage and the pleasurable feeling attending satisfaction of the nutritive instinct, and on the other between the closed slide and the unpleasant feeling of hunger and inhibited impulse. Now in her free life the spider had always employed her web in the service of the nutritive impulse. Associations had therefore grown up between the definite positions of her web and definite peculiarities of the objects to which it was attached, as well as changes which it produced in the positions of certain of these objects,—leaves, small twigs, etc. The impression of the falling slide, that is, called up by association the idea of other objects similarly moved which had been held in their places by threads properly spun; and finally there were connected with this association the other two of pleasure and raising, unpleasantness and closing, of the door. That was surely enough to rouse the prisoner to action. Any other intellectual or inventive activity is entirely unnecessary. If she had not had these associations at her disposal, she would certainly never have hit upon the plan she did.

§ I

THERE seems at first sight to be a very considerable difference between the expressions of mentality in the higher animals, more especially the more highly developed domestic animals, and the simple or complex associations which we ascribed in the previous lecture to certain of the invertebrates, such as spiders, ants, and other insects with very definite instincts. The perpetual intercourse of the domestic animals with man seems to bring them quite near to him on the mental side ; he exerts a determining influence upon the contents of their ideas, the direction of their associations, and their whole affective life. The dog shares the joys and sorrows of its master ; it reads anger, happiness, or despondency in his face. The trained poodle is made as happy as can be by its master's praises, and shows every sign of pride when entrusted with certain duties, as the carrying of a basket or a walking-stick. Now it is true enough that all this points to a great diversity of feeling and a considerable adaptability to the feelings of others. But the emotions expressed never belong to the sphere of intellectual feeling,—logical, æsthetical, etc. So that the only certain inference from the actions in which the animal appears to resemble man so closely is that it is endowed with a very active associational mechanism.

There can be no doubt that the behaviour of our more highly developed domestic animals indicates this activity of association. As soon as you have taken your hat and stick, your dog knows that you are going out, and shows by tokens of pleasure and

A A

other unmistakable gestures that it wishes to go with you. A poodle of my own used to be subjected to a thorough washing every Saturday, and disliked it very much. Various preparations that were going on in the house would remind him that Saturday had come; and he not infrequently disappeared early in the morning, and only returned late in the evening, when all fear of the cold water was over. On these occasions he usually spent the day on the square in front of the house, looking longingly up at the windows and obviously divided between the wish to return, and aversion to the fate that awaited him. He was all the happier when Sunday morning arrived. On that day my brother, who was living in a neighbouring town, was accustomed to come over and see me. The dog was more attached to him than to any of the inmates of the house. He never failed to keep watch behind the front door from the very first thing in the morning, welcoming approaching footsteps with a hopeful wag of his tail, and dropping his head despondently if they disappointed him. At last, when my brother really appeared, he was greeted with every manifestation of the most extreme joy. Experiences of this kind not only show that the mechanical operation of association may extend over a considerable time, as is proved by the recognition of an individual after a number of days, but also,—and it is this which distinguishes the present instance from a case of simple recognition,—that an animal is itself able to associate events which coincide in space or time, and to extend these associations over a relatively long period. The poodle knew, of course, that it was Saturday simply by the special preparations for cleaning the house. With that idea was inseparably associated the very unpleasant idea of his own washing. This association was not only strong enough to keep him away from the house for the whole day, but was further connected with the other and far more complicated association of the following day and the arrival of my brother. Of course, the regularity of the visits furthered the formation of the association. At the same time, we have here a development of temporal ideation reaching far beyond the connection of directly simultaneous or successive events. It would be utterly wrong, though quite in the manner of current animal psychology, to make the dog reason in this way : 'Yesterday was house-cleaning and washing-day ; my friend usually arrives on the day after that ; therefore he will come to-day.' The simpler, and therefore the only justifiable, explanation is, that the experience of previous weeks had made the succession of these events a stable contiguity-association in the animal's mind ; and that its expectation of my brother's arrival after the preceding occurrences was of the same kind as its expectation that it would get something to eat after the filling of its platter. The only difference between the two cases is that the former association extended over a longer time and embraced a greater number of events than the latter.

A whole host of observations, which are usually interpreted in terms of intelligent action in the strict sense of the word, are more or less like these examples of my own. I will only cite the following, which I take from Romanes' book on *Animal Intelligence* (p. 418). The servants of a house had been accustomed during a frost to throw the crumbs remaining from the breakfast-table to the birds. A cat took advantage of this, and now and again obtained a hearty meal from one or two of the assembled guests. The practice of feeding the birds was therefore discontinued. The cat, however, scattered crumbs on the grass itself with the obvious intention of enticing them. Romanes supplements this by another story, in which crumbs were sprinkled on the garden-walk, and the cat lay in ambush to catch the sparrows. In this case, however, the cat used to conceal itself from the walk in a neighbouring shrubbery, and there await the coming of the birds. But the sparrows showed themselves more wide-awake than the cat : they waited on a wall, never venturing to fly down till their enemy wearied and went away. Romanes credits the two cats with the same form of inference. In the latter case the animal reasoned : ' crumbs attract birds ; therefore I will wait for birds when crumbs are scattered.' The first cat proceeded a stage further : ' therefore I will scatter crumbs to attract birds.' There can be no doubt that the two cases are analogous, and that the simpler is altogether a matter of ordinary association of the sort which is constantly directing animal action. But there can be no doubt either that the cat did not make the author's inference. When once the association between scattering crumbs and attracting birds had been formed, the feeling of expectation was sufficient to cause it to lie in wait for prey as it was accustomed to do in other cases. This association was precisely like that which the sparrows had formed, even in its effect upon the will; except that the cat was led to undertake a particular action, and the sparrows to refrain from it. We must admit that the case is somewhat different when the cat itself scatters crumbs. The tale is so unlikely, in view of what we know of the general behaviour of the animal, that we may possibly refer it to self-deception on the part of the observer, or conjecture that some chance incident has been wrongly interpreted. The same story told of monkeys would sound more probable, on account of the high development of the imitative impulse in those animals. There is nothing to take the place of this in the cat except the carnivorous instinct to secure prey. But even if we grant the correctness of the observation, the action might be explained by associative processes. The stable association of scattering crumbs and attracting birds, taken together with the impulse to catch them, might have enabled the cat to supply one term of the association-series for itself. But if we leave out this element of spontaneous activity connected with the specific carnivorous instinct, we see that the range of association in the cases cited is not nearly so great as in the above-quoted recollection and recognition by my dog.

That the action of animals may be determined by memorial ideas, as well as by the corresponding sense-impressions, can be very readily shown. I often made the following amusing experiment with my own poodle. I had taught him to spring over a stick which I held out at the word ' Jump!' One day I called the word out to him without presenting the stick. At first he looked at me in surprise, and then, as I repeated the command, barked impatiently. At last, after I had given the order several times with a stern face, he decided to make a spring into the air, but barked loudly at me afterwards, as though to complain of the absolute absurdity of my command. When I had repeated the experiment a number of times, the animal came to respond at once by springing into the air, but never failed to protest by growling and barking. The word of command aroused the memorial idea, and this was sufficient to arouse the action produced by the actual presentation of the stick ; while the feeling of contrast between idea and object, and of the purposelessness of the act gave rise to unpleasurable emotions conflicting with the dog's habitual obedience.

§ II

The criterion of 'intelligent' associative action and of intelligent action proper can only be this,—that the effect of association does not go beyond the connection of particular ideas, whether directly excited by sense-impressions or only reproduced by them ; while intellectual activity in the narrower sense of the word presupposes a demonstrable formation of concepts, judgments, and inferences, or an activity of the constructive imagination. If Pliny's story of the elephant practising dancing by moonlight were true, it would undoubtedly be a case of imaginative activity in this sense. On the other hand, the jumping of the dog over an imaginary stick at the word of command implies memory, but not imagination; *i.e.*, it depends not upon the spontaneous arousal of associations, but upon their discharge by external impressions.

The *play* of man and the animals differs in the same way as their 'intelligence.' We regard certain actions of the higher animals as playful when they take the form of imitations of purposive voluntary actions. We know that they are imitations because the end pursued is only a fictitious end,—the real end being excitation of joyous emotions similar to those which follow as secondary effects from genuine purposive action. That means, you see, that the play of animals is for all practical purposes identical with play among mankind. Our own play, at least in its simpler forms,—*e.g.*, in the play of children,—is merely an imitation of the actions of every-day life stripped of its original purpose, and resulting in pleasurable emotion. The play of animals bears the same relation to the play of man as animal life does to human life. The impossibility of transcending a certain circle of ready-made associations characterises the play of animals (even of the most highly developed), as it characterises their mental life in general. Over against the countless varieties of the play of children, reflecting all conceivable relations of life, stands the single form of mock fighting among the animals. (Trained animals do not, of course, concern us ; their performances are not real play.) Dogs, cats, and monkeys, even when they are playing with their young, show their affection by pretending to fight with them. And though it is true that play is an indication of high mental development, and brings the animal nearer to ourselves than any other activity, it is rather the fact that it plays than the nature of the play itself which is the important point. Only those animals can play which reproduce in memory pleasurable experiences, and are able to modify them so that only their pleasant side comes to consciousness, and their unpleasant aspect disappears. At the same time, any comparatively complex associative and affective activity in the mind of an animal is a sufficient reason for the origination of playful actions. Animal play never shows any inventiveness, any regular and orderly working out of some general idea. And only where this is found can play be re-garded as the expression of really imaginative activity. The simple and original play of the animals is, if the expression may be allowed, a purely associational play. When a dog sees another dog, it does not necessarily feel any hostility towards it, but still has an inclination to exercise its strength in a mock fight, and so to gain the pleasure which it has experienced in real ones. If it obeys this impulse while its mood is friendly, or at least not hostile, the result is a mutual trial of strength in a playful contest ; ending often enough, as you have probably observed in dogs and monkeys, in the arousal of the real joy of battle and a fight in bitter earnest. In many animals, how-ever, and especially in those which, like the dog, have been domesticated for a long time, the inherited instinct appears in a moderated form from the first ; the connate fighting impulse seems to become a connate impulse to play.

§ III

We must conclude, then, that those animal actions which remind us most forcibly of imaginative activity do not show any of the specific characteristics which distinguish imagination from memory. There is no purposive and comprehensive connection of ideas, none of the inventiveness which that implies. In the same way the animal actions which border most closely on the realm of human understanding give us no warrant for inferring the existence of true concepts, judgments, and inferences. That writers upon animal psychology have so often asserted the contrary is due to the interpretation of comparatively simple associative processes as apperceptive intellectual operations.

Romanes gives an account of a series of tests of the intelligence of an elephant sent him by one of his correspondents (p. 401). The story goes to show that 'elephants possess abstract ideas.' Even if we do not give the phrase 'abstract ideas' its philosophical meaning, but simply understand by it general experiential concepts, we must still admit that the facts recorded do not prove their existence, but merely indicate a fairly highly developed associational activity. An elephant was made to lift various objects with its trunk,—bundles of clothes ; tree-trunks ; heavy pieces of metal. It was noticed that the animal gradually 'took in a knowledge of the nature of the things it was required to lift' ; light objects were thrown up easily and quickly, heavy ones slowly and with obvious muscular preparation, cutting instruments with a certain degree of caution. The observer concludes ' that elephants recognise such qualities as hardness, sharpness, and weight.' You will probably agree that nothing more is necessary for such actions than the formation of definite associations between the visual impressions of an object and its tactual properties. Such associations would in any case have been necessary to produce the observed result, even though the elephant really possessed the general concepts of light, heavy, cutting, etc. But if once the associations were formed, they would be altogether sufficient to determine the 'intelligent' lifting of such objects, and there would be no need at all of the further formation of general concepts.

I spent a great deal of time in trying to discover some certain indication in various actions of my own poodle of the presence or absence of general experiential concepts. I was never able to demonstrate them ; but I made many observations which rendered its power to form them exceedingly improbable.

I taught the dog to close an open door in the usual way by pressing with the fore feet when the command 'Shut the door!' was given. He learned the trick first of all on a particular door in my study. One day I wished him to repeat it on another door in the same room; but he looked at me in astonishment and did nothing. It was with considerable trouble that I persuaded him to repeat his trick under the altered circumstances. But after that he obeyed the word of command without hesitation at any other door which was at all like these two. It is obvious that when the animal had learned the trick of closing the door for the first time he had formed no general concept of 'door'; otherwise he would have found no difficulty in shutting others. His action plainly depended upon a quite individual association. Some force had to be used to form this, as is always the case with such performances. I stood the dog up on his hind legs against the open door, while I gave the word of command, until he responded to the order by voluntary movement. But you will say: might not the further step necessary to the formation of a concept have been taken when the animal learned to close the second door? After that he was able to close others for which he had not been specially trained. I do not believe that the assumption of the formation of a concept is necessary even here. It is evident that when the association between word of command, movement, and closing of the door had been formed at several doors the more special association between the particular door and the action of closure must have become obscured. The association of particular ideas had developed into a true similarity-association. There is not the slightest indication of the presence in the dog's mind of the principal characteristic of the formation of concepts,—the consciousness that the particular object vicariously represents a whole category of objects. It had throughout only a very confused idea of the 'nature' of a door. When I ordered it to shut a door which opened from the outside, it made just the same movement,—opened the door, that is, instead of closing,—and though I impatiently repeated the command, it could not be brought to do anything else, although it was obviously very unhappy at the ill-success of its efforts. Only after I had on several occasions let it out of the room, and made it shut the

door from outside, did the inability to close it from inside
decide it to go out and repeat the attempt from the other side ;
and then it would at once begin to scratch at the closed door
to beg for readmission.

What holds of concepts holds equally of the alleged judgments and inferences of our more intelligent domestic animals ;
on closer consideration they may be all resolved into obvious
associations, and they fail at the point where the sphere of externally suggested association touches that of intellection proper.
I used to take my poodle on summer evenings into a garden
lying across a stream, to which we had access by boat. The
banks were very crowded, and boats were always plying to and
fro. One day the poodle had stayed behind with some other
dogs, and came to the bank too late, so that the boat was at
some distance from the shore when he reached it. There was
nothing left for him but to swim the stream ; and since it was
fairly broad, and he had little love for water, that was very unpleasant. Some days afterwards the same misadventure befell
him. He ran up and down the bank whining dismally, and
obviously very averse to repeating his bath. At that moment
a boat-load of passengers pushed off; he sprang in and came
dry-footed and extremely happy to the other bank. Henceforth he not infrequently crossed in the same way. Now what
was going on in his mind during this particular action? Did he
think : 'my master has crossed the stream; this boat is also
about to cross: therefore I shall overtake him if I step in ' ? Of
course we can translate the animal's action into this chain of
reasoning. Only we are not then dealing with the process in
its mind, but with a logical interpretation of that process. But
are not the interpretation and the process equivalent? And
since the chain of reasoning would lead to the same result, how
do we know that it did not actually take place in the dog's
consciousness? This instance is in truth a good illustration of
the fact that the outcome of a logical reflection is not so completely identical with that of a mere ideational association.
The two processes present characteristic differences which acquire especial significance in cases like this, where we are determining the nature of internal experiences. If the dog had sprung .
at once into a boat lying upon the shore and waiting to take

in passengers, we should still perhaps hardly have been justified in crediting it with logical reflection ; the association between the boats and their passage over the stream might have been formed in its mind through its frequent journeys across. Nevertheless this association would have been considerably more complicated ; and its inclusion of a number of memorial elements, in the absence of the corresponding impressions, would have brought it very near the intellectual operations. But it never occurred to the animal to enter an empty boat and wait. Not till the boat pushed off did it associate this new crossing with previous journeys. So that the action bears every mark of a simple association. If on later occasions,—and I am not sure about this,—the dog did spring into an empty boat, or one just filling up, we need not refer this action to the sudden birth of logical reflection, but can explain it as a gradual extension of the associational series, resulting from practice. Many accounts of animal intelligence which are given without falsification or exaggeration in works on animal psychology would, we may be sure, oftentimes admit of a similar interpretation, if we were informed of all the stages in the animal's conduct. You could hardly imagine a better example of intelligence than the action of a dog which should cross a stream of his own accord with a number of passengers who were total strangers to him in order to visit a garden on the other side. And yet the act becomes simple enough when we can follow its gradual development in detail.

§ IV

All these manifestations of animal 'intelligence' may be adequately explained as relatively simple associations. Whenever we are in a position to investigate the nature of the connection of conscious processes involved, we look in vain for traces of logical reflection or real imaginative activity. We can now understand how it is that the animals lack one function which is characteristic of the intellectual processes, at the same time that it is their invariable concomitant,—language. Animals can express emotion ; the more highly developed of them can to a limited extent give evidence of the connection of ideas with their emotions. But the expressive movements of animals never

show that regular articulation, that reflection in organic structure of the nature of intellectual ideation, which is characteristic
of language proper.   Animals possess certain elements of
language, just as they possess certain elements of consciousness
which might serve as the basis of intellectual function, but they
do not possess language itself.   So that the mere absence of
this external mark would justify us in inferring the absence of
those mental functions of which it is the mark.   As a rule, it is
not any physical obstacle, as is so often thought, which prevents
animals from talking.   In very many animals the development
of the organs of speech has gone far enough to enable them to
clothe thought in words, if the thought were there to clothe.
The question why the animals do not talk is most correctly
answered in the old way : because they have nothing to say.
Only we must add that certain movements and sounds characteristic of feelings and ideas seem to be the forerunners of
language ; and that animals give signs that in this connection,
as in others, their mental life is the immediate precursor of our
own.

Indeed, the importance of association for the animal consciousness recalls what we have already said of its value for the human
mind.   When we began our consideration of the mental life of
animals, we condemned the tendency of animal psychology to
translate every manifestation of ' intelligence ' into an intellectual
operation.   The same reproach could be made against certain
more or less popular views of our own mentality.   The old
metaphysical prejudice that man ' always thinks ' has not yet
entirely disappeared.   I myself am inclined to hold that man
really thinks very little and very seldom.   Many an action which
looks like a manifestation of intelligence most surely originates
in association.   Besides this, man is constantly translating acts
of logical thought back again into customary associations, and
so increasing the sphere and the intellectual consequences of the
associational processes.   By practice we can reduce anything to
association.   Trains of thought which at first involved considerable intellectual labour are completed with increasing certainty
and mechanical facility the oftener they are repeated.   We can
hardly overestimate the facilitation of logical operation and constructive imagination which this tendency brings with it.   All

the work spared by associative practice can be employed in new intellectual achievements. For this reason, thought proper is continually engaged with permanent mental associations at the same time that it is making new ideational connections. It is a process compounded of logical and associative connections of ideas. We may rightly give the name of thought to a train of ideas whose associations are manipulated by the attention for definite intellectual ends, and are only allowed to have weight within the bounds set by those ends. The best confirmation of this is given by the expression of thought in language. Though the general content of the expression is the result of an intellectual process, still the ready-made thought-forms furnished by association play no small part in the whole process.

The fruitfulness of this interaction of association and intellection helps us to understand how it is that many psychologists, so far from translating all mental activity into logical reflection, prefer, on the contrary, to regard thought and imagination as forms of association. We saw earlier that there are external and internal characteristics which clearly differentiate these processes. The associationist psychology can give no account of them. It ignores them, identifying imagination with memory and referring logical thought to logic as distinct from psychology, as though the forms discovered by logic must not in the last resort rest upon psychological facts and laws. According to this school, the creations of imagination and intellect keep even pace with the activity of association. Dreaming and mental disturbance furnish, of course, a direct refutation of such a view; but the associationist theory makes light of that. It pays so little regard to the laws of ideation that it does not need to spend any time in considering exceptions.

§ V

In spite of these essential differences between the intellectual processes and pure associations, there is an intimate correlation and mutual furtherance of the two in our own consciousness. That is possible simply because they are both stages in the same development. The duty of association is to form those multifarious connections among the conscious elements which enable us to comprehend a whole number of previous experiences into a resultant mental force, such as that which is employed in each separate act of voluntary attention, and on the basis of which the activity of association may be directed to the achievement of definite intellectual ends. So that intelligence springs from association, and then turns round again to enrich it by new connections which will facilitate the employment of thought in the future.

It is this relation between association and intelligence which must decide our answer to the final question which arises out of our investigation of the intellectual capacities of animals : are we to think that the gulf which separates the animals from man, which is oh its mental side the gulf between association and intelligence, can ever be bridged ?

In view of the facts of individual human development, we can hardly do anything else than reply with an unconditioned affirmative. The boundary line between the purely associative processes which simulate intelligence and really intellectual actions may be crossed, because in actual fact it is crossed in the life-history of every one of us. From the stock of associations which we begin to lay in from our earliest childhood, there gradually develops the collective mental force of the individual personality, which manifests itself in self-consciousness, in active attention, and in voluntary control of ideation. And in this last especially we can clearly trace the influence of the increasing store of stable associations and the corresponding enrichment of affective and conative mentality which comes with it.

But in its application to the relation of man to the animals our question falls into two special inquiries : is it probable that species or individuals of animals existing to-day will ever cross the boundary line ? and is it probable that man himself at some period of his development bridged the gulf which to-day separates him from the animals ?

The first of these questions may be as certainly answered in the negative as the latter may in the affirmative. The step from association to intelligence proper is undoubtedly the longest ever taken in the course of mental evolution. When once the mind has reached the level of logical thought and constructive imagination, it has before it that prospect of unlimited advance which must inevitably at some point in time be realised in civilisation and history. That any species among the higher animals can make this tremendous progress is rendered altogether improbable by the general nature of their psychophysical organisation. Besides which this organisation appears to be so far determinate that further modification cannot transcend very narrow limits. And perhaps the struggle for existence in the organic world would prevent any large number of quite differently constituted beings from rising by their own efforts to the height of civilisation and historical existence on a single planet.

It is otherwise with the second question. The laws of physical development put it beyond all doubt that man passed through lower forms of life before he reached his present stage of organisation. And the laws of mental development make the same conclusion at least exceedingly probable. Just as every one in the course of his own individual development takes the step from association to that intellectual conscious activity which arises from it, so must mankind at large have done at some point in the world's history. It was the first step from savagery to civilisation. And surely it is no disparagement of the value of mental development to regard it as having been from the beginning what we see it to-day,—an evolution of mind from itself, proceeding under the conditions set by the environment in accordance with the universal laws of mental life.

§ I. Connection of Affective States in Consciousness. § II. Sensory Accompaniments of Compound Feelings. § III. Emotion. § IV. Intellectual Feelings.

§ I

OUR original plan in these lectures was to begin with the examination of *particular* mental processes,—sensations and ideas, feelings and voluntary actions,—and then to turn our attention to the interconnection of all these processes in consciousness. But when we came to analyse complex states of mind, it seemed better to take their components separately and examine them as we examined the simpler phenomena of mind, although their isolation was no more than an abstraction. We have accordingly spoken in the preceding lectures only of the ideational side of consciousness. We could not leave feeling and will altogether out of account, because of their importance for the apperception and association of ideas. But we said nothing of the relation of feeling and internal voluntary action to the other affective states of consciousness and to external voluntary action. We will now glance briefly at the more complex affective processes.

The sense-substrate of all the more complex affective states we have found to be the *common feeling*. Like the particular sense-feelings which give rise to it, this is either a pleasurable or an unpleasurable feeling. Indefinite as these categories may seem, they are characteristic for the mental nature of affective activity in general. The physical excitations underlying the sensations to which affection attaches differ only in intensity and in the nature of the stimulus which arouses them. Sensations as such, therefore, present only intensive and qualitative differences. But when the physical excitation also gives rise to an affective process, we find these two polar opposites of pleasure and pain, the principal determinant of which is the intensity of stimulus. As we intensify any sense-impression, pleasurable feeling passes gradually into unpleasurable, and that into pain. While the intensity of the physical excitation increases continuously, its affective concomitant changes from one quality into its exact opposite.

At the same time the conditions of this qualitative change are really present in the physical excitation. The effect of stimulation upon the course of the physiological processes in the organism must be either favourable or inhibitory. Moderate stimulation is indispensable for the continuance of normal function. Organs which are not used for a long time degenerate morphologically. Stimuli whose intensity is regulated by the normal capacity of the bodily organs themselves are therefore most beneficial for the maintenance of life : they call for a uniform expenditure of energy which is never pushed to the limit of exhaustion, but always leaves some small surplus of force in reserve. These normal internal stimuli also excite feelings of moderate intensity which unite to form the common feeling of comfort. When, on the other hand, the intensity of stimulus becomes so great that the organs are in danger of exhaustion or of entire destruction, there follows either a general disturbance or a temporary arrest of the vital mechanism. This excessive stimulation conditions not only a morphological and physiological derangement of the bodily organs, but also feelings of unpleasantness or pain.

The reason for these differences in the affective character of the excitation,—differences of pleasure and pain,—is given, therefore, with the physical attributes of stimuli and the changes which they set up in the organs of sense. But the differences themselves are altogether mental in nature, manifesting themselves simply as modifications of the general state of mind. Though originally produced by sense-stimuli, they contain no necessary reference to physical processes, so that we may have feelings which bear the same relation to the simple sense-feelings as does a train of logical thought to a particular sensation. Pleasure and pain mean alteration of our general psychical condition. They do not, therefore, require a sense-stimulus for their origination ; they may be called up by the mere idea of a sense-stimulus or by the intellectual content of a single idea or an ideational association.

At the same time, the purely mental feelings always rest upon a sensory substrate ; just as on the intellectual side of mind the most abstract concept is still so far dependent upon sensation that it cannot be thought without translation into a vicarious sense-idea.

This connection with sense can be demonstrated just as plainly for feeling as it can for idea. Feelings which attach to the most complicated ideational series have the same names as the simple organic feelings. 'Pain' may signify either the excessive excitation of a sense-organ, or the higher degrees of general mental discomfort. We speak of the 'pain' caused by the loss of a friend, or by the disappointment of our hopes, just as we do of the 'pain' of a wound or of a diseased organ. We talk of 'burning' love, of 'pressing' care, of 'gnawing' remorse. In a word, our names for all kinds of affective states which have no direct connection with sense-impressions are derived either from peripheral sensations or from the feelings to which they give rise.

We may, of course, term this secondary meaning of the words purely figurative. We speak metaphorically of being 'torn' by conflicting emotions. So 'agony' of mind, the 'weight' of care, and similar expressions which are used both for bodily and mental conditions would be metaphors which we easily overlook because we are always using them and have no others at our disposal. Nevertheless there must be some psychological reason for this figurative characterisation of the affective consciousness ; there must be some sort of relation between the sense-feeling from which the name is taken and the intellectual feeling to which it is applied. The most obvious relation would be a temporal association of the two processes.

The question then arises whether mental pain is associated with bodily, and whether when pain 'presses' and remorse 'gnaws' the sense-feeling of pressure or of gnawing pain is really present.

•

§ II

If we look closely at the intellectual feelings, especially in their more intensive stages, we can hardly doubt that they are invariably accompanied by sense-feelings. These concomitant feelings often attain an intensity equal to that of the sense-feelings aroused by direct external or internal stimulation. Sometimes they may be even localised with some degree of definiteness. They always evince a determinate quality which varies with the general affective condition, and which is reflected in the expressions which we employ to describe this. All excessive feeling is attended by physical pain, whether diffused over the body generally, or restricted to a particular organ. Moderate excitations also affect the sense-feelings though less strongly, and are more definitely localised. References to this localisation of the sense-stimulus in particular affective states are found in ancient literature. Every passion was supposed to be seated in a particular organ ; and it must be admitted that where observation was wanting imagination took its place. Anger was placed in the liver, envy in the spleen, the higher emotions in the organs of the breast. Even to-day the heart is the seat of the most various affective states. Care and disappointed hope bring on heartache ; despair dies of a broken heart ; love through all its changes and chances has its source and centre in the heart, and ' faint heart never won fair lady.'

There is really good reason for this relation of the heart's activity to the state of feeling ; for the cardiac nerves are those most easily excited by changes in our affective condition. Every affective excitation manifests itself in a weaker or stronger, quicker or slower, heart-beat. Joy and hope make the pulse quick and strong ; care and anxiety render it weak and slow ; terror arrests it altogether. And there are many indications that other organs react to affective changes. It has often been noticed that violent anger results in a return of the bile to the blood, which means a derangement of the function of the liver. The lachrymal glands are very easily excited by the feeling of sorrow. And we should undoubtedly discover other similar interrelations were it not that they have no external symptoms. Besides the particular organ which is especially concerned in a particular affective state, there are always other organs more or less affected ; and it is the common feeling resulting from the sum total of these sensations that constitutes the sense-substrate of the total affective condition. The muscles, *e.g.*, are almost always involved in this secondary excitation. We have a direct measure of the energy and tension, or of the exhaustion and relaxation of our muscles in sensation of movement ; and our general affective state is altogether different according as the limbs are motile and elastic or are heavy loads to weigh the body down. The feeling of the moment is here of the greatest importance. A feeling of joy and excitation makes movement easy and prompt ; a depressing feeling renders it slow and heavy.

We must try to explain how it is that sense-excitations are always present in greater or less degree in affective processes. The view of the ancients that the excited organ is the direct seat of the feeling cannot, of course, be maintained. We know quite certainly that the parts of the body most closely related to mental activity are the central organs of the nervous system. It is here that the impulse must arise which has the sense-feeling as one part of its result. The symptoms observable in the peripheral organs simply indicate that this central excitation has a different seat at different times, and consequently produces different organic changes,—now altering the pulse, now deranging the liver, now affecting the muscular system. In fact, we have before us a phenomenon which presents a certain analogy to reflex action : only that its source is not to be looked for in external stimulation, but in an activity of the central organs.

These peripheral symptoms are of especial importance, as proving that there is no complete separation of mental process and bodily function in the sphere of feeling any more than in that of ideation, but that the two are intimately interconnected.

§ III

The affective processes which attach in consciousness to ideational connections are generally designated *emotions*. They are among the most important mental phenomena, exerting a marked influence both upon ideation and voluntary action. They resemble feelings in that they are subjective processes not directly referred to external objects ; they differ from them in that they involve change in ideation and reactions in the organs of movement. Feelings, that is, are not accessible to external observation, or at least only become so when they pass over into emotions. Then they are reflected in certain *expressive move-ments.* These are further connected with reactions of the heart, the blood-vessels, the respiratory muscles, and certain secretory organs, which take on a special characteristic form in each par-ticular emotion.

This twofold relation of emotion to feeling and ideation has led to a diversity of view as to its nature. It has been regarded both as an intensive feeling, and as a feeling originating from the train of ideas. Neither of these definitions does it full justice. The typical emotion has three stages : an initial feeling ; a subsequent change in the train of ideas, intensifying and quali-tatively modifying the initial feeling ; and (always supposing that the emotion is distinct and well defined) a final feeling, of greater or less duration, which may possibly give rise to a new emotion of which it forms the initial feeling. The principal difference between feeling and emotion, that is, consists in the second stage : the alteration in the train of ideas. The presence of this alteration enables us to divide emotions into two classes, excitant and inhibitory. Instances of the former are joy and anger ; of the latter, terror and fear. At the same time all very intensive emotions are inhibitory in character, and it is only when they have run some part of their course that their excitant side comes to consciousness. On the physical side, the effect of emotion upon the train of ideas is accurately reflected in ex-ternal movement. The excitant emotion quickens ideation, and involves heightened mimetic and pantomimic movement, increase of cardiac activity, and dilatation of the blood-vessels ; the in-hibitory emotion paralyses, or at least relaxes, the muscles, slows the heart-beat, and contracts the vessels. All these physio-logical effects have their accompaniment of sense-feeling which intensifies the affective elements in the emotion.

Less intensive degrees of emotion are called *moods.* It is a general rule that the duration of emotion varies inversely with its intensity ; so that moods are more permanent states of mind than emotions proper. Violent emotions are sometimes termed *passions*. The name indicates that strongly affective states, which oscillate between the feelings of pleasure and pain, tend invariably towards the side of the latter. 'Passion' also implies that a particular emotion has become habitual. Hence the word is often used to denote a permanent condition which finds its expression in frequent outbursts of emotion.

The most indefinite emotions are joy and sorrow. All the rest may be regarded as forms of one or the other of these two fundamental moods. When, *e.g.*, sorrow is directed upon the external object which excites it, we call it care. We can only be careful about others ; and if we wish to express the fact that an object arouses no interest in us, we say that we do not 'care' about it. The subjective opposite of care is melancholy. The melancholiac is centred in himself; he withdraws from the world to brood in solitude over his own pain. Care and melancholy become anxiety and dejection when they pass from emotions to permanent moods. Intermediate between these objective and subjective forms of sorrow, stand gloom and depression. We may be gloomy as to our fate in the world and depressed about a loss we have suffered, or we may be gloomy or depressed without any external reason simply because our mood will have it so.

Joy, like sorrow, assumes different forms according to the direction which it takes. But we have not nearly so many words to express joyous emotion as we have to express sorrow. A joyous mood we call cheerfulness, or in its higher stages hilarity. But we cannot tabulate the joyous emotions as objective and subjective, as we could their opposites. It may be that our poverty of words points to a distinction in the facts of our affective life. The joyous emotions appear to be more uniform, less variously coloured, than the sorrowful.

The emotions of joy and sorrow, whether their reference is mainly external or to the feeling subject, are always subjective in character ; the affective excitation of our own minds is always the principal thing. A mood, on the other hand, may be objectified by our putting our own feelings into the external objects which excite them. If joy and sorrow are the expressions of an internal harmony and disharmony, these objective emotions are the result of some external harmonious or inharmonious impression. Like and dislike are the most general forms of objective emotion, corresponding to joy and sorrow on the subjective side. They further imply a movement to or from the object; what we like attracts us: what we dislike repels us. And this movement finds its expression in the various particular forms in which the general emotions occur. The attraction which a pleasing object has for us we call charm; a thing is 'charming' which both pleases and attracts us. The opposite of charm is repulsion, a violent dislike which makes us turn away from an object in displeasure. Repulsion becomes aversion, and at a still higher stage anger, when it is turned directly upon the repellent object; it becomes chagrin and mortification if the unpleasant mood can find no outlet. The extreme degree of anger is rage; the extreme of mortification is exasperation. The opposite of chagrin is contentment; when pleasantly con-. cerned with external objects it becomes delight, when quietly occupied with its own affairs, happiness.

The two opposite processes of charm and repulsion find a meeting point in indifference. Indifference has a tendency in the direction of unpleasantness; when sense or thought is sated with the indifferent or perhaps originally attractive object, it passes over at once into repugnance. Repugnance is as much sense-feeling as emotion. In the latter shape, it has an objective form, antipathy, and a subjective, discontentedness. If the emotion becomes a permanent mood, we have weariness and dissatisfaction.

In all these cases, emotion and mood are at once distinguishable from sense-feeling by their connection with a train of strongly affective ideas. When we feel joy or sorrow, our mood is the result of some pleasant or painful experience which may be resolved into a number of ideas. If we are mourning the death of a friend, our consciousness is filled by affectionate memories, more or less clear or distinct, which co-operate to produce the emotion. If we are made angry by some insolent remark, our first feeling is one of violent displeasure; then our mind is flooded by a torrent of ideas connected with ourselves, the personality of our assailant, and the more immediate circumstances of the insult. Most of them will not attain to any degree of clearness, but all are held together by the feeling of displeasure, which in its turn is intensified by the sense-feelings accompanying our expressive movements.

A simple sense-idea which has no special relation to our past mental history will, therefore, hardly be able to excite an emotion, though it may call up quite intensive sense-feelings. Where an emotion appears we may assume the presence of memorial ideas, of experiences in which a similar sense-impression was somehow concerned. The full and harmonious tone of a peal of bells sounds holiday-like to us, because we have been accustomed from childhood to interpret the chimes as harbingers of holidays and religious festivals; the blare of the trumpet reminds us of war and arms; the blast of the horn brings up the green wood and the tumult of the chase; the call of the cuckoo tells us that spring has come; the chords of the organ suggest a congregation assembled for devotion.

It is probably memory again which determines our affective reaction to colour-impressions, although in their case the ideas aroused are not so clear or distinct. Why is white the colour of innocence and festivity, black the colour of mourning and severity? Why do we choose blood-red to express energy and spirit, or purple to express dignity and solemnity? Why do we call green the colour of hope? It would be difficult to trace the mood to its original source in each particular case. In many cases it probably arises from an obscure association of the colour with the occasions when custom prescribes its use. Purple has been the royal colour since time began; and black is almost everywhere the colour of the mourner's garments.

It is true that this association does not fully explain the connection between the sense-impression and the mood which it arouses. There must be some original reason for the choice of one particular sense-stimulus, and no other, as the expression of an affective state. It is perhaps justifiable to look for this reason in the relationship between the sense-feeling and the affective character of particular emotions. The sensation as such could then originally excite only a feeling; but this might become emotion as soon as consciousness had at its disposal affectively efficient memorial ideas, into which the sensation naturally entered as a normal constituent.

Emotions exhibit peculiar modifications when their affective character is not determined, as in the cases hitherto considered, by impressions and ideas belonging to the present and thought as present; but by ideas which refer to the *future*, whether in the way that an occurrence is definitely expected, or that some indefinite idea of the future gives rise to a feeling, and through it to an emotion.

The most general of these expectations of the future is expectation itself. In it we outrun the impressions of the present, and anticipate those which the future will bring. We look forward to its realisation; and if this realisation is postponed, it becomes what we call strained expectation; the bodily feeling of strain accompanies the emotion. In expectation the muscles are tense like those of a runner awaiting the signal for the race, although very possibly the expected impression demands no motor response whatsoever. Expectation becomes watching if the expected event may happen at any moment, and our sensory attention is wide awake to prevent its passing unnoticed. The tension is relaxed with the appearance of the expected impression. If the consequent perception fulfils our expectation, we have the emotion of satisfaction; if not, that of disappointment. Satisfaction and disappointment constitute sudden relaxations of expectant attention. If expectation is prolonged, its tension will gradually disappear of itself: for, as you know, every emotion weakens with time.

The opposite of disappointment is surprise. Surprise is the result of an unexpected event. In it we have ideas suddenly aroused by external impressions, and interrupting the current train of thought in a way which we did not anticipate, and which at the same time strongly attracts our attention. Surprise may be in quality pleasurable, painful, or altogether indifferent. A special form of it is astonishment. Here the event is not only unexpected at the moment, but unintelligible for some time afterwards. Astonishment is therefore a kind of continued surprise. If it passes into a still more permanent mood, it becomes wonder.

The feeling of rhythm, which is the single psychological

motive in dancing, and ranks with harmony and disharmony as a psychological motive in musical composition, contains the elements both of expectation and satisfaction. The regular repetition in rhythmical sense-excitation makes us expect every succeeding stimulation, and the expectation is immediately followed by satisfaction. Rhythm therefore never involves strain, or if it does, it is simply bad rhythm. · In pleasant rhythms satisfaction follows expectation as quickly as possible. Every impression arouses the expectation of another, and at the same time satisfies the expectation aroused by its predecessor, whose temporal relations it reproduces. Rhythm, that is, is an emotion compounded of the emotions of expectation and satisfaction. A broken rhythm is emotionally identical with disappointment.

Hope and fear may be regarded as special forms of expectation. Expectation is indefinite. It may refer to a desirable or undesirable, or perhaps to a relatively indifferent, event. Hope and fear definitise expectation : hope is the expectation of a desirable result, fear the expectation of something undesirable. It is hardly correct to call hope a future joy, and fear a future sorrow. The feelings can as little penetrate into the future as the senses. Hope and fear are the expectation of future joy and future sorrow, but not joy and sorrow themselves. Either of them may be unrealised, just as expectation may lead to satisfaction or disappointment.

Fear of some immediate disagreeableness is called alarm. Fright bears the same relation to alarm as does expectation to surprise. Fright is the surprise occasioned by some sudden terrifying occurrence. It becomes consternation when the occurrence physically paralyses the individual experiencing it ; and it is called terror when he stands amazed before the event. Consternation is therefore the more subjective side of fright, and terror its objective side. If fear is continued, it becomes uneasiness. The uneasy mind is always afraid ; every occurrence alarms it. In other words, the emotion has become permanent, but at the same time somewhat less intensive.

§ IV

The emotions both of the present and future assume the most varied forms according as the ideational content of the moment changes. Especially important are those attaching to certain intellectual processes and originating in the peculiar feelings which accompany them. We can distinguish four kinds of *intellectual feelings :* the logical, ethical, religious, and æsthetic. Attaching as they do to very complicated ideational connections, they almost invariably pass over into emotions, and in that form exert upon our mental life an influence which far exceeds that of any other affective process. Their analysis belongs, of course, to the special sciences from which they take their name. But we will devote a few words to the logical emotions ; first because they are often overlooked altogether, and secondly because their relationship to the emotions of the future enables us to use them as illustrations of the passage of emotion in general into the particular forms of intellectual emotion.

It is well known that the rapidity of the course of thought exerts a considerable influence upon our general affective condition. It is not indifferent to us whether our ideas succeed one another at their normal rate, or proceed slowly with many inhibitions and interruptions, or pour in upon us in perplexing confusion. Each of these cases may be realised whether from internal or external causes. Our state of mind at the moment, the topic of our current thought, and external sense-impressions may all be of determining influence. The traveller in a new country is well content when his carriage takes him quickly from one impression to another,—not so quickly that he cannot assimilate what he sees, but not so slowly that he is always wishing himself farther on amid new scenes. He is not so satisfied if he is lumbering along in a heavy waggon, passing for days together through the same scenery when he longs to be at his journey's end, or is curiously anticipating novel experiences. Nor is he quite happy when the railway takes him swift as an arrow through a country rich in historical association, and he tries in vain with deafened ears and tired eyes to fix some of its features in his memory.

This general result can be produced by internal causes just as well as by the variation of external impressions. If you have to solve a mathematical problem in a short time, your thoughts trip each other up ; you are in a hurry to get on, but are obliged to go back, because you have been following out a second thought before you had brought the first to its conclusion. And it is not less disagreeable to be arrested in the middle of your task because your thought halts, and you cannot answer the next question. On the other hand, work becomes a recreation when one result leads certainly and easily to another.

We have therefore the three emotions of confused, inhibited, and unimpeded thought. The two last are closely related to the emotions of effort and facility. Correlated with these are the sense-feelings attaching to ease and difficulty of muscular action. They are generally present to some degree in the corresponding emotions, even when the causes of these are wholly mental. The feeling of effort is a weight which presses upon the affective condition, and whose removal is accompanied by a sudden feeling of pleasure. This characteristic feeling of relief affects us mainly by way of contrast to our previous mood.

Special forms of the emotions of unimpeded and inhibited thought are those of enjoyment and tedium. In enjoyment our time is so well filled by external or internal stimuli to ideational activity that we hardly notice its passage, if we do at all. The nature of tedium is indicated by its name. Our time is unoccupied, and passes slowly because we have nothing else to think of. Tedium, therefore, has a certain affinity to expectation, but it is an expectation that has remained indefinite. It does not expect or anticipate any particular occurrence, but simply waits for new events of whatever kind they may be. A long continued expectation always passes into tedium, and an intensive tedium is hardly distinguishable from strained expectation.

Related to the feelings of effort and facility are those of failure and success. Investigation and discovery are attended by feelings which show a close resemblance to those of effort and facility. The feelings of agreement and contradiction are somewhat different. They originate in the comparison of simultaneous ideas, which in the one case are accordant, and in the other refuse to be connected.

Doubt, which we have already discussed under the heading

of oscillatory feelings (Lecture XIV., p. 219), is not the same as contradiction. The doubter cannot decide which of two alternatives is the correct one ; he is in contradiction with himself. The conflicting ideas are nothing real, but simply products of his own thought, so that there is always the possibility that the contradiction in doubt may be resolved by experience or more mature consideration ; and so far doubt is related to the emotions ·of the future. This relationship becomes still more apparent in a special form of doubt,—the feeling of indecision. When we are undecided, we are in contradiction with ourselves as to which of different roads we shall follow, or which of different actions we shall choose. Indecision is therefore a doubt implying reference to action and resolved by it.

§ I. Expression of the Emotions. § II. Impulsive and Volitional Action. § III. Instinctive Action. § IV. Theories of
Instinct.

§ I

WE have seen that the movement among ideas which is
characteristic of emotion in general is always attended
by physiological movements, which exhibit specific differences
according to the intensity and quality of the particular emotion.
These *expressions of the emotions* have more than a symptomatic
interest : they are _genetically_ important. It is through them
that we are able to understand the relation of emotion to the
development of _external voluntary action_. Emotion bears the
same relation to this as feeling does to the *internal* will-process.
The transition from volition to external voluntary action runs
parallel with that from feeling to emotion. But just as not
every feeling develops into a volition, so emotion need not
necessarily or invariably lead to a voluntary act. To take a
special instance, the control of emotion which is natural to the
morally and intellectually mature consciousness consists for the
most part in its inhibition at the boundary line which separates
it from external voluntary action. In the savage and the animal
any emotion that is at all intensive passes over irresistibly into
action. And even where the inhibition is effective, the internal
tension always finds relief in movements whose only *differentia*
is that they are not intended to bring about any determinate
result. In this way arise the 'pure' expressions of emotion,
which are simply symptomatic of a particular internal affective
state. They are *rudiments of true voluntary actions.*

Among the regular expressions of emotion are the _mimetic_
movements. They are the most characteristic of any for the
nature of the particular emotion. Physiologically considered,

they correspond to definite reflex movements in the facial sense-organs. Thus the mimetic movements of the mouth, which are so important for the expression of the affective state, resemble the reflexes set up by the action of taste-stimuli (acid, bitter, sweet, etc.). When a man 'looks sour,' the lips are drawn out laterally, so that there is more space between them and the sides of the tongue, which are especially sensitive to acid. In the 'bitter' expression the posterior portions of the tongue and palate, the parts most sensitive to bitter, are held apart. 'Sour' and 'bitter looks,' *i.e.*, depend on reflex movements which serve to prevent the contact of certain ill-tasting substances with the portions of the organ most sensitive to them. With the 'sweet' expression it is just the reverse. The tip of the tongue is the part most sensitive to sweet. The expression consists in a sucking movement, calculated to bring the tip of the tongue into as complete contact as possible with the sweet substance. We may imagine that all these movements depend upon a uniform connection of certain nerve-fibres and nerve-cells, the reflex movement being gradually restricted by that process of regulation which we discussed above (Lecture VIII., p. 128). Direct evidence for this latter supposition is furnished by the fact that in early life the mimetic movements are more diffuse and indefinite than they become later on ; the movements of the mouth, *e.g.*, are invariably accompanied by general facial contortions, and often by movements of other parts of the body.

But the mimetic movements appear not only as the response to special sense-stimuli, for which they are teleological reflexes, but also as the expression of internal emotion. Unpleasant excitations, of whatever kind, will manifest themselves in 'sour' and 'bitter' looks. The 'bitter' expression varies with the different degrees of contempt, abhorrence, and loathing ; the 'sour,' which culminates in weeping, may denote mental as well as physical pain and affective disturbance. So that the facial expression becomes *symbolic*, so to say ; it is the sensible index of a mental condition. This assumes, of course, that the sensible expression and the sense-excitation producing it are more or less closely related to the emotion. And that is the case. All emotions, you remember, are accompanied by sense-feelings, though these may only become clearly perceptible when the emotion is very intensive. Now these mimetic movements mean movement-sensations from the muscles, and they in turn give rise to sense-feelings which call up clearly enough the peripherally excited sensations to which they correspond. When we are looking 'sour' or 'bitter' or 'sweet,' we think that we are actually tasting some acid or bitter or sweet substance ; because whenever these stimuli affect us the reflex movement follows, and so the sensation of the mimetic movement is fused with the sensation of taste proper.

The process by which these movements develope, then, will be somewhat as follows. Every affective excitation is attended by bodily movements. Some of these gradually obtain an advantage over the rest, those, *i.e.*, whose affective tone is similar to that of the emotion. This is a process of restriction of movement, completely analogous to that of the gradual restriction of reflex movement discussed above. It is true that the mimetic movements and the sense-feelings attaching to them are few in number as compared with the infinite diversity of emotion and mood. They can do no more than indicate the general class to which a particular affective state belongs. Still they admit of a certain amount of variation, as different facial expressions are combined or modified in detail. But the mimicry becomes more and more indefinite and equivocal as the emotion grows more intensive.

Those mimetic movements which serve as a means for the expression of emotion and mood cannot obviously be regarded as true reflexes, whose invariable antecedent is the operation of sense-stimuli. They may with better right be termed *impulsive* movements, if we understand by 'impulse' the effort of consciousness to induce the physical condition appropriate to a given psychical condition. The reflex need not involve any conscious process at all ; in impulse some such process appears as a necessary condition, either antecedent to the external movement, or at least simultaneous with it. Do not misunderstand the ascription of a *symbolic* meaning to impulsive movements as compared with the same mimetic movements in their purely reflex function. We do not mean that they once were simply reflexes, and that the symbolic meaning has gradually been developed from their former significance. Observation decisively negatives any such view. We have, on the contrary, every reason to suppose that the movements were *first impulsive and later became reflex.* The new-born child, which has never tasted acid or bitter or sweet, makes the corresponding mimetic movements quite unmistakably. When it cries, the 'sour' and 'bitter' expressions appear, alternately or in combination. Before its lips have ever closed on its mother's breast, it makes sucking movements, and so 'looks sweet.' In the course of some weeks there develops the mimetic movement of laughing, the index of pleasurable mental excitation.

These phenomena indicate quite clearly that the human child when it first comes into the world possesses feelings and emotions ; and that even at this early stage of life the emotions find expression in movements whose affective character is related to that of the emotions themselves. There is presupposed either a previous mental development, or a connate adaptation of bodily movement to mental state. There has obviously been no such development in the course of the individual life. We must, therefore, assume a connection which for the individual is original, *i.e.,* connate.

How is this to be explained? The most obvious thing to do is to derive the association from the organic interconnection of nerve-fibres and nerve-cells. We may assume that the great majority of the sensing organs are intimately connected within the central nervous system with the motor fibres running to the mimetic muscles. But there is always the possibility that these connections are further developed in the course of the individual life ; movements which at first were diffuse and indefinite gradually becoming restricted. And observation raises the possibility to certainty : we find a continuous and continually increasing restriction of the mimetic movements. At the same time, we found ourselves obliged, in dealing with the general theory of the reflex process, to assume the existence of a certain disposition or tendency due, to the original interconnexion of fibres in the central nervous system. The theory explained the increasing limitation of the reflex response in the life of the individual by the supposition that the connexion of sensory and motor nerves is the most direct possible, *i.e.,* that it represents the path most usually followed by an excitation-process.

But when we said just now that the connection of the mimetic movements with emotion was 'original for the individual,' there was implied the possibility of pushing our investigation beyond the limits of the individual life. The question now becomes a problem in evolution.

§ II

You know that Darwin based his hypothesis of the origin of species by 'natural selection' upon two principles,—the principle of variability and the principle of inheritance of individual characteristics. It is surely evident that these cannot be meant as really explanatory principles, but only as general rubrics, under each of which are included a whole number of problems to be solved. For our present purpose, however, it is enough to remark that, whatever their ultimate causes may be, they are undoubtedly as valid for mind as they are for body. Suppose that both conditions, variation and inheritance, have been at work for an indefinite time, and that the physical peculiarities of the organic world have differentiated it more and more ; there will also be constant differences to be found in mental disposition or tendency. The perfection and differentiation of species as regards body and as regards mind constitute, that is, two parallel processes of development. When certain nerves, muscles, and central organs habitually function in response to psychophysical impulses, their physical development must necessarily follow suit ; while, on the other hand, furtherance of physical development means increase of mental function.

If we apply this hypothesis in our special case, it seems quite adequate to explain the appearance of connate impulsive actions observed in the new-born child. There is no reason why in the course of many generations certain nervous fibres and nervous cells should not advance in development and others recede, new ones be produced and old ones disappear. Even as between different individuals of the same species the number of these elements may differ very considerably. And the differences of family, race, and species arise through the summation of these individual variations by inheritance ; while upon the development of the separate parts of the nervous system and its terminal organs depends further the capacity of a simultaneous excitation of different parts of the body, *i.e.*, the inclination towards combined movements of some particular kind.

We may therefore regard the conditions of the development of impulsive movements as at once *physical* and *psychical*. Let us suppose that there exists an organism with the very simplest nervous system, consisting, say, of a few cells and connecting fibres. The impulsive movements called forth in such a creature by sense-stimuli will be altogether irregular. But very soon particular sensory fibres, which, owing to their position or for some other reason, are more frequently stimulated from without than their neighbours, will begin to develop more strongly. The immediate result will be a corresponding development of the motor fibres most directly connected with them. In this way a connexion will be formed, which may be perpetuated ; *i.e.*, which will be present in the individual's descendants from the very first. Regarded from the psychical side, this process appears as a gradual restriction of the effect of emotion to those actions which call forth feelings similar to the emotion in affective tone, and which thus enter into intimate association with it. The association cannot, of course, be inherited as such. But since the corresponding physical connection within the nervous system is transmitted from one generation to another, the impulsive movement in the individual is just as reflexly certain a response to the central excitations underlying emotion, as it is to external sense-impressions whose effect for feeling is analogous. In this way the affective associations which have been gradually acquired in the course of a long generic development may be present from the first, and require but little further development by individual practice.

At this point it becomes evident that no hard and fast line can be drawn between impulsive and expressive movement. Every impulsive action is a consequence, and therefore an expression of emotion. The animal which, impelled by its desire for food, throws itself upon its prey, is thereby giving expression to a state of mind dominated by emotion just as certainly as the man who expresses his grief by tears. The only difference is, that in expressive movement in the narrower sense the external action has no special purpose ; it has no direct effect upon the satisfaction of the pleasurable or painful feeling connected with the emotion. In this sense the expressive movements are rudiments of impulsive movements. But very frequently the more active emotions, such as anger or the pleasure in a coveted object, pass directly over into impulses and impulsive actions proper. Anger, *e.g.*, becomes transformed into the instinct of revenge, and this finds its expression in movements which seek to satisfy the revengeful feeling by an injury done to the object of anger. Impulse, that is, bears the same relation to emotion in the internal experience as impulsive action bears to expressive movement. And just as in the development of mental life impulsive action is the earlier, and pure expressive movement,—a mere rudiment of it,—necessarily the later, so the universal animal impulses,—the impulses of nutrition, of sex, of revenge, of protection, etc.,—are indubitably the earliest forms of emotion. Or if we wish to express the same thought in somewhat different language : the emotions are impulses which have become complex, but which in proportion to their complexity have lost their characteristic of activity.

We are able, then, to distinguish impulsive action from pure expressive movement by the fact that the former has a definite purpose, which is consciously attained or at least attempted, while the latter, though it shows some faint indication of a purpose, does not imply the least consciousness of intention to attain it. In saying this, we are at the same time characterising impulsive action as *voluntary* action. This, and this alone, is the criterion of voluntary action : that the thought of the end to be realised accompanies or precedes it. Impulsive action, therefore, is simple voluntary action in the sense explained above (Lecture XV., p. 228).

Again, when a feeling is transformed into an emotion, it takes part in an ideational movement which is itself accompanied by feelings. Generally some particular idea stands out in this movement as the efficient cause of the process ; and arouses the appropriate impulsive action either simultaneously with its own appearance or directly afterwards. If at this stage a number of partial emotions combine to form one compound affective state, there may plainly be present together in consciousness a number of conflicting motives. And so there is developed, in natural order from the simple, the compound voluntary action, or *act of choice*. At any one of these stages voluntary action may be mechanised to reflex action : the process has been described above in the course of our description of the separate conscious processes (Lecture XV.). The pure expressive movements also fall under this law of mechanisation ; their accompaniment of the emotions has ceased to be a matter of consciousness and volition.

§ III

Movements which originally followed upon simple or compound voluntary acts, but which have become wholly or partly mechanised in the course of the individual life or of generic evolution, we term *instinctive* actions. 'Instinct' is derived from *instinguere*, to incite or impel; and in meaning, as by definition, it comes very near to impulse. The only difference between the two consists in this,—that 'impulse' is generally used to denote the simpler purposive movements; 'instinct' to denote the more complex impulsive actions, which presuppose a long course of individual or generic practice. Instinctive action, therefore, stands midway between reflex movement and pure voluntary action. Thus the mimetic movement which follows the application of an acid stimulus to the tongue will be counted as a reflex ; hardly regarded as instinctive. But the involuntary movement of defence that a man makes when a stone is thrown at him we shall be inclined to term an instinctive action. It is evident enough that it must often be difficult to draw any very definite line between movements which have become entirely mechanical and those that still contain the impulsive element. Under certain circumstances the mimetic reaction to acid may be impulsive. Indeed, this will happen fairly often, *i.e.*, whenever there is at once associated with the acid taste the impulse to keep the tongue away from the stimulus. On the other hand, the movement of defence may appear as a simple reflex, occurring before the impression of danger has come to consciousness at all. This uncertainty of definition, combined with the current psychological restriction of the concept of will to choice, explains how it is that the chapter on instinct is one of the most debated fields in the science, notwithstanding that the now universal recognition of the genetic view of animal life in general has removed the principal obstacle to the comprehension of the more complex animal instincts.

§ IV

Still even to-day the *theories* of instinct . form a regular museum of conflicting opinions. Some regard it as a purely mechanical result of the physical organisation, a compound reflex movement, only different from the simple in that the motor responses to particular stimuli are more complicated and extend over a longer period of time. Others look upon the instinctive action of animals as a manifestation of connate ideas. A third view considers it as voluntary action, involving consciousness of end or purpose, but characterised by diminution in the clearness of ideas. The two last hypotheses have in modern times been gradually superseded by a fourth and fifth, which have grown up under the influence of the theory of evolution. These, together with the first (the pure reflex hypotheses), may be regarded as the standard theories at the present day. The first of them makes instinctive actions 'mechanised rudiments of manifestations of intelligence.' It emphasises the opinion, especially with reference to animal instincts, that this mechanisation has been going on for countless generations. The second, of which Darwin is the representative, explains instinct as *inherited habit*, determined principally by the influences of the environment and the struggle for existence, but also to some slight extent by intelligence. Like all habit, instinct has been subject to change; but natural selection has brought it about that these changes have always been purposive, advantageous to the species.

We may reject at once as wholly untenable the hypothesis which derives animal instinct from an intelligence which, though not identical with that of man, is still, so to speak, of equal rank with it. At the same time we must admit that the adherents of an intellectual theory in a more general sense are right in ascribing a large number of the manifestations of mental life in animals not, indeed, to intelligence, as the intellectualists *sensu stricto* do, but to individual experiences, the mechanism of which can only be explained (as we saw above) in terms of *association*. The precautions which the spider takes in fastening the threads of her web, and in selecting a suitable spot for it, point quite decisively to associative mental activity. The same is true of the many alterations made by honey-bees in the ordinary structure of their comb when they are disturbed by pieces of glass or other objects introduced into the hive. Indeed, it is probably impossible to adduce a single instance of instinct in which the actions of the animal do not afford evidence of some amount of individual experience. At the same time, there is another and parallel class of actions to be taken account of, which, although wholly purposive, cannot either be interpreted as the outcome of teleological reflection, or be explained from impressions and associations experienced during the individual life. When the bird builds her nest, or the spider spins her web, or the bees construct their comb, these are distinctly purposive actions ; indeed, they are more purposive than the other actions of the same animals which are explicable in terms of individual experience. If it were really teleological reflexion that led the bird to build her nest, the spider to spin her web, and the bees to make their comb, we should be compelled to attribute to these animals a degree of intelligence which the experience of a single life could hardly be expected to develop even in man.

Another argument that makes against this explanation is the regularity with which the same actions are repeated by the different members of a single species in cases where no connexion can be demonstrated between the various individuals such as might possibly account for the uniformity. Of course, there is an intimate connexion existing between the inhabitants of the same hive or ant's nest, and between parents and young in the species in which the family holds together for some little time. But in numberless instances the animal begins its life in total independence of its fellows. When the caterpillar hatches out of the egg, its parents are long since dead ; nevertheless, it spins the cocoon that they did. And, lastly, to interpret instinctive action as intelligence would in very many cases be to predicate of it a prevision of the future. It is hard to suppose that this prevision is conscious when there are neither analogous experiences given in the previous life of the individual, nor any way by which they could be communicated to it. The night-flying *phalœna* covers the eggs which it has laid with a layer of fur made from its own hair, to protect them against the cold, before the winter has come. The caterpillar changes into a chrysalis without any experience of the metamorphosis which it is to undergo.

We cannot better demonstrate the impossibility of a derivation of instinctive action from conscious reflection than by quoting an illustration from an earlier author, in which all the contradictions which the theory involves are brought together into short compass. The caterpillar of the emperor moth spins at the upper extremity of its cocoon a double arch of stiff bristles, held together above only by a few fine threads. The cocoon, *i.e.*, opens at the very least pressure from within, but is able to resist quite strong pressure from without. Autenrieth writes of this in his *Ansichten über Natur- und Seelenleben :* ' If the caterpillar acted from reflection and with understanding, it must, on human analogy, have pursued the following train of thought : that it had reached its chrysalis stage, and would therefore be at the mercy of any unlucky accident, without possibility of escape, unless it took certain precautionary measures in advance ; that it would have to issue from its cocoon as imago without having organs or strength for breaking through the cover it had spun as caterpillar, and without possessing any secretion, like other insects, which would if emitted eat through the threads of silk ; and that consequently, unless it took care to provide as caterpillar a convenient exit from its cocoon, it must certainly come to a premature end in imprisonment. On the other hand, it must have clearly recognised during its work upon the cocoon that, in order to have free egress as imago, it would only be necessary to construct an arch which could resist attacks from without while opening easily from within ; and that these conditions would be fulfilled if the arch were made of stiff threads, inclined together in the median line, and with their ends left free. At the same time it must have realised that the plan could be carried out if the silk employed for the construction of the other parts of the cocoon were employed with special care and skill at the upper end. Yet it could have learnt nothing of all this from its parents : they were dead long before it had issued from the egg ; it had had no practice or experience, for the spinning of the cocoon happens only once in a lifetime ; it could not imitate a neighbour, for the species is not a social one. And during the whole of its existence as caterpillar its understanding could have been but very little developed : it crawled about on the branch where it first saw the light, devouring leaves, an occupation which required no consideration, since the food was there waiting for it ; it clung fast with its feet, perhaps, to avoid falling to the ground, and crept under a leaf to escape the rain ; it got rid of its old uncomfortable skin some few times by involuntary contractions of its entire body, but without making any cocoon : —and that was the whole of its life, the sum of its opportunities for the exercise of intelligence.'

Instinctive action, then, cannot be explained either in terms of conscious reflection or from individual associations : the hypothesis requires an amount of prevision on the part of the animal which is psychologically impossible. But the opposite theory, recently defended by Herbert Spencer, that instinctive action is simply compound reflex action, determined by the laws of the physical organisation, is equally untenable. That the caterpillar secretes silk, the spider the material for her web, and the bee wax, is just as much a matter of physical necessity as is the emission of any other secretion. But that these substances when secreted are worked up in such definite and artistic forms is altogether inexplicable from the facts of physical organisation. That accounts for the *material* which the animal has at its disposal, but not for the *form*, which is the real result of its work.

Worse still, if that is possible, is the view which stands midway between the intelligence and reflex theories, and which regards *connate ideas* as the motives of instinctive action. The bee is supposed to have in its mind from the first a pattern of its hexagonal cell, the spider a pattern of the meshes of her web, the caterpillar a picture of its cocoon, and the bird one of the nest it is to build ; and each of these animals must necessarily translate its idea into reality. The older philosophical idealism found in such a hypothesis a welcome support for the doctrine of innate ideas. But it contradicts everything that our analysis of the human consciousness has taught us. (It is impossible to prove the existence in our own minds of ideas which do not spring from the experience of the individual life.) The congenitally deaf has no knowledge of tone, the congenitally blind none of colour. And the probability that complex ideas can be innate is infinitely less. Besides, the observation of instinct does not by any means give unqualified support to the hypothesis. If there is so definite an image of the hexagonal cell in the bee's mind, how is it that all the cells of the hive are not made of the same size? You see, there must be present in its consciousness not the idea of a single cell, but that of the whole number of cells belonging to the colony, if its action is to become intelligible in every respect. The bird builds its nest of certain determinate materials, from which it never varies except in cases of necessity. Does the innate idea of the nest include the ideas of every twig and straw used in its construction? It is evident that this theory becomes entangled in difficulties no less grave than those which proved fatal to the hypothesis of intelligent action. It requires the assumption not of a single innate idea, but of a whole connected series, in a word of an innate activity of thought with a large store of experience behind it.

Only two hypotheses remain, therefore, as really arguable. One of them makes instinctive action a mechanised intelligent action, which has been in whole or part reduced to the level of the reflex; the other makes instinct a matter of inherited habit, gradually acquired and modified under the influence of the external environment in the course of numberless generations. There is obviously no necessary antagonism between these two views. Instincts may be actions originally conscious but now become mechanical, *and* they may be inherited habits. This compromise would have a great deal to recommend it, if we might slightly alter the first theory, and make instinct, according to it, *partly* a matter of mechanised volition and partly of action which is still determined by psychical motives. If we are ourselves to appeal to the facts for a decision in favour of one of the two views or for a verdict for or against a combination of both, we shall do well to keep in mind the rules laid

down above in connection with our consideration of the manifestations of animal 'intelligence' (Lecture XXIII.). Never have
they been so sadly sinned against as in this particular chapter
of psychology on the nature of animal instinct. The first ran,
you remember, that we must always set out from known facts
of the *human* consciousness ; the second, that simple principles
of explanation are always to be preferred to complex.

We must, therefore, go on in the following lecture to discuss
briefly the instinctive actions in man. When we have done
that, we may pause to look back once more upon the very
difficult phenomena presented by animal instinct.

§ I

BY an instinctive action we understand, as remarked above, something purposive, but involuntary, half impulsive and half reflex. It cannot be doubted that in this sense many human activities come under the category of instinctive action.

We laugh and weep, we make the most complex mimetic movements, without, or even against, our wish or our knowledge. Most of our movements are determined by emotion, and volition manifests itself quite as often in the moderation or inhibition of movements as in their independent initiation. Not seldom the will simply definitises the direction of a movement; its execution is left to instinct. When we walk, it is generally volition that prescribes the road; but step follows step instinctively. Many actions at first require practice and the exercise of voluntary effort, but when once they have become familiar, may be performed under almost exclusively instinctive control. The child learning to write will laboriously copy every stroke of the pen; the ready writer needs only the intention to write some particular word, and it stands before him on the paper. The novice at the piano must strain his attention upon every note, in order to find the appropriate key; the practised player translates the printed page mechanically into the proper movements. Any movement that has become altogether habitual is made instinctively. An impulse of will is, of course, necessary at the outset; but its effect extends to a whole series of actions, and each particular one takes place without effort and without knowledge: the series once started is continued to its end with the same unconscious certainty and purposiveness as the reflex. The voluntary movements of early childhood are uncertain and awkward ; practice has not had time to transform them into instinctive acts. And the same is true of the adult whenever he wishes to perform some as yet unaccustomed action, of however simple a character. Precision and grace of movement, then, depend upon certainty of instinct, not upon firmness of will.

This transformation of voluntary into instinctive activity is greatly furthered by the influence of the environment. From the first days of life we are surrounded by our fellow-men, and imitate their actions. And these mimetic movements are instinctive in character. As soon as the child's consciousness is aroused from its first sleepy passivity, it begins to perceive the expressions of others' emotions, and to respond to them by similar emotions with corresponding impulses. The continued imitation by which a child comes to learn the language that is spoken round it is impulsive, not voluntary. Even the peculiar word-formations of child-language are not, as is often wrongly held, invented by the child, but borrowed by it from its environment,—from the words of nurse and mother, who in their intercourse with it adapt themselves to its level of mental development and capacity of articulation. And with them, again, this formation of special baby-words and imitative sounds is to a very slight degree a matter of purposed invention ; for the most part the adaptation and imitation are themselves instinctive. Voluntary act and instinctive movement, suggested by environment and example, cross and recross in human conduct from the beginning to the end of life. And if the sum of action resulting from personal choice and intellectual reflection were laid in one scale, and that proceeding from instinct and imitation in the other, there can be little doubt that the beam would incline on the side of instinct. Suppose a bird were to become interested in zoological investigation ; he might well regard mankind as the richest of all creatures in instincts. Man shares with the birds the instinct to live in wedlock ; like the fox, he educates his children ; he has the beaver's impulse to build houses, and the bee's custom of founding states and sending forth colonies ; while he has in common with the ant a pleasure in war, in slave-making, and in the domesticating of useful animals.

There is, it is true, one immense difference. In man all these instincts, at least in the form which they have assumed in the course of history, are the fruits of a continuous intellectual development, not a trace of which is demonstrable among the animals. And a great gulf is set up also by the fact that within the limits of these general norms of life individual volition has ample space for the determination of its particular conformity to them. Still, if human conduct as a whole is divided into the two great departments of voluntary and involuntary action, there can be no doubt that for the vast majority of us the principal incentive to those very acts which constitute the universal criterion of the *genus homo* is not reflection and free-will, but instinctive imitation of our neighbours. Reflection and volition begin as a rule only when the general norms of life have to be applied in the particular case. How the individual builds his house, or where he lives, may be a matter of pro-tracted consideration for him. But that mankind at large build houses and seek shelter seems to him to be as natural and right as it probably does to the bee to construct its hexagonal cells. And even the question of the particular disposal of his own life, which is so tremendously important for the civilised man, generally troubles the savage but little. He builds his hut or pitches his tent as his fellows do, and as his forbears did before him. So that human life is permeated through and through with instinctive action, determined in part, however, by intelli-gence and volition. As for that, all forms of psychically con-ditioned action are mixed processes. It hardly ever happens that a fact of consciousness admits of complete subsumption to any of the categories that psychological abstraction enables us to set up. Like mental life in general, it contains a mixture of various elements.

§ II

Instincts which, so far as we can tell, have been developed in this way during the life of the individual, and in the absence of definite individual influences might have remained wholly un-developed, may be called *acquired* instincts. You can see from what has been already said that all and each of them,—from the instinctive finger-movements of the practised pianist down to the instinct to build a shelter and wear clothes for protection against the weather,—spring from two conditions, one physiological and the other psychological. The former consists in the property of our nervous organisation gradually to *mechanise* complex voluntary movements ; the second, in the operation of the *mimetic impulse,* which is probably natural to all animals that live in any kind of society, but is especially powerful in man. This impulse is itself an instinct; mimetic movements are, as a rule, impulsive, and not volitional. But it is at the same time the fountain-head of many other instincts, and especially of those whose development is furthered by a social mode of life. It is a necessary corollary from these remarks that the first of the two conditions will be effective, even if the second be absent, in the case of the acquired instincts *sensu stricto,*—instincts developed during the individual life as a result of individual practice, such as the instinctive movements of the skilled pianist. These are purely matters of physiological practice ; so that it is not difficult to understand that the movements may occasionally become quite reflexive. The hypothesis which is most nearly adequate to this special case, then, will be the fourth of those which we reviewed above as professing to account for instinct in general :—that of the passage of intelligent into reflex action. I say ' most nearly ' : for the expression ' intelligent action ' is not admissible in the present instance, any more than in the other contexts in which we have discussed it. In most cases there are no acts of intelligence involved at all, but only associations ; and in any case intelligent action must have been reduced to association before it could become mechanised. The piano-player has first of all to form a stable association between the printed note and the movement of touch. But this association gradually lapses from consciousness, and the interconnection of movements becomes purely mechanical.

The operation of the second condition, the psychological impulse toward imitation, is to be seen,—often enough in combination with the physiological factor,—in the case of the social instincts. The fact which lies at the root of the imitative impulse is this,—that as a rule any action resulting from psychical motives excites in all individuals of the same species an emotion similar to that experienced by the agent himself. And similarity of emotion means similarity in its external expression. The simplest manifestations of the imitative impulse, then, will be found in the different forms of violent emotional expression. The passionate gestures of a speaker are reflected in the involuntary movements of his audience. As we look at a terrified or sorrowful face, our own features assume a cast in keeping with the feelings it expresses. In all these cases the imitative movements are purely instinctive. On the other hand, if the strangeness of the presentation is such as to evoke an act of will on our part, the instinctive reaction passes over into some less simple form of action. This is seen in all the human social instincts, where the sphere of instinct borders on that of custom. The phenomena here are of a kind so mixed and complex, that their instinctive element is usually entirely overlooked.

§ III

To be distinguished from these acquired human instincts are others, which are *connate*. They are, perhaps, more modified in man than in the other animals by civilisation and education ; but they are still indispensable for the origin of the most important vital functions. There are in particular two fundamental instincts of organised nature,—the impulses of sex and of nutrition,—which appear unchanged in man, as connate instincts. The investigation of the conditions of connate instinct in general is exceedingly difficult. But that, of course, is so much the more reason for starting out from the facts of the human consciousness, which furnish the only directly accessible observational material.

Do connate instincts spring from connate ideas, or do they depend upon intellectual processes? You will see at once that such hypotheses as those could never have been set up had not mankind been left out of account in their formulation. Or is the impulse to imitation in some way or another a factor in their constitution, as in that of the acquired social instincts? To this question also we may return a negative answer, without more words. Are we, then, to look on these manifestations of original

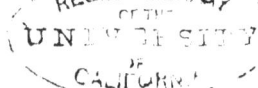

instinct as something analogous to the mechanised voluntary actions that now resemble reflexes? Certainly, if you observe the first sucking movements of a new-born mammal, those that appear before it has satisfied its hunger by actual sucking of milk, you will not find much to object to in the term 'reflex.' But none the less it is impossible to suppose that these reflexes have originated in a similar way to the mechanised movements (say those of the pianist) that have come about by practice. No! so long as we confine ourselves to the life of the individual, there can be no question that they are original, and not acquired. It looks, therefore, as though we had found an exceptional case to support the reflex theory, which has proved untenable everywhere else.

But we must not decide in its favour too hastily; we must go to observation for refutation or confirmation. The reflex theory assumes that the sucking movements of the new-born mammal are not only involuntary, but unconscious. Like reflex movement in general, they are purely physiological in nature: they show an entire absence of psychical motives. Now, though such an assumption might look reasonable enough on the study-table, it is hardly a theory that any one would hold who had ever really seen the movements of a hungry infant. Every feature and every gesture betoken the presence of unpleasant feelings. Plain enough to read in its crying and movements is the inarticulate complaint: 'I am hungry.' Give it anything that can be sucked, a finger or a corner of its pillow. All movement ceases; sucking, and only sucking, is the business of the moment. It is not long, of course, before the restlessness comes back again, only to be finally overcome by the satisfaction of hunger.

It is wholly impossible that all this is a matter of purely physiological reflexes. If emotional expressions have any significance at all, the infant's movements can only be interpreted as psychically conditioned actions, *i.e.*, manifestations of impulse. No doubt we must suppose that in these first impulsive movements there is not present a shadow of the idea of the end towards which the impulse is directed. But that is not at all necessary for the origin of emotional and impulsive expression. Sensations, with the feelings attaching to them, are altogether adequate to the result. And they are given in the sensation of hunger, which is physiologically conditioned, and the unpleasant feeling connected with it.

At the same time, there is one part of the effect that these causes do not suffice to explain,—the very phenomenon which gives to these impulsive movements their character of purposiveness, and renders it possible for them to attain their end :—namely, the sucking movement of the lips, which is in no sense a characteristic of unpleasant emotions in general. None the less we may regard this as a special emotional expression, inseparably associated in the human infant with the intensively toned sensation of hunger. And if the movement is one of expression, its purposiveness becomes intelligible. For while expressive movements are the means of expression of individual emotions, their general nature and in particular their characteristic of purposiveness result from a process of development extending beyond the individual : their physiological conditions are inherited or, what in this connection is the same thing, were acquired in the course of earlier generations, reaching back into an unlimited past. And this shows us the grain of truth that is contained in the reflex theory. The sucking movements of the new-born child are reflexes, in the sense in which expressive movements in general are reflexes. Their purposiveness, like that of the reflexes, is due to an organisation acquired in the course of generic, not individual, evolution. But they differ from the reflexes proper in this, that they are accompanied by emotions in the mind, and that their performance is regulated by these emotions. It is just the combination of these two characteristics that constitutes the peculiarity of the *connate impulse.* It stands midway between the reflex and the acquired impulsive action : related to the former in that its ultimate basis is physiological, and to the latter in that it springs directly from psychological conditions, which may at any time interfere to modify its original character.

§ IV

If we survey all those phenomena of human conduct which are referable to instinct, we see that the simplest conditions of instinctive action in general are to be found in the cases where it is the result of individual practice. Here the action simply indicates a disposition of the physical organisation, which has been induced by movements often repeated in the past. The performance of a definite complex act and its connexion with an adequate sense-stimulus have become more and more matters of course; till at last they are rendered completely mechanical. In the second place come the acquired social instincts, whose conditions are complicated by the development of the social emotions and the corresponding mimetic actions. Lastly, the connate instincts oblige us to assume that the disposition of the physical organisation *plus* the mechanisation of complex movements correlated with it, if induced through a number of generations, leaves behind it permanent physical effects, common to all individuals alike; so that certain impulsive movements, subserving the elementary necessities of life, take on the reflex form. They may then constitute a starting-point for fresh developments, through which the impulse can arrive at a special degree of perfection in special individuals.

The effect of 'practice' and 'habit' can only be due to after-effects of excitation, of the kind assumed by us for the explanation of instinctive movements. And since the expressions of instinct are *par excellence* 'customary' or 'habitual' actions, their subsumption to the general law of practice needs no justification. That law runs as follows: the more frequently a voluntary action is repeated, the easier is it to perform, and the greater is the tendency of its constituents (if it is a complex act) to take on the reflex form, *i.e.*, to arrange themselves in a connected series of movements, which runs on mechanically when once initiated by an adequate stimulus.

The formulation of this law shows us at once that its basis must be physiological. The goal attained by the process of practice is simply the mechanisation of movements which were originally dependent upon psychical antecedents. That must mean that mechanical, *i.e.*, physiological, alterations of the nervous system are at the bottom of the whole matter. We are still so much in the dark as regards the real nature of nervous processes, that we need not be surprised to find the exact physical and chemical character of these alterations quite unknown. If we know nothing more about them, we are at least certain that they exist ; the witness of the actual results of practice cannot be called in question. There is hardly any movement of the human body, however difficult, which we cannot, by continued practice and repetition, reduce to a mechanical certainty so complete, that it will be performed, even without any intention on our part, as the necessary reaction to certain sense-stimuli. Very remarkable instances of this mechanisation of complex actions by practice occur now and again in the conduct of 'absent-minded' persons. It is quite a common experience to begin a customary action at a time which is altogether unsuited for it,—the stimulus having been given by some familiar impression. We may intend to pass by our own house or the place of our daily business, but suddenly discover that we have mechanically followed our usual route, and entered the building without in the least meaning to do so. Some years ago I was occupied with certain physiological experiments on the frog, each of which involved the performance of a fairly complicated operation. It happened one day that I had taken up a frog for the purpose of making a quite different experiment. I suddenly found, to my great astonishment, that, instead of making the intended experiment, I had performed the customary operation. Now we certainly cannot regard acts of this kind as pure reflexes. The impressions are not only physiological stimuli, but psychological motives as well. But the reaction to them is impulsive : the familiar visual impression calls up the sensations, feelings, and movements associated with it. The movement could not, however, become instinctive in that way unless the succession of movements had been thoroughly practised physiologically. The greater the extent of this practice, the more effective is the inhibition of the conscious realisation of what we are doing, which puts a stop to the unintended action.

There are experiences of the most different kinds, then, which put the physiological effects of practice beyond the shadow of a doubt. But there is yet another proof of their reality in the functional properties of the nervous elements. If you excite a motor nerve by a stimulus so weak that it only just occasions a contraction of the muscle attached to it, and continue to apply this same stimulus at intervals just sufficient to avoid exhaustion, you will find (especially if the nerve is in good condition) that the contraction gradually increases in amount. This increase of excitability by stimulation can be best seen in the reflex movements that follow the stimulation of a sensory nerve connected with the cord,— supposing always that the experiment is made under conditions which preclude the adverse influence of fatigue. The molecular changes in the nervous elements on which the increase of excitability depends are, as was said above, still unknown. But we can get some idea of them by taking a few common illustrations of the facilitation of a movement by its repetition. As a carriage-wheel, for instance, turns round the axle, the rough surfaces are gradually worn smooth ; the frictional resistance is diminished. A watch, as you all know, goes better the more regularly it is wound up : and so on. Similarly, we may suppose, repetition facilitates the functioning of the nervous elements by removing all manner of obstructions and inhibitions. Now a complex muscular movement consists of a definitely arranged sum of simple movements, every one of which depends upon some elementary excitation-process. Each preceding excitation in such a series serves as the adequate stimulus for the succeeding one. This means that the effect of practice consists not only in the facilitation of every particular component of the complex process, but also in that of the definite combination of elementary movements which go to make it up.

You can easily see that this law of practice possesses a significance for the physical basis of our mental life which extends far beyond the sphere of instinctive action. Not only the combinations of certain movements, but the associations of sensations and ideas in general, are rendered stable by practice. Contiguity- and similarity-associations alike bear witness to its influence. The former are directly correlated with the habitualness of certain excitatory processes in a sense-

centre, the effect of which is to facilitate the genesis of sensation when the same impression is repeated ; the latter depend
upon our habituation to a particular connection of simultaneous or successive excitations. Regarded from this point of
view, that is, instinct appears as an extension of association
to the motor sphere.

These laws of practice suffice for the explanation of the
acquired instincts. The occurrence of connate instincts renders
a subsidiary hypothesis necessary. We must suppose that the
physical changes which the nervous elements undergo can be
transmitted from father to son. Later generations will then
be affected in two ways : they will from the first acquire
familiarity with certain complex movements more easily, owing
to connate dispositions of the nervous system ; and they will
react to particular stimuli by reflex movements of mechanical
certainty, owing to particular nervous dispositions of a more
definite and clearly marked kind. The assumption of the
inheritance of acquired dispositions or tendencies is inevitable,
if there is to be any continuity of evolution at all. We may
be in doubt as to the extent of this inheritance : we cannot
question the fact itself. It is in particular the inherited reflexes
of the human infant, so important for the development of its
instinct of nutrition, that belong to those constituents of original disposition which reach far back to the beginnings of
generic development. But more individual gifts,—the transmissibility of certain talents is unquestionable,—also appear to
lend probability to the view that the propagation of definite
dispositions takes place, at least within certain limits. Dispositions of this kind, however, are not the products of any very
long development ; and are probably to be looked on rather
as dispositions facilitating the practice of new functions than as
ready-made systems of reflex arcs. It is of great importance,
by the way, in this matter of the transmissibility of more
special gifts, that the disposition of associations and the direction of instincts be in complete agreement. A connate talent,
especially if its field of exercise is internal rather than external,
depends at least as much upon the disposition to form certain
associative connexions as upon the facility of certain complex
forms of movement. But in every case the point to remember

is that it is the *disposition*, not the actual functional capability, which is connate. Every instinctive action, however original it may be,—the taking of food by the infant, *e.g.*,—must to a certain extent be acquired afresh by the individual. Far more practice, then, will be required for the realisation of the connate talent, which has so short a period of development behind it. Readiness of movement and many-sidedness of ideational connexion are the promise of the connate disposition; the fulfilment of the promise comes later in life. Ideas cannot be inherited any more than complex volitional actions. Talent and instinct alike are latent until external stimulation calls them into actual life.

§ V

We have now reviewed the conditions of origin of the human instincts. How does the matter stand with the analogous phenomena presented by the animal kingdom? Are they deducible from the same conditions,—perhaps with the difference that the various factors are concerned in different amounts? Or must we look for other and peculiar explanatory reasons?

Different in expression as animal and human instincts are, their fundamental similarity can hardly be doubted. The first question must be answered in the affirmative, the second in the negative, though the negation cannot be absolute: for the conditions of human life are such that in it certain influences tend to disappear, and may accordingly be left out of account, which acquire a very considerable importance in the life of animals. To see the necessity of this admission, we have only to cast a glance at instincts which enable an animal,—*e.g.*, a caterpillar,—to provide not only for itself, but also for its larval condition and even for a still later condition, that of the imago, without the aid of the example of other animals or of any previous experiences of its own. Can the principles which our explanation has adopted explain this,—that a caterpillar living in a pomegranate cuts a way out of the fruit just before its transformation, and then makes this particular part of its home fast with silk thread to the nearest branch, that it may not fall to the ground before the transformation is complete? Many similar instances are quoted by Darwin in a posthumous essay on instinct, published as an appendix to Romanes's work *Mental Evolution in Animals.* Here belongs, too, the case of the caterpillar of the emperor moth, which we employed as an argument against the intellectualistic hypothesis (p. 391). These are all connate instincts ; so that the closest analogy to them in our own experience would be furnished by the sucking movements of the hungry infant. But these are sufficiently simple to be referred to a ready-prepared reflex mechanism. Can the same be said of the complicated animal actions which conform so wonderfully in the different species to the special conditions of life ? And, granted that it can, how far will the previous life-history of the species enable us to explain the origin of the particular reflex mechanism ?

We do not know the details of this life-history. And therefore we must give up any hope of a real genetic explanation of instinct. All that we can do is, first, to test the general question of the possibility of the origin of reflexes, which do not simply involve a definite and unchangeable co-ordination of movement and stimulus, but a co-ordination which may vary with variation of its special conditions ; and, secondly, to inquire whether the term 'reflex' is really applicable to the facts as stated. Now it is true that brainless animals exhibit reflexes of the kind that varies with its special conditions. A frog the whole of whose brain has been removed, with the exception of the optic lobes, not only tries to escape when its skin is stimulated, but avoids obstacles placed in its way. But in other respects the movement has all the characteristics of a reflex. Apply this to the present case. There can be no doubt that here, too, variations occur in accordance with special conditions ; if only for the reason that (as our previous discussion shows) the movements, like those of the hungry infant, are not purely reflexive, but expressive of emotions,—the expression being mediated by preformed purposive connections within the nervous centres. So that, strange as the instinctive action of an animal like the caterpillar may at first sight appear, it yet differs only in degree from the action of the human infant, which we have found comparatively easy of explanation. [1]

---

[1] For a description of the various animal instincts, *cf.* G. H. Schneider, *Der thierische Wille* (1880). Unlike so many works upon the same subject,

There still remains one point which requires further elucidation. Hitherto, relying upon the facts of human experience, we have been bridging the difference between connate and acquired instincts in this way : we have supposed that father can transmit to son the physiological dispositions that he has acquired by practice during his own life, and that in the course of generations these inherited dispositions are strengthened and definitised by summation. But is it possible to conceive of any specific life-history into which there could be crowded such a multitude of tendencies as should finally give rise to a succession of instinctive actions so complicated as those of the caterpillar, of the emperor moth, or even of the bird of passage, which flies south in winter without precept or example ? Surely the analogy of the practised pianist fails us here. But is it really applicable even to the connate instincts of man ? Do not they imply besides the action of will, which is required to introduce what afterwards becomes habit, a compelling force residing in the external conditions of life ? We do not know from what beginnings the functions of nutrition have been developed in man, except in so far as the facts of structural evolution admit of functional inference. But his general mental attributes enable us to assume quite definitely that the earliest development and consolidation of habit took place under the conjoint and unfailing influence both of external circumstances and of voluntary actions proceeding from feelings.

And this leads us to the principle which Darwin enunciated as of prime importance for the development of instinct and for the course of evolution in general,—the principle of *adaptation to environment*. There can be no doubt that this adaptation and *voluntary action* constitute the two universal determinants of the development of animal impulses. The first supplements the second ; volition must have an object towards which it is directed. The converse, of course, need not necessarily be the case. In the vegetable kingdom specific alterations are gradually effected by the sole operation of the environment, influencing the functions of growth or favouring certain peculiarities which are thus more readily and certainly perpetuated. And this

this volume may be recommended as giving an impartial and accurate account of observed facts.

passive adaptation will naturally be found among animals as well, since they share with plants all the physiological functions which are capable of modification by it. But Darwin's explanation of the development of instinct as being mainly the result of passive adaptation seems to·contradict the facts. Instinctive action is impulsive, that is voluntary action ; and however far back we may go, we shall not find anything to derive it from except similar, if simpler, acts of will. The development of any sort of animal instinct, that is to say, is altogether impossible unless there exists from the first that interaction of external stimulus with affective and voluntary response which constitutes the real nature of instinct at all stages of organic evolution. We may possibly succeed in deriving a complicated form of instinct from a more simple one ; but we can never explain instinct in terms of something which is as yet neither instinct nor impulse.

External conditions of life and voluntary reactions upon them, then, are the two factors operative in the evolution of instinct. But they operate in different degrees. The general development of mentality is always tending to modify instinct in some way or another. And so it comes about that of the two associated principles the first,—adaptation to environment,—predominates at the lower stages of life ; the second,—voluntary activity,—at the higher. This is the great difference between the instincts of man and those of the animals. Human instincts are habits, acquired or inherited from previous generations ; animal instincts are purposive adaptations of voluntary action to the conditions of life. And a second difference follows from the first : that the vast majority of human instincts are acquired : while animals,— apart, of course, from the results of training, which do not concern us here,—are restricted to connate instincts, with a very limited range of variation. This makes it to a certain extent intelligible that the older psychology, failing to see the close connection of habit and practice with instinct, usually ascribed instinctive action to the animals alone and denied it to man. The corollary from that connexion really is, that animal instinct is more predominantly reflexive, more exclusively constituted of purposive movements given with the connate physical organisation. If the complexity of a number of instinctive actions in animals seems to contradict that view, you must remember that throughout the animal kingdom they remain relatively uniform. We might almost say that the whole organisation of the central nervous system seems in many cases to be determined by certain associations that have been established by instinct.

LECTURE XXVIII

§ I

NEW and peculiar conditions for the development of instincts are to be found in the common life of animals. This is, of course, a product of *social* instincts; but it reacts in the most various ways upon the original impulses which occasioned it.

At the very lowest stages of animal life we see every creature seeking its like. Many of the medusæ and molluscs, many insects and fish, unite temporarily in swarms or schools. In all such cases it is not the individuals, but the species, which know one another. At the same time, the origin of the social impulse can only be looked for in a *feeling of inclination*, however primitive, which attracts animals of the same species towards each other by the intermediation of certain sense-impressions,—perhaps of smell or sight. At a higher stage of development this feeling of inclination shows itself as an *individual* attraction of animal to animal. But this is not found until we come to the higher birds and mammals. Dogs, as you of course know, manifest very pronounced likes and dislikes. If two poodles, *e.g.*, are kept in the same house, there may spring up between them a kind of friendship; the survivor mourns the loss of his comrade. Horses from the same stable become similarly attached to one another. Most remarkable are the friendships which arise between animals of different species as a result of living together. Even a dog and a cat may become friends. The inclination in all these cases is purely individual. The dog will pick out his comrades from a score of other dogs; and however gracious his behaviour to a particular cat, will chase all her companions with true canine hostility.

§ II

When the feeling of individual inclination combines with the sexual impulse, we have the phenomena of *animal marriage*. We can only speak of marriage when the union of male and female for the fulfilment of the sexual functions is a permanent one, demonstrably based upon individual inclination. There is no trace of it among the invertebrates or the lowest vertebrates. Although the 'insect-states' are really extended families, there is no proof that their individual members know one another as such, or are held together by any permanent mutual inclination; indeed, the facts enumerated in Lecture XXIII. make such a hypothesis exceedingly improbable.

On the other hand, marriage is a very common phenomenon among birds and mammals. That our domestic animals furnish so many exceptions to the rule is probably the result of domestication. By close association with man, the animal loses touch with its kind. Most animals are monogamous, although polygamy is a well-known institution among birds. Polyandry does not appear to have been observed in animals; it is confined to certain savage tribes.

We have many reliable observations to show that with many birds the marriage contract is a matter of free choice. Males and females that are kept together in a cage by no means always pair. There are preferences and aversions shown, for reasons often inexplicable to ourselves. Male song-birds contend for the females in song. The birds of paradise are said to spread their gorgeous plumes till the female chooses the wooer that pleases her best. Animals of a fiercer disposition do not get through their wooing so peaceably; the males generally have to fight pretty vigorously for the object of their choice. Lions and tigers wage bloody war for the possession of a spouse; and stags will wound one another to the death in their struggle for a doe. The males of polygamous species are especially ferocious in this quarrel for wives: you know that two cocks in the same yard are impossible. In this matter of choosing a mate, peaceably or otherwise, recourse is always had to the special weapons and peculiar ornamentation that so frequently characterise the male animal: think of the antlers of the deer, the spur of the gamecock, the tusk of the boar, the lion's mane, the varied plumage of many birds. All the birds that breed with us,— pies, storks, swallows, sparrows, doves, and what not,—are monogamous. The nest is nearly always the family residence ; male and female build it together, and share in the tending of eggs and young. Only the swallows have separate nests for male and female. Besides the common fowl, the ostrich and the cassowary are polygamous.

The marriage-relation among animals takes on a different form in monogamy and polygamy. The cock looks after his hens, seeks food for them ; they follow his call. But the hen does nothing for the cock except obey him. On the other hand, she watches, feeds, and protects the chicken, while the cock does not trouble himself about them. In monogamy it is generally otherwise. A pair of pigeons share all there is to do between them. Male and female take turns in sitting, and both alike see to the feeding of their young. It is obvious that these differences depend upon differences in the feelings of individual attraction ; and we shall therefore be right in explaining by their aid the difference between the monogamous and polygamous form of the marriage-relation.

The stability of animal marriage seems in general to be proportional to affection for the young. And this again becomes stronger the more careful and lasting attention the brood requires. A secondary reason for the continuance of the union after the young have ceased to require care is the need of mutual aid and protection. This holds especially of animals which construct nests or lairs, or which live in holes. So far, therefore, animal marriage is intimately connected with the conditions of specific physical organisation. But it would be unwarrantable to ascribe all its phenomena to this source. Individual inclination certainly determines choice in animals as well as in man ; accidental contiguity is a second and different cause. If there is an intimate connection between the fulfilment of a mental impulse and the satisfaction of a physical necessity, that is no more than we find everywhere else in our investigations of life and mind.

§ III

In marriage individuals are held together by the feeling that they belong to one another.   If this feeling is extended so far as to embrace a large number of animals, we have an *animal society*.   Most birds and mammals tend to unite in flocks or herds.   Domestication may overcome this tendency ; but it is seldom absent in the wild or feral state.   Even dogs that have run wild not infrequently get together into a pack.   Our oxen and sheep have retained the impulse to social life even under domestication.   Many animals herd only for a particular purpose, especially for plunder or food ; but even then the herd is very often composed of a definite group of individuals.   Migrant birds flock only when about to migrate ; the passage is made by thousands together.   At its conclusion the individuals separate, to reunite in the following autumn.   In the meantime the only connexion that the members of the same flock have with one another is that of locality ; they generally settle down near one another.   A flock of daws likes to settle, if possible, all together in the same ruin ; storks of the same flight nest in neighbouring villages.   There seems to be evidence in all these cases that the primitive feeling of inclination which brings the members of a species together at the very lowest stages of animal life is reinforced by individual inclinations, though there may be but few in every flock or herd that are held together by them.

We are taken a step further by animals which construct interconnected lairs or holes, destined to contain not one family only, but the progeny of a whole colony.   This inclination is a direct outgrowth from the impulse of the individual to build. The otter tends to settle down in the neighbourhood of other otters.   The same is true of the hamster and the beaver.   Sometimes the partitions between nest and nest are broken down, and the whole system thrown open ; as is the case with the hiding-places of rats and mice.

These forms of social union are common enough.   A special place among them must be assigned to the so-called insect-*states ;* 'so-called,' for they are not really 'states.'   The expression applied to these animal communities has done more to mislead than to explain.   It has led to the assumption that all their phenomena are to be interpreted in terms of those of human governments and institutions ; which in its turn tempted observers to parallel the division of labour in these societies, conditioned by the facts of physical organisation, with class distinctions in human society, and so to explain their observations by reading into them their own thoughts and feelings. We have already had illustrations of this method of procedure in our review of ant-life (Lecture XXIII., p. 343).

Insect-states are really extended families. The dwelling-places of the colonies are nests of a more or less complicated structure, according to the size and composition of the society. In most cases those orders of animals which contain the species that live in states include others which have not carried their social life farther than the stage of simple nest-building. In some species of wasps,—*e.g.*, the digger-wasps and solitary wasps, —the males and females live separately, though the female digs a hole in the mortar or wood of a wall, in which she lays her egg, putting in small caterpillars along with it, to serve as food for the newly hatched larvæ. The nest of the common wasp is more extensive. In the spring the female builds in a tree, on a roof, or in the ground a few hexagonal cells of vegetable material, lays an egg in each of them, and feeds the newly hatched larvæ until they crawl out. After this the young assist in the work of building, and a nest is gradually constructed, the female depositing an egg in every new cell. The females that develope at this period are themselves incapable of laying eggs ; their whole energy is consumed in the business of nest-building, and their sexual organs remain immature. These sexually undeveloped females are accordingly called workers. Not till towards the end of the summer are eggs laid which produce males and perfect females. These males fertilise the females in the autumn. When cold weather sets in they and the workers die ; but the females survive the winter, and at the return of spring begin to make nests and lay eggs. The female solitary wasp generally begins her work in some narrow hole in a wall, into which she had crept for the winter ; and the colony issues forth later, when the space has grown too small for it, to build a larger nest. What is true for the wasp holds also for the humble-bee (*bombus*), a relative of the ordinary honey-bee. The female is fertilised in the autumn, survives the winter, and begins an underground nest in the spring, in the building of which she is assisted by the females or workers which are first hatched. Towards the end of the summer sexually mature insects make their appearance ; and with the coming of winter the whole colony perishes, with the exception of the females which seek shelter underground.

There are two features in particular of these communities of wasps and humble-beés which for a long time defied any attempt at interpretation : the presence of sexless workers among perfect males and females, and the constancy with which the appearance of the latter two forms was restricted to the end of the summer. The first problem was solved so soon as it became known that the workers were not really sexless, as had been supposed, but simply immature females. This arrest of development could easily be explained from the expenditure of force necessary for building the nest ; and experiment showed that a more abundant supply of food did really suffice to change the workers into ordinary females. The second problem was answered by the discovery, first made in the case of the honey-bee, that the laying of male or female eggs depends entirely on the nature of their fructification by the female herself. After fertilisation by the male, the female retains the injected seed in a small pocket opening into the canal by which the eggs escape. This arrangement is of immense importance, because in these insects all eggs, even those which are not fertilised by the male, are capable of development. The fertilised eggs produce female, the unfertilised male insects. It is now plain why the humble-bee and the wasp at the beginning of summer lay only eggs which develop into females : the female fructifies her eggs as long as she retains any of the seed which she received from the male the autumn before. When this store is exhausted, the eggs can only produce males. But even of the fructified eggs it is only those last laid that can become perfect females, because it is only after the completion of the nest and the production of a sufficient number of workers that the larvæ are well enough fed to attain to their complete development. So that what looked at first sight like a preconceived design in these simplest insect-states proves to be the necessary result of physical organisation and of the relatively simple instincts which accompany it.

Taking the wasp as our guide, we shall not, perhaps, find it so very difficult to explain the organisation of the bee-state. The female bee, the ' queen' as she is called, also lays fertilised and unfertilised eggs. But she lays both kinds from the first, and distributes them among the cells of the hive, which the workers have built from wax of their own production. The cells are of two kinds,—wide and narrow. The wide are for the unfertilised eggs, which develope into males or drones; the narrow for the fertilised, which develope into workers. Besides this, the queen lays a few fertilised eggs in specially wide cells. The larvæ from these are fed more abundantly than the rest ; they become perfect females or queens. Sometimes the workers will take a larva from an ordinary cell into a royal cell which is not quite finished; then by means of good nourishment it becomes a queen. In the spring, as soon as the brood of queens is beginning to approach maturity, the hive becomes restless ; and on the first fine day a part of its inhabitants swarms out in quest of a new abode. This first swarm is quickly followed by others ; so that a single hive may found several colonies in the course of the summer. The old queen always goes with the first swarm, leaving the hive before the brood of new queens are out of their cells. The first of these latter to appear remains queen of the hive ; the others fly off with a portion of the workers to found other colonies. If two queens of the new brood make their appearance simultaneously, they fight till one or the other is overcome and killed, unless she avoids her danger by swarming out in time. The hive, therefore, never contains more than one sexually mature female, though the number of drones is very various, ranging from none at all to nearly a thousand. The drones are not confined to the limits of their hive. In the spring they fly out on every warm day and meet the young queens. But in the autumn, as soon as provisions become scarcer, they are expelled by the workers, and perish on the first cold night.

What distinguishes the bee's hive from the societies of wasps, hornets, and humble-bees, therefore, is a more hard and fast division of labour. In containing only one female, the hive resembles the nests of these other insects. But it is essentially different from them in its mode of origin. The wasp's nest is begun by a female, so that her solitariness is a matter of course. But the bee's hive is a society from its foundation, a society which grows, but without undergoing any radical change. The solitariness of its queen is partly the result of force. But it is this very interconnection of bee-states, the fact that each is a colony from some pre-existing one, that enables us to under-stand the mode of origin of the bee-societies and their difference from the associations of related insects. The natural history of the building of every wasp's nest is just a repetition of the same processes. But the bee-state stands in connection with its parent state, that again with its own, and so on. It has, in other words, a historical relation to the past and the future. If we suppose that there occur in such a society expressions of the universal im-pulse towards imitation, it is a necessary corollary that a colony will not have to begin its life from the beginning, but carries to its new home the customs acquired by previous generations, whether these are transmitted in inherited organic dispositions or are perpetuated by being directly handed down from the older insects to the younger. But it would be altogether gratu-itous to assume that the organisation of the hive has always been what we find it to-day. We know from experience that the habits of animals may change. One can wean the domestic insects from swarming and founding colonies by enlarging their hive as circumstances require. Populous bee-states will now and then give up the work of collecting honey, and take to plundering the smaller hives in their neighbourhood. And if we see changes like these in the habits of animals going on under our eyes, there is nothing to prevent our concluding that the peculiarities of the bee-society have arisen gradually and slowly, and that its customs have been fixed and settled both by inherited physical dispositions and by imitation. This conclusion is all the more probable in that the mode of origin of the beehive of to-day contains indications that its primitive mode of origin was something different. The first socia union of the insects, you see, cannot possibly have branched off from any pre-existing society. How could it have been brought about?

The question is answered by the condition in which we still find certain of the insects that are most closely related to the honey-bee. Every female wasp founds her own family ; every female bee must originally have founded her own family in the same way. Worker and queen at once, she prepared by herself the first cells for her brood. Now an alteration in these conditions may have been brought about by the greater length of life of the bee-communities. When more than one female had appeared in a single hive, jealousy made any peaceable common life impossible ; death and exile were the only alternatives for the weaker faction, and the latter would have been already suggested whenever the crowded condition of the nest prevented any further increase of population. So far everything is intelligible. But how is it that the queen voluntarily deposits drone-eggs in the wide cells, and worker-eggs in the narrow ones, and that the workers kill the royal larvæ if the weather is unfavourable to swarming ? These customs, too, we have every reason to suppose, are matters of gradual development, products of the natural evolution of instinct. The size of the cells in which the larvæ develop, *e.g.*, would have to be settled by reference to their wants. At first all the cells might have been made of one size. It would soon be found that the more poorly nourished larvæ, which were destined to become workers, required less space than those which turned into queens or drones. When once the right size had been hit upon, it might be adhered to in the future, since the bee-state is in touch with past traditions which lay down rules of conduct for its members. The younger generation had only to follow the example set them by their elders. For this reason the bee-state never needs to return to the primitive stage, and model its organisation from the very beginning. There is no exaggeration in saying that it is based, like our own civilised states, upon the work of all preceding generations.

Ant-communities differ from those of bees chiefly in the number of females which they support. For the greater part of their lives the males and females are winged ; and they are larger than the wingless neuters which make up the great bulk of the population. These neuters, like the worker-bees, are immature females. With the ants, too, the division of labour seems at times to extend even to the workers : especially with the termites, or white ants of Africa and Southern Asia. These ants build hills which often attain the height of several feet. Their workers are of two classes,—the workers proper, to which the peaceful avocations of the colony are entrusted; and the soldiers, whose duty it is to attack strange nests or defend their own from attack. This difference in instinct is probably correlated with difference in the physical strength of the individuals. And everything that we know of the intellectual capacities of these insects would lead us to suppose that the division of labour is not consciously agreed upon. A very similar instinct is displayed by the Amazon ant, which carries off the larvæ from the nests of weaker species, and makes workers or 'slaves' of them. This instinct is rooted in the general aversion that the different species of ants manifest towards one another, and has been gradually developed from the mob-fights in which the feeling of mutual dislike often culminates. Another specific instinct of ants is the custom of keeping plant-lice as 'domestic animals,' for the purpose of feeding themselves and their larvæ from the liquid secreted in the abdomen of these animals. There is nothing strange in such an expression of the nutritive impulse : the plant-louse, being one source of food among others, would naturally be carried with the rest into the ant-hill.

The phenomena presented by these animal states can only be seen in their proper light, if we keep in mind at the same time the mental capacities of the individuals which compose them. I have already pointed out to you that the exaggerated ideas of the early bee and ant naturalists as to insect intelligence must be considerably modified in view of the results of observations made under careful experimental conditions. The members of a bee- or ant-community do not know one another individually. And the feeling of inclination which holds them together is of a collective, indefinite nature, standing on a far lower developmental plane than the analogous feelings of birds and mammals, which lead to marriage or to the formation of less extended associations. The power of communication is also extremely limited, confined in all probability to certain manifestations of the imitative impulse. Numerous proofs of the comparatively low development of the mental life of the individual in these insect-communities have been collected by Sir John Lubbock, to whose work upon ants, bees, and wasps

I must refer you, in the absence of any observations of my own.[1]
His investigations show clearly enough the immense advantage
of experiment in this field over simple observation.   Lubbock approached every single question with pre-conceived ideas, derived
from observation of the general results of instinct and naturally
tending towards an overestimation of the intellectual capacities
of the insects.   But experimental tests always gave the same
result,—that the impulses of the common instinct left hardly any
place at all for the exercise of individual intelligence or the expression of individual feelings of inclination.   And even
Lubbock's conclusions require one further limitation ; this concept of intelligence still plays far too great a part in his
pages.   The very modest performances which he ascribes to
intelligence are entirely explicable in terms of comparatively
simple associations.   And that implies that the feelings and
impulses operative in the instinctive actions of the insects in
question are of an extremely primitive kind.   So that when
we talk of their having feelings of inclination and aversion, or of
their impulse towards imitation, we must be careful not to
regard these feelings and impulses as identical with the analogous
processes of our own consciousness, still less with these processes *plus* the products of our reflection upon them.   We have
before us no more than the first obscure movings of feelings and
emotions, which we do not find in their clearly conscious form
till we reach the higher animals or even man, but which for
that very reason act with all the greater certainty and uniformity
at this low level of development.   We are fatally inclined to
make the same mistake with regard to the elementary psychical
factors, feelings and impulses, which lead to the formation of
animal societies that we make in foisting our own point of
view upon their complex results,—these communities themselves.
We talk of the organisation of insects into a state, of queens and
workers, of soldiers and slaves, even of the rearing of domestic
animals.   And so we tend to read into their loves and hates,
acts of succour and of imitation, conscious processes completely
analogous to those which the terms call up in our own minds.   We

[1] *Ants, Bees, and Wasps: a Record of Observations on the Habits of
the Social Hymenoptera*, by Sir John Lubbock, Bart., M.P. (Int. Sci.
Series).

must remember that we are really in face of very primitive forms of mentality, which may be every bit as different from its more highly developed stages as is a single cell from a complex organism.

But if we are always obliged to measure the animal mind by the standard of our own consciousness, applying this as best we can where the conditions are so different, the other side of the matter is not less important. We must look into these facts of animal psychology for light upon the phenomena of the human mind. Another fatal tendency on the part of the psychologist is to measure every human action by the highest standard applicable to it. We look at it from the standpoint of intellectual reflection, and then make this reflection,—our own affair entirely,—the condition of its origin. Man lives in wedlock ; he combines with his fellows to form a community ; he founds states. All this as he does it presupposes an immense sum of intellectual work, accumulated through countless generations and implying the development of the higher feelings. In every particular case of human action this accumulated store is drawn upon. But it is surely wrong, in the light of the instances which the animal kingdom furnishes of the manifestation of social impulses, that a part played in the constitution of human society by original, natural impulse should be so entirely overlooked as it not infrequently is. Why, even in man it is only the special development which the phenomena have undergone, not their existence or their origin, which is the result of civilisation. The witness of animal psychology tells with all possible directness for the naturalness of the first beginnings of human social life. The investigation of the interaction of the two factors, nature and civilisation, in their gradual development forms the subject-matter of other disciplines upon which we cannot enter,—social psychology and social science.

§ I

WHEN we were considering will in its significance as an elementary psychical phenomenon, we found that the facts comprehended under the term constituted the links in a chain of development. (The lower stages of this development, simple voluntary acts, were classed together as manifestations of *impulse*) (the higher stages, acts of choice, as those of *volition* proper.) In reviewing the expressions of instinct we have become familiar with a whole number of phenomena whose invariable mental condition is some impulsive act, while at the same time the peculiarities of the physical organisation exercise a determining influence upon their development. It now remains to consider briefly the second and higher form of voluntary activity, volition proper, in its relation to the entirety of conscious processes.

We took our best examples of instinctive action from the animal kingdom. In the present investigation of volition, on the other hand, we are exclusively restricted to the human consciousness, although it is certain enough that instances of volitional action are not infrequent in the animal world, and especially among its more highly organised members. But the problem of volition, or, as it is generally called in consequence of the popular restriction of the concept of will to the sphere of choice, the problem of will, is practically confined to man, for this reason,—that the one question which is of decisive importance for our understanding of the nature of voluntary action and its relation to the other facts of our inner experience, a question which has long divided psychologists and philosophers alike into two hostile camps, is one that must be answered by an appeal to our own minds. It is the question of the *causality of will.*

§ II

An impulsive action is one, as we have seen, which is univocally conditioned ; there is only one motive present in consciousness. *Volitional* action arises from the choice between different motives, clearly or obscurely conscious. In impulse, therefore, the feeling of *our own activity* is less developed than in volition ; whilst, since this latter involves a decision as between various conflicting motives, the feeling of our own activity rises in it to that of *freedom.*

But if freedom is a result of the possession of will, of the choosing or selective will, how does it come about that the relation of the two is so often transposed ? Instead of saying, ' I am free, for I can will,' we are apt to say, ' I can will, for I am free.' Is not this a confusion of cause and effect ? It is plain enough that our consciousness of freedom can only have its source in the power of willing. The prisoner is not free, because his will is without effect. He would gladly be out of prison : but that is wishing, not willing. A firm belief in our power to do is an indispensable condition of willing, which is just the decision to act. How, then, are we to explain the fact that the consciousness of freedom, whose root is in the will, thus denies its origin, and makes itself out to be the cause of that from which it has really resulted ?

We know that we are free when we act of our own power, unimpeded by any external obstacle. Action by our own power we term *volitional* action, and regard as the consequence of our freedom. But what do we suppose to be the cause of this freedom ? There appears to be at this point a sudden break in the chain of cause and effect. We say the very *concept* of freedom excludes any idea of causality. For if it were dependent upon some cause or other, it would cease to be what it is,—freedom. Freedom and necessity mutually exclude each other.

Notice now the steps by which we have arrived at this conclusion. We should not be justified in saying the very concept of *will* excludes any idea of causality. For the fact that we do not know all the causes of a volition cannot be regarded as necessarily implied in the concept. What is done, then, is this. Freedom, the concept of which excludes causality, is interpolated as a middle term, on the one side of which volition is subject to causality, while on the other it is independent of it. For it is now subjected to a special causality, the causality of freedom, while made independent of general causality, the causality of natural processes.

It is this view of the matter which has given rise to the conflict between ordinary *determinism*, which maintains the universal validity of the law of causation, and *indeterminism*, which postulates freedom. 'The will cannot be free,' says the determinist, 'for a free will would not accord with the actual causal connexion of world-processes. Natural law would be replaced by miracle. No! every action, however free it seems, must have its cause. It is a necessary occurrence, and the agent cannot help himself.' 'The will is free,' replies the indeterminist, 'for we have an immediate consciousness of its freedom. Natural necessity and personal freedom are opposites. But the latter is vouched for by the inner voice of conscience, requiring from the agent responsibility for his every action.'

The opponents of the freedom of the will, that is, assert that its assumption is nonsense ; its advocates maintain that it is necessary. Which party in the dispute is right ?

We must insist, in the first place, that all the *ethical* arguments which have been brought to bear upon the question of the freedom of the will are out of place. They may *move* us ; they may incline us to the hypothesis of the freedom of human volition : they cannot *prove* anything. Even if a denial of the freedom of the will imperilled the validity of conscience and shook the foundations of our whole ethical system, still, if clear proof could be adduced that the will is *not* free, science would have to take its course. But happily that is not the case. Whichever theory holds the field, practice may stay quietly at home. You may remember what Kant said : 'Every being who can act only under the idea of freedom is in his action really free ; that is, he is governed by all the laws which freedom would necessarily bring with it, just as really as though his will were proved to be free to the satisfaction of theoretical philosophy.' The undeniable fact that we have a *consciousness* of freedom makes fatalism impossible, unless, indeed, this consciousness itself be regarded as included in the universal causal nexus.† For this consciousness of freedom tells us that we have the power to act without being consciously impelled by any constraining force, external or internal ; it does not tell us that we act without a cause. The defenders and the opponents of the freedom of the will have not seldom been at one in their confusion of constraining force and cause. Really, the two are wholly disparate concepts. We cannot say the earth is constrained to move, but we can say man is constrained to die. Only a being who knows that he is free can be constrained. The fatalist makes the mistake of destroying freedom and putting constraint in its place, constraint being in actual fact a condition which arose out of freedom, and cannot be conceived of without it.

So that if we take the concept of freedom in its proper sense, we shall say, ' The will is free,' for everything that stands in the way of a purposed voluntary action is felt by consciousness as constraint, while will seems to it the very opposite of this constraint. Freedom and constraint are reciprocal concepts ; they are both necessarily connected with *consciousness* ; outside of consciousness they are both imaginary concepts, which only a mythologising imagination could relate to things. If we say, ' The earth is subject to constraint because it moves round the sun,' we might just as well go on to assert that the sun is free, because it moves the planets.

Herbart remarks somewhere: ' If we regard ourselves as not free, we are really not free ; but if we ascribe freedom to ourselves, it by no means follows that we are so in reality.' We may say with equal justice : ' If we know the cause of a phenomenon, it necessarily follows that this really has a cause ; but if we do not know the cause of a phenomenon, it by no means follows that it has no cause.' But it is this last and erroneous inference which the adherents of absolute indeterminism draw when they conclude, from the premise that we cannot discover in consciousness all the causes which determine the will, that the will itself is the first cause of our actions.

It was attempted to support this negative proof from con-

† *Corrected series*

sciousness by a further positive argument. In nature, we are told, every occurrence presupposes a previous condition of things of which it is the inevitable consequence. This previous condition must itself have a predecessor, and so forth. But for the beginning of this infinite series we must postulate a primary, spontaneous impulse, if the origin of the world is to become intelligible at all. Now if it is once shown that *one* point stands outside of the universal causal nexus, there is no difficulty in conceiving of any number of causally connected series arising in the progress of the world's development, and each possessing its own particular beginning. If I now undertake the performance of some voluntary action, this fact, with all its consequences, means the beginning of a new series, each term of which is determined by natural causes except the first one, which is beyond their reach.

There are two weak points in this argumentation. In the first place, the assumption of a first beginning of things seems to be impossible for consciousness, whether pictorially or conceptually represented ; and secondly, even if a first beginning of the world had to be assumed, the hypothesis that similar beginnings could take place in the midst of the course of the world's development would be an analogical inference, destitute of all positive foundation.

The fundamental error in these and other arguments for or against the freedom of the will goes deeper. It consists in considering the entire question simply under the concept of *natural causation*. The very first requisite is a treatment of it as a question of psychological experience. If we regard it from this point of view, we see at once that the *psychical* causes, whether of a voluntary act or of any other manifestation of consciousness, are never wholly discoverable, for two reasons : first, because they lie outside consciousness, and belong to an inaccessible series of past experiences ; and secondly, because they form part of a more general conscious nexus, of which the individual mind constitutes only one link.) The general direction of the individual will is, you see, determined by the *collective will of the community* in which its possessor lives. And it is particularly in this connection that we find reason for the belief that the causality of our mental life cannot be subsumed without more ado under the familiar laws of natural causation, such as that of the equivalence of cause and effect.

§ III.

An attempt to construct the history of a nation or of mankind at large in terms of the laws of natural causation would not only be vain in practice: it would be wrong in principle. If the individual can say that, in place of acting as he did in some particular case, he might have acted otherwise, we must also be able to say of every event in history that it might have happened differently.   In both cases the *necessity* of natural causation is wanting.   For historical events and for the voluntary actions of an individual we can only adduce determining *motives* ; we cannot prove constraining *reasons*.   In this regard the concepts of historical occurrence and of voluntary action are exactly equivalent.   The only difference is, that one refers to a community, the other to an individual.

The general will of a community consists simply in the expressions of the wills of a large number of individuals.   The individual and his voluntary action are enclosed within concentric circles of more and more general volition.   First comes the general will of the little community in which he most immediately belongs ; then he, with this will, is subject to the will of a larger community; with this, again, to a still more comprehensive will ; and so on.   The relations in which the individual is thus placed are the principal determinants of his voluntary actions. But the general will of a community is usually in its turn determined by the wills of the more energetic individuals, which are acquiesced in by the individual wills of the majority.

It is a rule written upon the face of history that the frequency of expressions of volition is inversely proportional to the magnitude of their effects.   National action by which the course of history is suddenly changed is a matter of rare occurrence. Events which we can refer to the action of the general will of considerable communities constitute, as it were, the milestones of history.   In the intervals between them the general will is for the most part inactive ; though there are changes occurring within the community, oscillations in this direction or in that, they are not of vital import : they are like the variations of an individual will in obedience to the impulses and emotions to which a man's manner of life exposes him. The determination of the general will by those of a few prominent individuals has given place to its direction by a crowd of hardly noticeable influences, affecting each and all alike, directly or indirectly, by way of external condition or internal modification.

(The principal determinant of the individual will is, as we have seen, the will of the community.) In stirring times the course of events carries the individual with it, while in those periods of history when the general will is inactive the community remains in what we may call a state of equilibrium. But the social condition resulting from previous history, from external natural causes, and from the intervention of particularly strong individual wills in the ordinary progress of things, must, of course, itself contain motives of determining influence upon the voluntary action of the individual ; so that it is only to be expected that in the long intervals elapsing between historical events of the first magnitude the practically constant condition of society will bring with it a certain uniformity in the voluntary actions of the individuals composing it.

This general influence is confirmed by statistical facts. We find that the annual number of crimes, suicides, and marriages may remain constant for decades together, in civilised countries where the condition of society resulting from their past history is also approximately constant. Quetelet showed that the number of marriages every year is more regular even than the number of deaths, to which, of course,—except in cases of suicide, —the will has nothing to say. The same statistician proved also that so long as the course of justice, the prosecution and punishment of crime, remain unaltered in any nation, the crimes committed show a marvellous constancy in number, character, and distribution with regard to age and sex. And the same regularity obtains for suicide. It extends even to the manner of death chosen. Every year approximately the same number of men hang, shoot, poison, and drown themselves. From all this constancy we cannot but conclude that the historically determined social condition of a people is a dominant influence in the voluntary actions of the individual citizen.

And our conclusion finds still further confirmation in observations of a different nature, which afford us the means of isolating certain of the factors which combine to constitute the state of a society. If we compare the slight deviations from absolute regularity which the statistical tables show with the relations which help to determine that state, we are able in some measure to trace them to their causes. Thus it is demonstrable that famine increases the number of crimes against property and decreases the number of marriages. Violent epidemics, like cholera, bring with them a temporary decrease in the number of marriages, followed shortly after their disappearance by a still more marked increase. This latter phenomenon is to be ascribed to the increased mortality occasioned by the epidemic. Society seems to be hastening all unconsciously to fill up the gaps that death has made in its ranks. However irregular the actions of the individual, those of the community present the completest uniformity. But this regularity appears as the product of a blind necessity. Actions of every kind follow a definite numerical law, which no volition of the individual can avail to change.

But if in this summation of individual actions there is no trace of anything that could be ascribed to the influence of an individual will, are we not bound to conclude that this influence is illusory ? Is not the exception to natural law only an apparent one, which disappears when our observations extend over a sufficiently wide field ? Yes ; this conclusion has been drawn. The statistical figures prove, it has been said, that voluntary actions are dependent in measurable degree upon a series of external factors. Will within us, that is, corresponds to accident in the natural world without. Neither is a phenomenon without laws ; but both are phenomena whose laws cannot be deduced from the particular instance. In this way, it has been thought, the problem of the freedom of the will is solved by appeal to experience ; and the solution is—determinism.

But there is nothing in the facts of statistics to warrant such a conclusion, in the remotest degree. They simply show that the influence exerted by the condition of society constitutes *one* of the causes which determine the will. Whether it is the only cause, or whether there is not a whole number of co-ordinate causes to be found elsewhere,—on those questions they have not a word to say.

In extending our observations from the individual to a large community, we eliminate all the causes which condition the individual alone or only some small section of the community. It is the same procedure as is employed in physics. To eliminate chance influences which might vitiate the result of an observation, a large number of observations are taken. The more observations there are, the more probable it becomes that the separate sources of disturbance, which work in both directions as *plus* and *minus*, will compensate one another ; so that the average of the whole number will give us a result in accordance with the real fact under observation. But when we argue that, because statistics enable us to cancel out the influences that are restricted to the individual, therefore these influences do not exist, that is as bad as it would be to say in physics that the accidental errors, eliminated in the total number of observations were not present in the particular case. The physicist can afford to neglect them, simply because they possess no significance for him ; the psychologist cannot. The question before him is whether there exist, in addition to the influence exerted by the social state of a community, further determinants of volition of a more individual character. He must *not* neglect the deviations from the norm shown in the particular case ; for their presence constitutes the proof that such secondary determinants really exist.

Statistics itself teaches us that the effect of individual conditions determining voluntary action can really be traced in different degrees in the different circles of a community. The number of crimes, suicides, and marriages varies with age, sex, income, profession, etc. As soon, that is, as statistics goes more into details, it points to influences of a more special kind, depending upon the special nature of the state of society in that particular circle of the community. Still the utmost that statistics could do,—and this will really never be possible for it for many reasons,—would be to follow its investigations out till it reached such circles as stood in all external respects under identical influences, circles the age, sex, profession, etc., of whose members were absolutely similar. Statistics would furnish us with regular figures for the voluntary actions even of these narrowest circles, and we could calculate from them the force, as it were, with which each individual is attracted to a particular voluntary action by the conditions under which he lives. But so long as there remain individuals who resist this force we shall be obliged to take into account a *personal* factor if we are to understand the causality of the particular voluntary action.

§ IV

The determinants of volition which have their source in the social condition of a people, and the existence of which is demonstrated by statistics, come within the causal nexus of natural and historical processes. They serve, then, to prove once and for all that the will is not undetermined. But statistics can do no more than discover the *external* causes of voluntary action ; as to its *internal* causes, we are left wholly in the dark. These internal causes constitute the *personal* factor, which from its very nature must elude any statistical observation. Whether it operates by way of cause and effect, and if so what the form of this causation is, are questions which the rough averages of statistical examination cannot of course decide.

This personal factor conflicts in various ways with the other factors determining volition. Thus the general will furnishes a reason for the determination of the individual will, but it remains for the personal factor to decide whether the result aimed at by the general will shall also be the object of the volition of the individual. In the same way a determining influence is continually exercised by the state of society in the whole community and in the professional circle to which the individual belongs ; but here again the separate act of will is never performed without the decisive co-operation of the personal factor.

Now what is this personal factor, which of all the determinants of volition proves to be so indispensable ? When we have taken account of every one of the external reasons that go to determine action, we still find the will undetermined. We must therefore term these external conditions not causes, but *motives*, of volition. And between cause and motive there is a very great difference. A cause necessarily produces its effect ; not so a motive. A cause may, it is true, be rendered ineffective, or its effect be changed, by the presence of a second and contrary cause, but even then the result shows the traces of it, and that in measurable form. But a motive may either determine volition or not determine it ; and if the latter is the case, then exerts no demonstrable effect.

The uncertainty of the connection of motive and volition is due, and due only, to the existence of the personal factor. In consequence of this, all motives are seen to be insufficient for the complete explanation of a voluntary action ; they can never be constraining causes, but remain as partial determinants. And the motives of volition are insufficient for its explanation, simply because the nature of the personal factor itself and the manner of its co-operation with external factors are wholly unknown. At the same time the fact that an ineffectual motive leaves no trace upon the completed volition points towards the inference that external motive and internal factor do not co-operate as does a plurality of causes in nature, but that personality is the only immediate cause of action, *i.e.*, that the only direct effect of a motive is exerted upon the personality. Properly speaking, therefore, we may not talk of a 'personal factor,' since that expression implies the simultaneous co-operation of other factors. Rather, since all the immediate causes of voluntary action proceed from personality, we must look for the origin of volition in the inmost nature of personality,—in *character*.

Character is the *sole immediate cause* of voluntary actions. Motives are only mediate causes of them. Between the motivisation and the causality of character there is this essential difference,—that motives are immediately given or are at least determinable by a close examination of the external conditions of an action, while the ultimate grounds of causation remain unknown to us, opening out as they do into the infinite series of the psychological conditions of the development of the individual mind.

We estimate a man according to the reaction of his character to external motives. That is, we judge of his character from his voluntary actions ; we determine character from its effects, and can never define it in any other way than by reference to these effects. The real nature of personality, therefore, is always a riddle. Now, whenever we come to the limits of our solution of the problems of philosophy, there remains the final problem, a riddle which we cannot read. But in this case the knotty point seems to lie clearly before us in the midst of a series of cognisable causes and effects. The motives which determine the will are parts of the universal chain of natural causation. Nevertheless, the personal character, which alone can constitute volition, cannot be assigned a place in this causal nexus. We cannot therefore decide immediately and empirically that personality in its inmost nature, the source and origin of every difference that exists between individuals and between communities, is itself subject to natural causality.

It has been said that a man's character is a resultant of air and light, nurture and climate, education and destiny ; that it is predetermined by all these influences, like any other natural phenomenon. The assertion is undemonstrable. Character itself helrs to determine education and destiny ; the hypothesis makes an effect of what is to some extent also a cause. And the facts of psychological inheritance make it extremely probable that if our investigation could penetrate to the very beginnings of the individual life, we should find there the nucleus of an independent personality, not determinable from without, because prior to all external determination.

On the other hand, it is equally impossible to prove by an appeal to experience that character is not a product of the external influences at work to form it. Would two men the course of whose whole lives was absolutely identical show precisely the same peculiarities of character ? We cannot say ; the case has never been realised in experience. So far as the deficiencies in experience admit of any answer at all being returned to the question, we should conjecture that the truth lay somewhere between the two extremes : character is partly the result of conditions of life, partly an original possession of the personality. But the further question of the causality of character is not settled by that answer, for the beginnings of it, not caused in the individual life, may still be terms in some more universal causal nexus.

However that may be, personal character is the ultimate cause of volition. And this statement contains the immediate answer to another question, a question which can be met independently of any dispute as to the freedom of the will,—whether, namely, the individual is *responsible* for his actions or not. Punishment cannot, is not intended to, affect the external occasion of a crime, but the criminal, *i.e.*, the criminal character, as acting of its own initiative, in terms of its own causation. This character, you see, is placed in a more or less external community, and finds there a causality which is foreign to it. But, to estimate adequately the right of punishing it, we must look at the whole matter from the point of view of this wider community. Surely it must be conceded to every society as an inalienable right that it may defend itself against the attacks of its own members. For the general will stands in this respect above the individual just as unconditionally as the latter stands above the organs which obey the behests of the personal self.

The individual, then, brings with him into the world the germ of his future character. Two hypotheses are possible for the explanation of the existence and nature of this original endowment : we may say the germ of character in every individual is a special creation, or we may regard it as a resultant of the conditions embodied in previous generations. Our choice between these alternatives will be determined by our general metaphysical theory. If we look upon every form of life as an original creation, we shall find no difficulty in supposing that the birth of the individual involves a creative act, producing this or that bodily or mental force from nothing. If we believe in a developmental continuity, we shall choose the second path. There can be no doubt that the very earliest developmental stages of the individual contain in them rudiments of all his bodily and mental capacities. But we can neither demonstrate with certainty what the contents of these rudiments is, nor tabulate with any completeness the influences that come to work in the course of the individual life. What principally inclines us to admit no gap in the chain of processes connecting the special constitution of the individual with the general nature of the community into which he is born is the realisation of our endeavour to obtain a single theory applicable alike to the mentality of the individual and of society. If personal character grows out of a causal nexus extending far beyond personal existence, the determination of volition must also be sought outside of and beyond the individual life, and will prove incalculable from the factors which influence it. Every cause that stands behind the existence of the individual is itself the outgrowth of a still more remote chain of causation, and to follow this link by link to the end would be to trace out the causality of the universe.   Herein is to be found the justification for the view of religion which in its symbolic way makes will the gift of God.

But if character takes its origin in a causal nexus that extends beyond the individual life, it follows that the innermost causation of volition not only is unknown, but must necessarily remain unknown   And this gives us the distinction between *volition* and *chance*, which determinism is so fond of comparing. Chance depends on a defect of our knowledge which can possibly be made up ; volition depends upon a necessary and irremediable defect of knowledge.   That is why we are so apt to regard a chance occurrence in external nature as only an apparent exception to the causal law, while we look on volition as an actual exception.   The real reason for this difference is just what we have been saying,—that character, of which every voluntary act is an expression, has its origin outside of and beyond the individual life and consciousness, in the infinite continuum of mental development.   The more complete the determination of character by personal experience, the greater is the confidence of our prediction that it will act so or so in a particular case.   So it happens that the more the will matures, the farther it travels from its original inherent determination, the more certain does its direction become, and the closer the approximation of its external manifestations to a mental series necessarily and causally related.

LECTURE XXX

§ I. Concluding Remarks; the Question of Immortality. § II.
The Principle of Psychophysical Parallelism. § III. Old
and New Phrenology. § IV. The Empirical Significance
of the Principle of Parallelism. § V. The Nature of
Mind.

§ I

AT the beginning of these lectures upon the mental life of
man and the animals we declined to base our considerations from the outset upon any hard and fast conception
of the nature of mind, and to force the facts of experience into
agreement with that conception, in the way of the metaphysical
psychologists. On the contrary, we regarded it as our primary
duty to acquaint ourselves with the facts, and then, without the
aid of any other assumptions than those suggested by introspection and supported by experimentation and objective
observation, to try and establish laws under which the phenomena of mind might be subsumed.

But, now that we have come to the end of our task, it becomes
imperative for us to cast a glance over the body of facts that
we have collected, and to consider what answer is to be given
to the ultimate questions of psychology. The path that we
have travelled was not lighted by any metaphysical guiding
star. What is the result? Do these questions refuse to be
answered? do they transcend the limits of human knowledge?
Or has experimental psychology something to say about them,
something which may be believed and accepted as the issue of
an unprejudiced appeal to experience?

There is, indeed, one problem of speculative psychology which
we must exclude from the first as insoluble. Not only does it
transcend the limits of the empirical doctrine of mind: it does
not stand upon the plane of scientific knowledge at all. It is

the question of the condition of the mind before or after this conscious life of ours, a question which has really as little place in psychology as that of the 'creation' of the world has in physics or astronomy. The hope of constructing from the materials of our knowledge of the universe a conceptual edifice in which the objects of a supersensuous world are transformed into objects of knowledge,—that hope has always and again proved to be one of those fatal illusions from which neither belief nor knowledge has anything to gain.

If you need confirmation of this, look for a moment at the question of immortality, one of the principal problems of metaphysical psychology. It was necessary to put the imperishability of the individual mind beyond all doubt. That necessitated the continual emphasising of its substantial simplicity. And that led in the last instance to the logical extreme of the Herbartian metaphysic, in which we have a mental atom of simple quality with an unalterable content comparable,— these are Herbart's own words,—to a simple sensational quality, like 'blue' or 'red.' How does the imperishability of this mind-substance differ from, say, the imperishability of a material atom? Is it anything better?

The one aim of empirical psychology is to explain the interconnection of the phenomena of our mental life. It must decline once and for all to furnish any information regarding a supersensuous mental existence. At the same time, the question may with some right be raised whether it is not at least indirectly concerned in this problem. We cannot deny to *philosophy* either the privilege or the duty of passing beyond the mere explanation of facts of actual life, on the basis of the total sum of knowledge amassed by the several sciences. The actual character of the world-process renders it inevitable that the solution of this our first problem should be followed by the presentation of a second. Facts are given us in the form of continuous developmental series which in experience terminate at this point or that. Philosophy must go beyond experience, and strive to attain the ideal goal of all science,—a coherent theory of the universe. Now our mental life in particular is presented in the form of a whole number of developmental series, all directly or indirectly interconnected and all together tending towards the same end, which, indeed, is inaccessible to our immediate experience, but the nature of which we may infer, if we are allowed to assume that the developments beginning in experience are continued on the same lines beyond the bounds of experience. It is the aim of philosophy to *supplement* the world of experience in this way. In doing so she is only carrying to its logical conclusion a method of procedure which is begun in every one of the separate sciences, and which is rendered necessary both by the character of the experiential developments and the impulse to fill out incomplete systems of knowledge. Now the mental life of the individual stands at the centre of this plurality of mental developments. The individual, with all his actions and impulses, is placed in mental communities of wide and of narrow radius. As a member of such communities, he contributes his share in the last resort to the sum of the achievements and creations of the human mind. What is the ultimate goal of all this mighty current of mental development? Experience alone cannot answer ; while the ideal completion of experience, which philosophy tries to discover, can have no other foundations than the developments given in experience. It is here that psychology finds a place ; it is one of the first witnesses called upon by philosophy for information which shall aid in her ideal construction. And this must never be opposed at any point to established psychological fact.

Now, if we recognise the existence of this problem of an ideal completion of reality, we have also recognised a *continuance of mentality* in the widest sense,—*i.e.*, a persistence of the mental developments beyond every experiential limit wherever and whenever attained. For the hypothesis that mental development might somewhere come to an end, to be replaced by simply nothing, would, of course, imply a recognition of the invalidity of any ideal completion. More than that, the whole of the mental content of the universe would cease to have any significance. For what meaning could we read into mental life in general other than that of a great and lamentable illusion, the growing store of man's mental possessions confirming him more and more strongly in his justifiable expectation of further development, while the end of all things should still be nothingness?

There can be no doubt that it was this philosophical notion of purposiveness, and not any particular speculation as to the nature of the individual mind, which ultimately gave rise to the idea of immortality, and has empowered it to. resist throughout all times the attacks of philosophic doubt and the force of opposing philosophic argument. But mankind inclines to look at things *sub specie individualitatis* rather than *sub specie æternitatis*, and has therefore transformed this general conviction of the imperishability of mental development into a belief in the imperishability of each individual mind, with all its sensuous contents,—a contents that could only have been acquired under the special conditions of this present sensuous life.

Psychology proves that not only our sense-perceptions, but the memorial images that renew them, depend for their origin upon the functioning of the organs of sense and movement, of the nervous system, and ultimately of the total mechanism of the living body. A continuance of this sensuous consciousness must appear to her irreconcilable with the facts of her own experience. And surely we may well doubt whether such a continuance is an ethical requisite : more, whether the fulfilment of a wish for it, if possible, were not an intolerable destiny. But when we turn away from this, the idea of immortality in a bygone mythology, and return to its true philosophic foundation, empirical psychology has nothing to urge against it. For the mental development of the individual is a necessary constituent of the development of the universal mind, and points not less unequivocally than this to something lying beyond it.

§ II

Besides this first question, which has taken us from psychology into philosophy, and into the most difficult and uncertain part of philosophy, there are two others of general import to which we may be required to give a final answer on the basis of the facts which we have been discussing. The first is that of the *relation of mental to bodily processes* ; the second, that of the *nature of mind*, as inferable from our survey of the whole range of mental experience. Our only way to furnish an answer to either is, of course, to put together the results of our various investigations.

We emphasised the fact, at the very beginning of these lectures, that mental phenomena could not be referred to bodily as effect to cause. It is an inevitable presupposition of the natural sciences that the processes of nature constitute a straitly closed circle of movements of unchangeable elements, governed by the general laws of mechanics. Nothing can ever be derived from motion except another motion. In other words, the circle of these natural processes which are presented to our objective observation can never lead to anything beyond itself. Recognising this, we recognised the necessity of deriving every mental process from some other, the more complicated from the simpler, and of making it our business as psychologists to discover the mental laws of this interconnection. And at every stage upon the road which we have travelled we have found confirmation of this general position. Every well-established case of a connexion of mental phenomena has proved capable of a psychological interpretation ; more, we have always seen that no other interpretatory method could throw light on the specifically psychological character of the process under investigation. Thus the fundamental law of the doctrine of sensation, Weber's law, was shown to be a mathematical expression of the principle of relativity of mental states. And the different modes of ideational connection in sense-perception and in the temporal and spatial combinations of memorial images were referable to the laws of association, which themselves, when analysed into the two elementary processes of connexion by likeness and connexion by contiguity, appeared as directly dependent upon psychological conditions. Further, the laws of apperception, with their corollaries of the composition and disintegration of general ideas which underlie the intellectual processes, are only capable of psychological interpretation. Finally, the feelings, with their classification,— again, only psychologically intelligible,—as pleasurable and unpleasurable mental reactions, and the excitations of volition, took their places as terms in a developmental series, extending from the simplest forms of impulse to the most complicated expressions of self-initiated, voluntary activity. It may very well be that we have not yet discovered the simplest and best formulation for many of these causal connexions; and it cannot be doubted that many important laws of mental life still await discovery. But neither does it admit of any doubt that psychical can only be adequately explained from psychical, just as motion can only be derived from motion, and never from a mental process, of whatever kind.

At the same time, we found it to be a truth of equal universality that mental processes are connected with definite physical processes within the body, and especially in the brain ; there is a uniform co-ordination of the two. How are we to conceive of this connexion, if, as we have stated to be the case, it is not to be thought as that of cause and effect ? The answer to this question has been given in detail in the preceding pages of the book. The connexion can only be regarded as a *parallelism* of two causal series existing side by side, but never directly interfering with each other in virtue of the incomparability of their terms. Wherever we have met with this principle, we have named it that of *psychophysical parallelism*. Its validity cannot be doubted even by those who may be of the opinion that there may still perhaps be some metaphysical bridge to take them from physical to psychical, or *vice versâ*. Even they must admit that it is the most obvious empirical expression of the connexion which we actually found to obtain between the bodily and mental series of vital processes. But the question of the extent of the validity of the principle is a different matter. It requires further consideration ; and only at the conclusion of this shall we be able to hazard a conjecture as to whether we are dealing with an ultimate principle of dualistic metaphysics, beyond which our knowledge cannot go, or whether the psychophysical facts which we have co-ordinated tend at all to justify the philosophical attempt to fuse these two parallel and independent causal series in the last resort in a higher metaphysical unity.

The question of the extent of the validity of this principle of psychophysical parallelism can be approached either from the physical or mental side. From the former point of view, our direct experience of the parallelism tells us in plain terms enough that its range is exceedingly limited. Of the whole number of physical processes, which constitute the course of the material universe, vital phenomena form but a narrow and circumscribed part ; and of vital phenomena themselves there are again but few in which mental processes can either be perceived or inferred from objective observation. This is undoubtedly one of the principal reasons upon which is based the materialistic view that psychophysical parallelism itself formulates a causal dependence of the mental upon the physical. Regarded as systems of processes in nature, the physical is wider than the psychical ; mind is bound up with certain definite connexions and attributes of matter. And so it seems an obvious assumption that mental activities are functions of certain highly organised substances. But such statements do not meet the requirements of a really causal explanation. It is surely inadmissible to suppose that mental existence suddenly appeared at some definite point in the developmental chronology of life. It is a far more justifiable hypothesis that that point merely serves to mark in a general way the limen of a more clearly conscious mental life. An isolated sensation, out of all connexion with other sensations or ideas, could not make itself known to us, whether subjectively or objectively, by any symptoms of consciousness. But since our analysis of ideas takes us back to sensations as their ultimate elements, we have every right to assume that primitive mentality was a state of simple feeling and sensing ; while the possibility that this state accompanies every material movement-process,—that is to say, that the principle of psychological parallelism, even when regarded from the physical side, is of universal validity,—though, like every ultimate assumption, incapable of proof, is still certainly not to be denied. At least, it looks very much more probable than the materialistic function-hypothesis, if we accept the dictum ' Ex nihilo nihil fit.' That the beginnings of mental life are to be found in the vegetable kingdom, and particularly in the protozoa, whose life represents the earliest stages of development both of plants and animals, is a theory, it is true ; but it is the only theory which can explain the phenomena of movement displayed by these primitive creatures.

If, on the other hand, we prefer to consider the principle of psychophysical parallelism from its second or mental side, we again find ourselves in some doubt as to the extent of the connexion between mind and body. The older spiritualistic psychology was inclined upon the whole to restrict it to sense-perceptions and external voluntary actions,—processes whose relation to physiological conditions could not well be over-looked. But in more recent times there has sprung up a tendency both in physiology and in psychology to look upon a considerable extension of the sphere of psychophysics as right and necessary. Every conscious content which possesses sensible attributes of whatever kind,—*i.e.*, which is to some extent constituted by sensations, however slight their intensity,—must be recognised at once as a psychical content with a physical substrate. There is, as you know, no certain characteristic by which to discriminate the sensational content of a memorial or fancy image from that of a sense-perception. The ordinary one, that of the different intensity of the sensations, does not furnish a valid criterion ; for the intensity of a peripherally stimulated sensation may be just as near the limen of notice-ability as that of a memorial image, while the strength of the latter, when it takes the form of a hallucination or an illusion, may rival that of any externally excited sensation. Since, moreover, as we have seen, the intensity of sensation stands in a uniform relation to the intensity of the physical excitation, there is not the slightest reason to suppose that the difference between memory-image and sense-perception consists on the physiological side in anything more than a difference in the intensity of the underlying excitation-processes.

But if all the mental processes whose contents involves the presence of sensation in any form may be thus subsumed to the principle of psychophysical parallelism, it becomes im-possible to make an exception in favour of the intellectual processes. Every concept requires an idea to serve as its symbol in consciousness ; and an idea without sensational con-tents is an absurdity. Conceptual thought will, therefore, be accompanied by an excitation-process in certain sensory centres. If thought is engaged upon the composition or analysis of concepts, there will always be effected an alteration in the contents of these, *i.e.*, in the sensational contents of their repre-sentative ideas. Corresponding to every process of thought there will be some physical excitation, varying with the variation of sensational elements. And we can go even farther. The apperception of an idea, the strain of attention upon an idea, is always attended by changes in the sensational content of that idea. Sharp as is the general distinction between the clearness or obscurity of an idea on the one hand and its strength or weakness on the other, still both alike depend upon the greater or less noticeability of its sensational constituents and attributes. So that if sensations themselves are accompanied by physical processes, the alterations in ideas connected with alterations in certain of their constituent sensations will also be accompanied by them. In the case of strained attention we must add to these alterations the associated muscle-sensations, which must, of course, follow the rule governing sensation in general. And, finally, if the apperception of ideas can be subsumed to the parallelistic principle, we must recognise that its intimate relation to volition cannot but involve the internal impulses of will in the same fate. Every volition as well implies an alteration in the ideational,—*i.e.*, also in the sensational,—contents of consciousness. So that the physical processes which attend the external voluntary movement are only a further expression of a relation in which the will has stood from the time of its very first beginnings.

§ III

The result of all these considerations, then, is to make it exceedingly probable that no mental process which contains sensational elements of any kind can occur without there being at the same time set up corresponding physical processes. The universal validity of the principle of psychophysical parallelism is given with the sensible nature of the foundations upon which our whole mental life rests. There is no concept so abstract, no notion so remote from the world of sense, that it must not be represented in thought by some kind of sensible idea. But it would for this very reason be wrong to regard this parallelism as though it implied an equivalence of the two series of processes. Physical and psychical are, as you know, wholly incomparable. And they differ more especially in this point,— that the criterion of *value* which is the ultimate standard of reference both for those of our conscious activities which affect the world outside us, and even to a greater degree for our appreciation of the phenomena of consciousness itself, is wholly inapplicable to physical processes, or, at least, can only be applied where they can be derived from some mental purpose, *i.e.*, are subsumed to the psychological point of view. Regarded as such, considered simply from the standpoint of natural science, every physical process is a link in the unbroken chain of movement-processes, of as much or as little value as any other link. A memorial image may hurry through consciousness as the transient reproduction of some past experience to which we are utterly indifferent ; or it may serve as a vicarious idea to embody a concept which expresses an important result of logical reflection. Within the circle of physical processes there will occur in both cases the same weak sense-excitation, connected, if you will, with very different antecedent and consequent motions, but giving not the least sign of the difference in mental value which attaches to it. If we could see every wheel in the physical mechanism whose working the mental processes are accompanying, we should still find no more than a chain of movements showing no trace whatsoever of their significance for mind. So that, despite the universality of the parallelistic principle, all that is valuable in our mental life still falls to the psychical side. And the fact of parallelism can affect this value just as little as the necessity of embodying an idea in a word or some other sensible symbol, if it is to be a permanent property of thought, or even thought at all, affects the value of the idea itself. The value of a work of art of imperishable beauty does not depend upon the material of which it is made. The material only becomes valuable because capable of giving expression to the conception of the artist. And it is only carrying this relation of mental conception to its objective realisation one step farther back to apply it to the less durable, but therefore all the more plastic, material of ideation, upon the varying content of which consciousness has to work. The artist could not call his thought to life in stone or bronze, in word or picture, if it had not already gained the potentiality of that life in his own mind as a work of the constructive imagination from the sensible material of ideas.

It need now hardly be said, that psychophysical parallelism is a principle whose application extends only to the elementary mental processes, to which definite movement-processes run parallel, not to the more complicated products of our mental life, the sensible material of which has been formed and shaped in consciousness, nor to the general intellectual powers which are the necessary presupposition of those products. Phrenology, as you may know, localised memory, imagination, understanding, and even such narrowly defined faculties as memory for things or words, sense of colour, love of children, and so forth, in particular parts of the brain. It assumed that the physical processes in those parts,—and it left their physiological character altogether undetermined,—run parallel to these complicated mental capacities and activities. These are the ideas of the crudest forms of materialism, and render any psychological understanding of our mental life altogether impossible.

The absurdity of the phrenological hypothesis is not greatly diminished in the more modern form of it. Starting out from the facts of cerebral localisation, it assumes that each single idea is deposited in some particular nerve-cell ; so that the excitation of this cell is synchronous with the appearance of its special idea. We can only account for such notions by supposing that observers who had absorbed the false doctrines of the older phrenology, when they came into contact with the modern discoveries of the histology and minute anatomy of the brain, felt it their duty to transfer the phrenological functions of lobe and convolution to the more elemental cell. To do this, it was necessary to get rid of memory, imagination, linguistic talent, etc., and to endow the morphological units with the separate ideas of which the complex mental faculties are constituted. Now we have seen how complicated, as a rule, those mental processes are which terminate in the formation of an idea, how many sensations taken from the most various departments of sense may be involved in them. It is impossible to suppose that the structural elements of the brain can be related to mental processes in any way differently from the structural elements of the external sense-organs. Each such element is adequate only to a very simple function ; but it can play a part in the most diverse and complicated functions. A single cell from the visual area of the cortex can no more be the seat of a definite idea,—say, of a house or of the face of a friend,—than can a single retinal rod or *opticus* fibril. The phrenological view has only to be carried to its logical extreme for its impossibility to become manifest. Suppose that we are in daily intercourse with a friend ; that we have seen him in numberless situations. We must assume that he takes up not one cell, but a whole number of cells, in our brain. If our next meeting with him takes place under ordinary circumstances, we can use one of our stock of ideas ; if not,—if he has a new hat on, perhaps,—this new idea will have to be stored away in some cell that happens to be empty at the time. Or suppose that we have learned a word of a foreign language. It is deposited in some cell of the central organ of speech. If we hear the same word with some slight change of pronunciation, this modified form must be laid up in a second cell, and so *ad infinitum.* It is evident at the first glance that the hypothesis of idea-cells gives no account of the manifold forms of ideational and sensational connexion. It would fall to pieces at the first attempt from the inherent impossibility of its effort. For, as a matter of fact, it is never ready-made and isolated ideas that combine, but ideational elements, or, better, elementary ideational processes, as we saw when analysing out the simple associative processes underlying the cognition and recognition of an object. The radical error of the phrenological hypothesis is, that it substitutes an *anatomical* for a *physiological* parallelism. It is a true scion of the old-time phrenological doctrines in this as well as in its extraordinarily naïve notions about psychology in general.[1]

§ IV

The principle of psychophysical parallelism, then, refers always to a parallelism of elementary physical and psychical processes, and not to any parallelism of complex activities on either side or of mental function and bodily structure. But this suggests

[1] For other proofs of the untenable character of the neo-phrenological localisation-hypothesis, drawn chiefly from the phenomena of normal and pathological disturbances of memory, I may refer the reader to my *Essays*, pp. 109 ff. (Leipzig, 1885).

a further question,—whether a principle which after all includes two utterly disparate principles, disparate and yet never out of relation to each other, can properly be regarded as an ultimate psychological postulate. Is not a dualistic principle like this in opposition to our justifiable endeavour after a monistic world-theory? And if we cannot doubt its validity, since psychological and physiological facts alike attest it, should we not still, perhaps, look upon it as provisional only?

Certainly we have reached the point where psychological assistance can avail us no more, and where we must appeal to metaphysics for an answer. It is, or it should be, the aim of metaphysics to satisfy this craving of the reason for final unification. The results gained in the separate spheres of scientific investigation are unable to do this. If, then, there is anything at all for metaphysics to do, it is to furnish the ultimate explanation of this parallelism, which physiology and psychology accept as bare fact. Physiology cannot be called on for this explanation. It restricts itself to the explanation of the physical manifestations of life; and though often and again it comes upon the signs of mental function, it is obliged to consider this as a department of knowledge with which it has no concern. The problem of pyschology, again, is the explanation of the interconnection of the psychical manifestations of life, which form another and a separate causal series. But the two sciences supplement each other; where certain links are wanting in the causal nexus of the one side, they may be given in that of the other. In these cases, of course, physiology must have recourse to psychological, psychology to physiological, connecting terms. But it is always understood that the interpolation does not carry with it any real completion of the broken chain of connected processes; it is simply the substitution for a term of one series of the parallel term of the other. We may speak in such instances, perhaps, of the influence of mind upon body, or *vice versâ*. But we always mean, if we do not say, that the word 'influence' is not to be taken *sensu stricto*: that, for instance, a direct causal influence cannot be exerted by psychical term upon physical, but only upon the psychical process which this physical represents by parallelism. Thus an external voluntary movement is not produced by the internal act of will, but by the cerebral processes correlated with it; an idea does not follow from the physiological excitations of the sensory centre, but from the processes, sensational and associative, which run parallel to them. We must even suppose, continuing this train of reasoning, that it is not the physical stimulus which occasions the sensation ; but that this latter arises from some elementary psychical processes, lying below the limen of consciousness and connecting our mental life with some more general complex of elementary psychical processes in the world outside us. But since we are utterly ignorant of all that belongs to these, we have no choice : at the beginning of the development of the empirical mental life, we must substitute a physiological first term for the psychological. But is psychology here so much worse off than physiology? Will it ever be able to demonstrate the physiological processes which correspond to the highest productions of psychical life?

In all its empirical investigations, then, psychology is obliged to take up the same position as regards the links in the chain of physiological causality as physiology must assume with regard to psychological phenomena. The severance of the spheres of the two sciences must, to be fruitful, go hand in hand with mutual recognition of these spheres. The only views of the nature of the bodily processes which are possible for psychology are, therefore, those current in physiology and the other natural sciences : it must assume an actually presented, absolutely constant, material substrate, unalterable save as regards the movements of its parts. Over against this stands the circle of the psychical phenomena of life, an equally independent sphere of investigation, not admitting of causal explanation in terms of the connection of motions of matter. So for psychology, as for physiology, the principle of psychophysical parallelism turns out to be an ultimate postulate, behind which it cannot go.

The attitude of metaphysics in this matter is, of course, a quite different one. The very nature of the objects with which psychology and natural science alike begin their analysis furnishes it with a sufficient reason for the inquiry after a higher unity in which the dualism of the parallelistic principle may be resolved. All that we know of the phenomena of nature comes to us in the form of ideas. The distinction of idea and object, upon which the division of the experiential sciences into those of nature and mind depends, is simply a result of the analytic activity of thought. In itself the idea is at the same time object; there are no objects which are not also ideas, or which must not be thought of in accordance with the laws governing the formation of ideas. But if it is thought which, by abstracting and distinguishing, has broken up the original unity of the worlds without and within, you can easily understand the mind's persistent impulse to restore that unity as the final act of its own development. Nay more, you will recognise the endeavour as just, and its fulfilment as a task for science. To point out means to this end is the business not of psychology, but of philosophy. Psychology can only indicate the path which leads to territories beyond her own, ruled by other laws than those to which her realm is subject.

§ V

These considerations have brought us to the last task which remains to be performed. We have learned all that we could of the interconnection of mental phenomena. What now is the *nature of mind*? The real answer to this question is contained in all that has been said before. Our mind is nothing else than the sum of our inner experiences, than our ideation, feeling, and willing collected together to a unity in consciousness, and rising in a series of developmental stages to culminate in self-conscious thought and a will that is morally free. At no point in our explanation of the interconnection of these inner experiences have we found occasion to apply this attribute of mentality to anything else than the concrete complex of idea, feeling, and will. The fiction of a transcendental substance, of which actual mental content is only the outward manifestation, a fleeting shadow-picture thrown by the still unknown reality of the mind,—such a theory misses the essential difference between the inner and the outer experience, and threatens to turn to mere empty show all that lends solid value and real significance to our mental life. Conscious experience is immediate experience. Being immediate, it can never require that distinction of a substrate, existing independently of our subjective appreciation, which is rendered necessary in natural science by its conception of nature as a sum-total of real things presented to us and persisting independently of us. Our mental experiences are as they are presented to us. The distinction between appearance and reality necessary for the apprehension of the world without, and culminating in the concept of a material substance as a secondary conceptual hypothesis which so far seems to do justice to the facts of experience, ceases to have any meaning when applied to the apprehension of the thinking subject by himself. You can understand, therefore, that when we are analysing our internal experiences we are never met by the contradictions between particular phenomena which in natural science furnish both incentive and means to the gradual developing and perfecting of the concept of matter, a concept which, destined as it is to remain for ever a hypothesis, can still hope to approximate to the truth by an infinite number of efforts towards it.

There is just one single group of empirical facts which have with some show of reason been adduced to prove the necessity of assuming a mental substrate analogous to material substance, —the facts of the *revival of previous experiences*. If we can call up some past idea, it is urged, it surely follows that some trace of that idea has remained in the mind during the meantime, else its reproduction would not be possible. Now we have seen, of course, that no idea, that no mental process whatsoever, can be called up again unchanged. Every remembered idea is really a new formation, composed of numerous elements of various past ideas. Nevertheless, it might be supposed that these very elements were the ideational traces left behind in the mind. But it is evident that even in this form the theory has presuppositions due simply to a transference of the permanent effects observed in the case of physical processes to the hypothetical mental substrate, in other words to an unconscious intermixture of materialistic views. A physical influence acting upon a body produces some more or less permanent alterations in it. Thus we have every right to suppose that a nervous excitation leaves an after-effect in the nervous organs, which is of significance for the physiology of the processes of practice and revival. Now in the theory of 'traces' these physical analogies are applied without more ado to the mind. Mind is conceived either as identical with brain, or as a substance localised somewhere in the brain, resembling it and other material substances in every essential attribute. But the physical excitation-process can only leave its after-effect upon the nerve, because it is itself a process of movement in or with a permanent substrate. And if mental processes are not phenomena, but actual immediate experiences, it is very hard to see how their after-effects can be psychologically conceived, except also in the form of directly presented mental processes. If we try to imagine an idea as persisting beneath the limen of consciousness, we can as a matter of fact only think of it as still an idea, *i.e.*, as the same process as that which it was so long as we were conscious of it, with the single difference that it is now no longer conscious. But this implies that psychological explanation has here reached a limit similar to that which confronts it in the question as to the ultimate origin of sensations. It is the limit beyond which one of the two causal series,—the physical,—can be continued, but where the other,—the psychical,—must end ; and where the attempt to push this latter farther must inevitably lead to the thinking of the psychical in physical,—*i.e.*, material,—terms.

We conclude, then, that the assumption of a mental substance different from the various manifestations of mental life involves the unjustifiable transference of a mode of thought necessary for the investigation of external nature to a sphere in which it is wholly inapplicable ; it implies a kind of unconscious materialism. The consequences of this transference follow at once from its nature ; the true value of our mental life is in jeopardy. For this value attaches simply and solely to the actual and concrete processes in mind. What can this 'substance' do for us, a substance devoid of will, of feeling, and of thought, and having no part in the constitution of our personality ? If you answer, as is sometimes done, that it is these very operations of mind that go to make up its nature, and that therefore mind cannot be thought or conceived without them, why, then the position is granted : the real nature of mind consists in nothing else than our mental life itself. The notion of 'operation' as applied to it can only mean, if it has any admissible meaning at all, that we are able to demonstrate how certain mental manifestations follow from, are the effects of the operation of certain other mental manifestations. Physical causality and psychical causality are polar opposites: the former implies always the postulate of a material substance; the latter never transcends the limits of what is immediately given in mental experience. 'Substance' is a metaphysical surplusage for which psychology has no use. And this accords with the fundamental character of mental life, which I would have you always bear in mind. It does not consist in the connexion of unalterable objects and varying conditions: in all its phases it is *process*; an *active*, not a passive, existence; *development*, not stagnation. The understanding of the basal laws of this development is the final goal of psychology.

Butler & Tanner, The Selwood Printing Works, Frome, and London.

www.ingramcontent.com/pod-product-compliance
Lightning Source LLC
Chambersburg PA
CBHW020905210326
41598CB00018B/1781